NETWORKS IN CELL BIOLOGY

The science of complex biological networks is transforming research in areas ranging from evolutionary biology to medicine. This is the first book on the subject, providing a comprehensive introduction to complex network science and its biological applications.

With contributions from key leaders in both network theory and modern cell biology, this book discusses the network science that is increasingly foundational for systems biology and for the quantitative understanding of living systems. It surveys studies in the quantitative structure and dynamics of genetic regulatory networks, molecular networks underlying cellular metabolism, and other fundamental biological processes. The book balances empirical studies and theory to give a unified overview of this interdisciplinary science. It is a key introductory text for graduate students and researchers in physics, biology and biochemistry, and presents ideas and techniques from fields outside the reader's own area of specialization.

MARK BUCHANAN is a physicist and independent author. He writes a monthly column for the journal *Nature Physics*.

GUIDO CALDARELLI is Associate Professor in the Centre for Statistical Mechanics, CNR-INFM, Rome, and collaborates with the Research Centre 'E. Fermi', the Laboratory LINKALAB and the London Institute for Mathematical Sciences.

PAOLO DE LOS RIOS is a Professor in the Laboratory of Statistical Biophysics, Institute of Theoretical Physics, Ecole Polytechnique Federale de Lausanne (EPFL).

FRANCESCO RAO researches in the Laboratoire de Chimie Biophysique at the University of Strasbourg and at the Research Centre 'E. Fermi', Rome.

MICHELE VENDRUSCOLO is a Reader in Theoretical Chemical Biology in the Department of Chemistry at the University of Cambridge.

Networks in Cell Biology

Edited by

M. BUCHANAN

G. CALDARELLI
CNR-INFM

P. DE LOS RIOS
Ecole Polytechnique Federale de Lausanne

F. RAO
University of Strasbourg

M. VENDRUSCOLO
University of Cambridge

Shaftesbury Road, Cambridge CB2 8EA, United Kingdom

One Liberty Plaza, 20th Floor, New York, NY 10006, USA

477 Williamstown Road, Port Melbourne, VIC 3207, Australia

314–321, 3rd Floor, Plot 3, Splendor Forum, Jasola District Centre, New Delhi – 110025, India

103 Penang Road, #05–06/07, Visioncrest Commercial, Singapore 238467

Cambridge University Press is part of Cambridge University Press & Assessment,
a department of the University of Cambridge.

We share the University's mission to contribute to society through the pursuit of
education, learning and research at the highest international levels of excellence.

www.cambridge.org
Information on this title: www.cambridge.org/9780521882736

First published 2010

A catalogue record for this publication is available from the British Library

Library of Congress Cataloging-in-Publication data
Modelling cell biology with networks / edited by M. Buchanan . . . [et al.].
p. cm.
ISBN 978-0-521-88273-6 (hardback)
1. Cellular control mechanisms. 2. Biological systems. 3. System analysis.
I. Buchanan, M. (Mark), 1961– II. Title.
QH604.M63 2010
571.6–dc22
2010000127

ISBN 978-0-521-88273-6 Hardback

Contents

Contributors

S. Ahnert
TCM Group Cavendish Laboratory
University of Cambridge
Cambridge
CB3 0HB
UK

M. Madan Babu
MRC-Laboratory of Molecular Biology
University of Cambridge
Hills Road
Cambridge
CB2 0QH
UK

Matteo Brilli
Biométrie et Biologie Evolutive
UMR CNRS 5558
Université Lyon 1
43, Bvd du 11 Novembre
69622 Villeurbanne Cedex
France

Mark Buchanan

Gerard Cagney
Room F067
Conway Institute
University College Dublin
Belfield
Dublin 4
Ireland

G. Caldarelli
Centro SMC, INFM-CNR
Dipartimento di Fisica
Università 'Sapienza'
Piazzale Aldo Moro 5
Rome
00185
Italy

Elisa Calistri
Centro Interdipartimentale per lo studio delle Dinamiche Complesse (CSDC)
Università di Firenze
Via G. Sansone 1
1-50022 Sesto Fiorentino
Firenze
Italy

P. De Los Rios
Laboratoire de Biophysique
Statistique, ITP SB EPFL
Lausanne
1015
Switzerland

Thomas M. A. Fink
Institute M. Curie
5 Rue d'Ulm 75005
Paris
France

A. Gabrielli
Centro SMC, INFM-CNR
Dipartimento di Fisica
Università 'Sapienza'
Piazzale Aldo Moro 5
Rome
00185
Italy

D. Garlaschelli
Dipartimento di Fisica
Università di Siena
Via Roma 43 Siena
Italy

S. C. Janga
MRC-Laboratory of Molecular Biology
University of Cambridge
Hills Road
Cambridge
CB2 0QH
UK

Pietro Lió
Computer Laboratory
University of Cambridge
Cambridge
UK

Seesandra V. Rajagopala
J Craig Venter Institute (JCVI),
formerly The Institute of Genomic
Research (TIGR)
9704 Medical Center Drive
Rockville
MD 20850
USA

Francesco Rao
Centro Studi e Ricerche e Museo della
Fisica 'E. Fermi'
Compendio Viminale
Rome
Italy

Erzsébet Ravasz Regan
Beth Israel Deaconess Medical Center
Harvard University
Boston
USA

Gian Paolo Rossini
Dipartimento di Scienze Biomediche
Università di Modena e Reggio Emilia
Via G. Campi 287
Modena
I-41100
Italy

Daniel Segrè
Bioinformatics program
Boston University
44 Cummington St.
Boston, MA 02215
USA

Björn Titz
Crump Institute for Molecular Imaging
Graeber Lab
University of California, Los Angeles
570 Westwood Plaza
Los Angeles
CA 90095-1770
USA

Peter Uetz
The J Craig Venter Institute (JCVI), formerly The Institute of Genomic Research (TIGR)
9712 Medical Center Drive
Rockville
MD 20850
USA

Michele Vendruscolo
Department of Chemistry
University of Cambridge
Lensfield Road
Cambridge
CB2 1EW
UK

Introduction

M. BUCHANAN, G. CALDARELLI, P. DE LOS RIOS, F. RAO
AND M. VENDRUSCOLO

Biologists now have access to a virtually complete map of all the genes in the human genome, and in the genomes of many other species. They are aggressively assembling a similarly detailed knowledge of the proteome, the full collection of proteins encoded by those genes, and the transcriptome, the diverse set of mRNA molecules that serve as templates for protein manufacture. We increasingly know the "parts list" of molecular biology. Yet we still lack a deep understanding of how all these parts work together to support the complex and coherent activity of the living cell; how cells and organisms manage the concurrent tasks of production and re-production, signalling and regulation, in fluctuating and often hostile environments.

Building a more holistic understanding of cell biology is the aim of the new discipline of systems biology, which views the living cell as a network of interacting processes and gives concrete form to the vision of François Jacob, one of the pioneers in the study of genetic regulatory mechanisms, who spoke in the 1960s of the "logic of life." Put simply, systems biologists regard the cell as a vastly complex biological "circuit board," which orchestrates diverse components and modules to achieve robust, reliable and predictable operation. Systems biology suggests that the mechanisms of cell biology can be related to the information sciences, to ideas about information flow and processing in de-centralized networks.

This view, of course, has long been implicit in the study of cell signalling and other key pathways of molecular bio-chemistry. It has long been clear that such pathways do not act through simple sequential action in chain-like reaction paths, but as a rule exhibit a much richer dynamics involving multiple pathways working in parallel, with interactions passing between, and feeding backward as well as forward. Yet with rapidly advancing technology, and new theoretical tools coming from physics, engineering and mathematics, systems biology is beginning to reach beyond the recognition that such systems exist to elucidate specific quantitative

Networks in Cell Biology, ed. M. Buchanan, G. Caldarelli, P. De Los Rios, F. Rao and M. Vendruscolo.
Published by Cambridge University Press.

regularities in biological information flows, often involving operations such as feedback, synchronization, amplification and error correction that are familiar to engineers.

Significantly, the potential power of this perspective is being multiplied by an explosion of recent work in physics and mathematics, both theoretical and empirical, showing that many of the world's complex networks have hidden structural regularities of great importance. These networks – ranging from social networks and food webs to the Internet, and including genetic regulatory or metabolic networks within the cell – have a surprisingly universal character, and can be fruitfully described with a unified conceptual language. This work has infused science with new analytical measures such as betweenness centrality, network dimension, degree distribution and so on, which reflect local or global network properties. These measures have been found to have direct relevance to a network's stability or resilience, information processing efficiency, or dynamical richness, offering hope that the network perspective will prove invaluable for understanding networks in cellular biology.

Networks in Cell Biology aims to fill a yawning gap in the modern literature of systems biology. While a number of excellent texts at the graduate level cover recent developments in cell biology, and others offer timely introductions to recent advances in complex network science, no single volume yet focuses explicitly on advances in our understanding of cellular networks. The present book offers an up-to-date snapshot of such work, aiming for a balance between empirical studies and theory, which are natural complements in furthering knowledge in almost any area of modern biology. What do we know about the qualitative and quantitative structure and dynamics of genetic regulatory networks, or the network biochemistry underlying cellular metabolism? What do empirical studies tell us about the large-scale architecture of protein–protein interactions, and what are the key weaknesses and gaps in such data? What are the prospects for building realistic dynamical models for metabolic or regulatory processes, possibly with predictive capability?

This volume covers these and other topics of current research in systems biology, with an emphasis on recent advances in complex networks science, and on how the massive data now available in biology can help test and inform such theories in their application. With contributions from key leaders in both network theory and modern cell biology, this volume makes available in one publication the naturally diverse range of studies supporting a unified view of biological networks that is likely to be increasingly important in the future.

The contributions also illustrate a timeless truism of science – that scientific progress is not linear or predictable, and that it is often a new idea emerging from an unexpected direction – a technique for gathering data of a new kind, or a small shift in perspective – that transforms and renews a field by framing old problems in new ways. The new idea in this case is an old one, the basic perspective of network science, but updated and vastly furthered by modern data and powerful concepts and techniques coming from statistical physics.

1

Network views of the cell

PAOLO DE LOS RIOS AND MICHELE VENDRUSCOLO

1.1 The network hypothesis

A cell is an enormously complex entity made up by myriad interacting molecular components that perform the biochemical reactions that maintain life. This book is about the network hypothesis, according to which it is possible to describe a cell through the set of interconnections between its component molecules. Hence, it becomes convenient to focus on these interactions rather than on the molecules themselves to describe the functioning of the cell.

The central dogma in molecular biology describes the way in which a cell processes the information required to produce the molecules necessary to sustain its existence and reproduction. It is also becoming increasingly clear, however, that in order to establish a more complete description of the manner in which a cell works we require a deeper understanding of the manner in which the sets of interconnections between these molecules are defining the identity of the cell itself. It is therefore important to investigate whether the genetic makeup of an organism does not only specify the rules for generating proteins, but also the way in which these proteins interact among themselves and with the other molecules in a cell.

Complex networks Networks are a way to represent an ensemble of objects together with their relations. Objects are described by means of vertices (sometimes also called nodes) and their relations by edges (sometimes also called links or connections) connecting them, which can be weighted to reflect their strength. A network is thus entirely characterized by the set of its connections, not by the way in which it is drawn. The actual distance between two vertices is given by the minimum number of edges that can be found between them. A key quantity to characterize networks is the degree: it is defined as the number of edges per vertex. Most networks present in nature have large fluctuations in the degree value. This feature has profound consequences both on the stability of the sytems represented by the network and on the dynamics of the processes defined on this structure.

Networks in Cell Biology, ed. M. Buchanan, G. Caldarelli, P. De Los Rios, F. Rao and M. Vendruscolo. Published by Cambridge University Press.

Networks provide a way to organize and regulate efficiently complex systems [12, 52, 88, 89, 98, 153, 436, 583, 636]. In an effective network different parts are linked by reducing at a minimum the number of the interconnections. This feat can be achieved at the cost of a fragility of the structure of connections. If one passage goes wrong, the whole structure falls down. A bit of redundancy is a good compromise for a fail-safe system without too much effort. A trade off between redundancy and cost is probably at the basis of the statistical properties of such objects and explains in part their ubiquitous presence in the various cell activities. From the point of view of the researcher a network is also a powerful method to represent the data in one object, and to enable the quantitative assessment of the fragility or robustness of the system. The first step is to establish the topology, the second is to establish the dynamics – that is how the topology changes with time. The study of these two aspects is at the heart of current research about biological networks and it constitutes the core of the following chapters in this book.

1.2 The central dogma and gene regulatory networks

The central dogma The central dogma, which was introduced in 1958 by F. Crick [131] and later restated in 1970 [132], defines the way in which information flows between DNA, RNA and proteins. At the most fundamental level, DNA passes information to itself (through replication) and to RNA (through transcription), and RNA passes it to proteins (through translation). We also now know, however, that additional information also follows other routes that taken together link DNA, RNA and proteins in a much interconnected network. This view of the central dogma defines our current core understanding of molecular biology.

The central dogma of molecular biology defines the manner in which the genetic information is exploited to generate proteins. In this process, DNA is first transcribed into messenger RNA (mRNA), which is then translated into proteins (Fig. 1.1a). This description, originally proposed by Francis Crick [131], is at the foundation of molecular biology.

Our understanding of the more subtle aspects and of the fundamental implications of the central dogma is becoming increasingly detailed. While initially the information was thought to flow in one direction from genes to proteins, it is now well established that there are much more articulate interactions between DNA, RNA and proteins. Gene transcription is a highly regulated process, in which a major role is played by transcription factors, which are proteins that bind

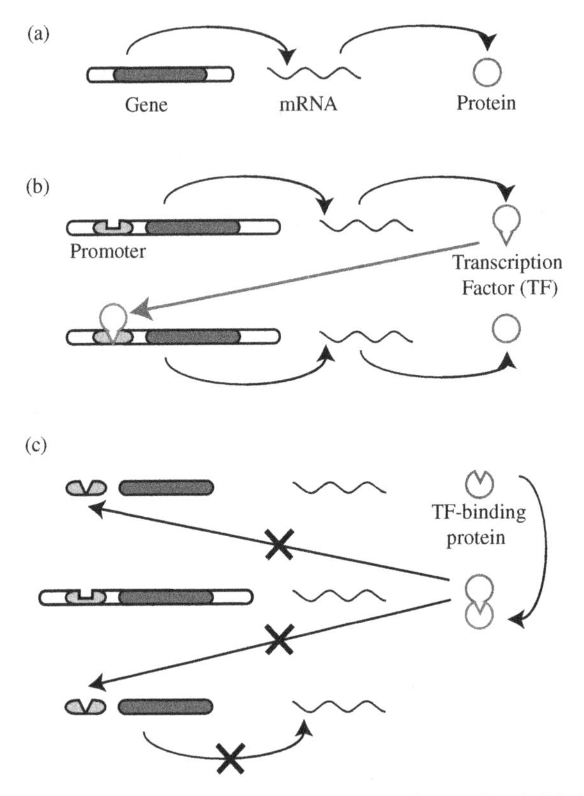

Fig. 1.1. The flow of information between genes and proteins is highly regulated in the cell by a network of interacting molecules. (a) In the simplest view, genes are transcribed into mRNA molecules, which are then translated into proteins. (b) The transcription of genes is controlled by transcription factors that bind to specific promoter regions that precede the genes themselves. (c) The activity of transcription factors can be regulated in a number of manners, including by binding with co-regulatory proteins that block their ability to bind to their corresponding promoters.

to specific promoter regions and control the transcription of the corresponding genes. As transcription factors are gene products themselves a backward flow of information from proteins to DNA is indeed present in the form of a gene-to-protein-to-gene control (Fig. 1.1b). For example, some genes encode proteins that bind to transcription factors, inactivating them. In this case the regulation implies a gene-to-protein-to-protein-to-gene control (Fig. 1.1c).

Furthermore, the cross-specificity between transcription factors and promoter regions is not extreme, allowing for some promiscuity. Indeed the same transcription factor can bind to several promoters and the same promoter can bind to several transcription factors. Moreover, promoter regions are often duplicated so that several genes might have the same promoters and are regulated by the same transcription factors (Fig. 1.1c).

Our understanding of the mechanism of regulation of gene expression is rapidly increasing. It has been clarified that RNA molecules have a variety of roles in the regulation of protein expression. For example, the recently discovered *small interfering RNA* (siRNA) and *microRNA* (miRNA) [513] molecules bind to the mRNA products of specific genes and inhibit or even enhance their further translation into proteins. This is a gene-to-RNA-to-gene form of control adding further layers of complexity to the way in which gene expression is regulated. It is thus impossible to represent such a rich pattern of relations as simple unidirectional flows. Rather, a network representation is particularly well suited to capture the feedback and feedforward regulation mechanisms that modulate gene and protein expression. Moreover, the network of gene regulation is highly dynamic in order to respond to the changing environment of a cell and to the different requirements through the cell cycle, and should therefore be considered in a time-dependent manner. Chapters 2 and 3 provide a description of the modes of regulation of gene expression, of their actuators and of the theoretical methods in use to reconstruct gene regulatory networks.

1.3 Protein–protein interaction networks

Protein–protein interaction network A protein–protein interaction network is a graph whose vertices are proteins and edges represent interactions between them, including for example those required to form macromolecular complexes or to establish signaling processes. Given this rather general definition, it is easy to understand why protein–protein interaction networks are becoming a very popular way to represent a variety of functions into the cell. Because of the disparate nature of these interactions, a variety of experimental techniques have been developed to detect them. These methods can be divided into physical (e.g. concentration, immunoprecipitation), library-based (e.g. protein probing, two-hybrid systems) and genetic methods (e.g. synthetic lethal effects) [474].

It is increasingly clear that the majority of proteins do not carry out their functions in isolation but by interacting together in complex manners [506]. There are two main ways to look at protein–protein interaction networks – physical and logical.

Physical interaction networks define interacting protein pairs according to physico-chemical principles. The characterization of this type of network is promoting strong links between structural biology and systems biology, and it is becoming possible to obtain such networks from experiments in which it is only known that a protein expressed by a given gene can interact with a certain set of other proteins, without the need of knowing the structures, or even the functions, of each of these proteins. The majority of such networks are currently *non-weighted*: only the presence or absence of specific associations are known, but not their strength.

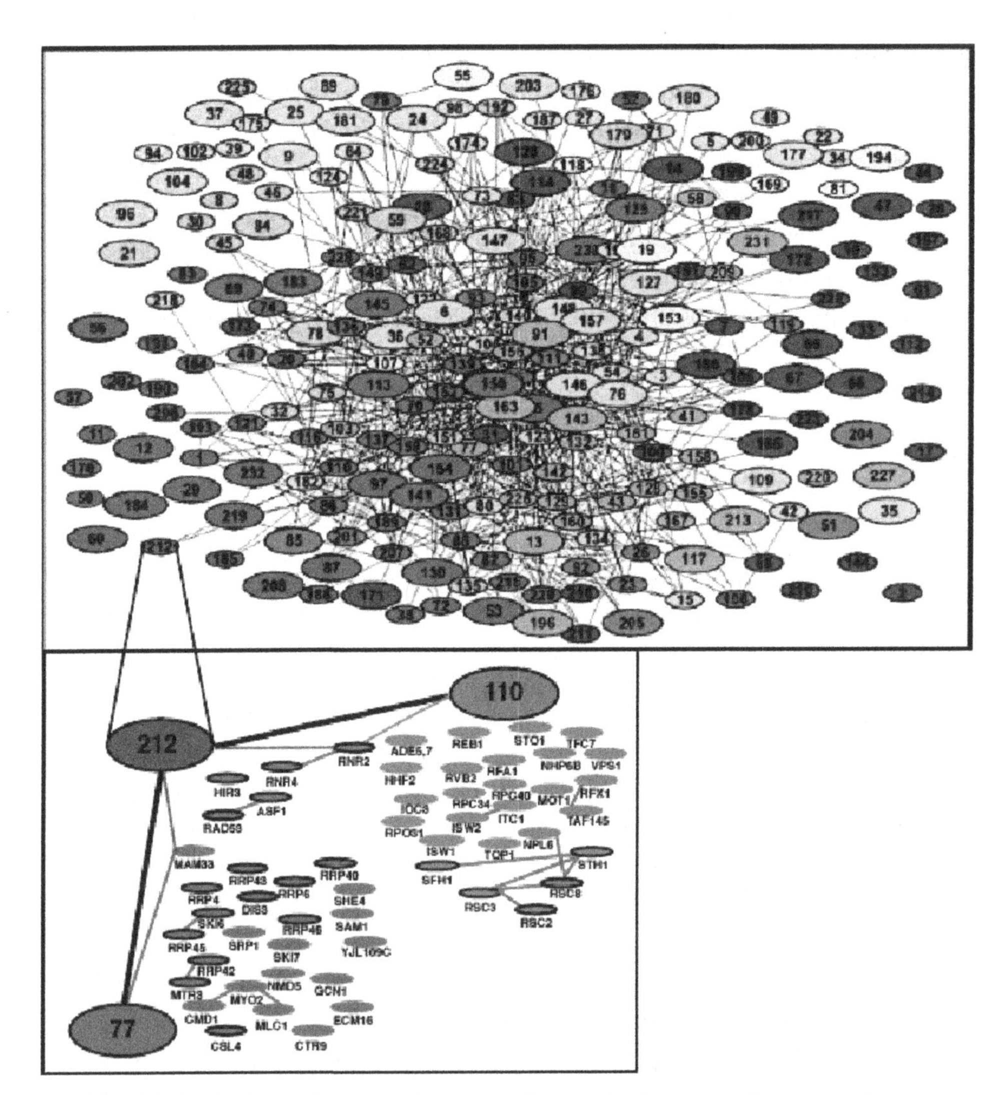

Fig. 1.2. Logical protein interaction network: proteins interact to form complexes, and different complexes may share the same proteins. When this is the case, network edges connect the complexes. Figure from [215].

Logical protein–protein interaction networks come into play once structures beyond the basic dimers become relevant. Indeed, the number of possible multimers grows geometrically and it is currently impossible to predict from basic principles all the possible assemblies. Mass spectrometry techniques allow to identify possible protein complexes, without strictly knowing which proteins are actually in contact with which other within the complex. Thus, these networks provide a glimpse of the logical organization of protein–protein interactions into functional collective units (Fig. 1.2).

Chapters 4 and 5 described how protein–protein interaction networks can be discovered by experiments and by theoretical inference.

1.4 Metabolic networks

Metabolism Metabolism refers to the ensemble of chemical processes through which living organisms transform resources taken from their environment in the molecules necessary for carrying out cellular functions. Since the products of a chemical reaction are often substrates (i.e. input molecules) of another one, the ensemble of metabolic processes can be conveniently organized as a network. The various chemical compounds present in the reactions are called *metabolites* and reactions are most often catalyzed by specific proteins called *enzymes*. In a metabolic network vertices represent metabolites and edges connect metabolites if they participate in the same reaction.

Simple metabolic processes consist of linear sequences (or pathways) of chemical transformations that take an initial set of molecules and transform them, by leaving by-products on the way, into different ones until the final products are obtained (Fig. 1.3a). A better exploitation of resources can be achieved by using the by-products of a pathway as inputs in other pathways (Fig. 1.3b). Metabolic pathways are indeed highly coupled, an organization that is facilitated by the fact that all the individual reactions take place within a confined space and therefore metabolites are readily exchangeable between different pathways (Fig. 1.4).

Since all reactions acting on the same substrates extract them from the same pool, a competition is permanently present among different pathways. This competition is regulated in a variety of ways, including a co-evolutionary fine-tuning of the enzymatic activities that allows all the reactions not to overexploit the resources to the detriment of the others; also, the different times at which different pathways might be active, thanks to the time-regulation of the process of enzyme expression,

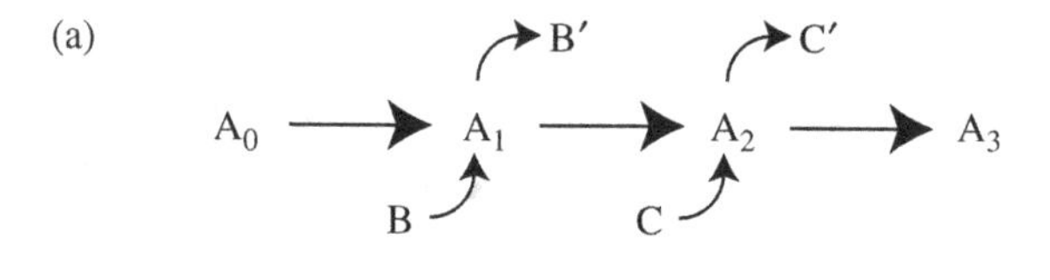

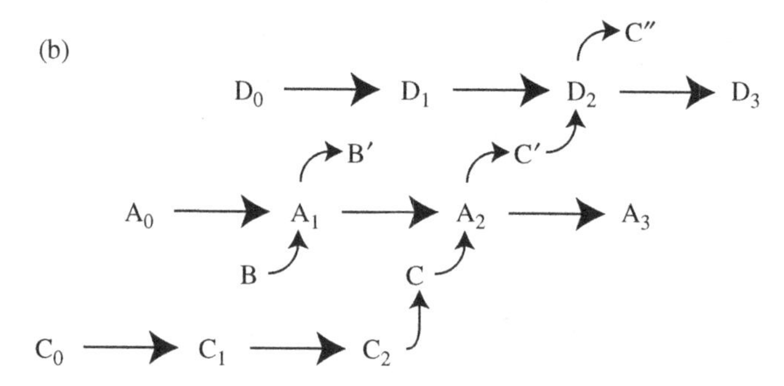

Fig. 1.3. Progressive intertwining of biochemical pathways. (a) A set of molecules A_0 is transformed at first into a second set A_1, through reactions that involve another set of molecules B, which in turn transforms into a new set B′. Then, A_1 is further transformed into a set A_2 through reactions with a set of reagents C, that is transformed into C′. After further reactions the end-product A_3 is achieved. (b) The set of molecules C that is the end-product of a reaction pathway, is coupled to the pathway leading to A_3, which can coupled to further pathways (e.g. leading from D_0 to D_3.

permit different reactions to take place at different times. Part of this fine-tuning is thus linked to the time regulation of gene expression outlined in the previous sections.

Moreover, metabolic networks must be flexible enough to allow even for large variations of their production rates, to cope with different, often significant, changes of the environmental conditions and hence of the needs of the cell. Also, they must be redundant, in order to make sure that the failure or degradation of a pathway is not, to as large a degree as possible, lethal to the cell.

Thus, there is a staggering number of requirements that evolution has had to satisfy while setting up metabolic networks, and it is likely that a full understanding of their rich structures and dynamical features will come only when most of such requirements will have been clarified and included in the theoretical models that we are developing.

Chapters 6 and 7 describe different theoretical techniques to analyze metabolic networks, to unveil their underlying hierarchical structure and to predict their behavior under different environmental conditions.

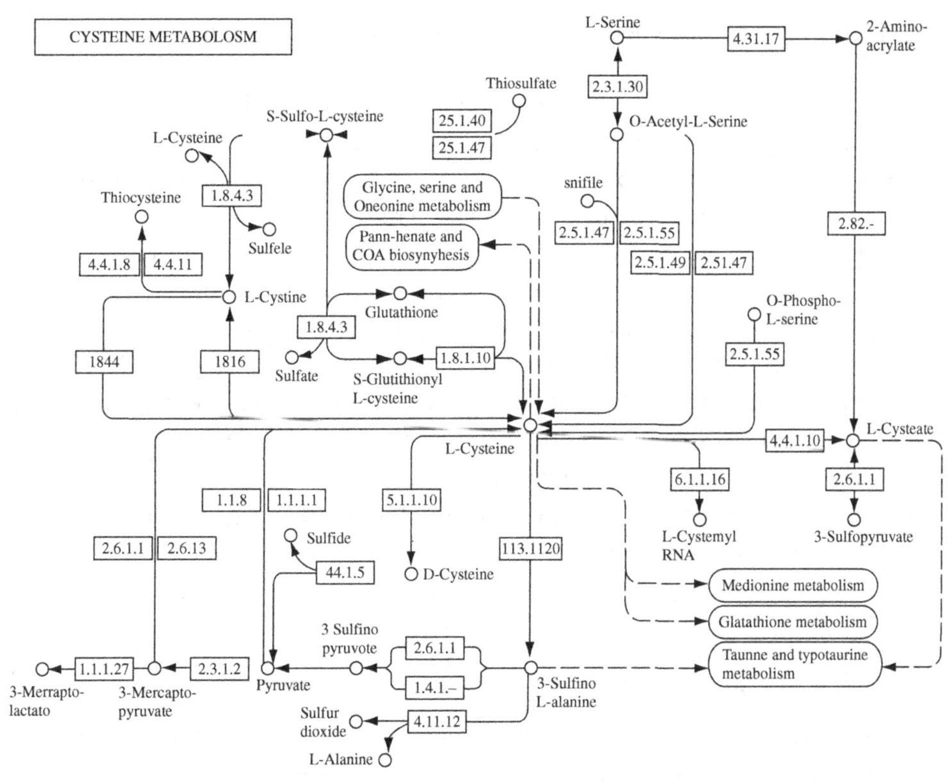

Fig. 1.4. Metabolic network for the synthesis of the amino acid cysteine (from the KEGG database, http://www.genome.jp/kegg/).

1.5 Signaling networks

> **Signaling networks** Signaling networks enable cells to sense changes in their environments and articulate the appropriate responses to them. These networks are made up by collections of interacting signaling pathways, which are cascades of biochemical reactions through which the signals corresponding to the external stimuli are transported to the repository of the genetic code, where they regulate the production of the specific proteins required to orchestrate the overall cellular response.

In order to survive, cells need to respond promptly to the challenges posed by their ever changing environments. They should therefore be able to sense the signals present in the environment and to react to them in the correct way. This response is achieved primarily by means of a series of sensor proteins located on the cell surface, which can alter their behavior upon changes in pH, temperature or salt concentrations, as well as upon binding to a variety of ligands.

After these signals have reached the cytoplasm, they must be collected and further reported where necessary, typically to genes that encode the proteins needed for the cellular response. For example, in many signaling pathways, sensor proteins are able to activate protein kinases, whose role is to phosphorylate (i.e. attach a phosphate group) other proteins, which as a consequence are activated or deactivated. The change of activity of such phosphorylated proteins ultimately results in the regulation of the expression of the genes responsible for generating the response of the cell to the specific external stimulus. Eventually a host of proteins are produced that not only provide the correct change of cell functionality, but also, down the line, shut off the signal, thus effectively down-regulating their own production once the cellular response is completed.

Chapter 8 is devoted to highlighting the different types of signal transduction in the cell and how they intertwine.

1.6 Networked networks and cell functionality

The networks that we have outlined do not work in isolation but, rather, in a highly concerted manner. Interactions between genes are mediated by RNA and protein molecules, whose building blocks (nucleic acids and amino acids, respectively) are the end-products of metabolic networks. In turn, the reactions present in metabolic networks (e.g. fatty acid synthesis [385]) are almost invariably catalyzed by enzymes that are end-products of gene regulatory networks, which involve a myriad of protein–protein interactions. Moreover, different modes of gene regulations are enacted depending on external and internal conditions (changes of temperature, nutrients, salts, etc.). The transduction of these signals depends on the regulation of protein–protein interactions, such as kinases with membrane proteins and with their targets, of complexes involving transcription factors and of the end-product proteins with the membrane proteins themselves to turn off the signal. Thus, gene expression networks are regulated by signaling networks and transduced via protein–protein interaction networks.

These are just a few examples of the seamless connection between the different classes of biological networks that are crucial to cell functionality. The ultimate goal of the network approach to cell biology is to understand the whole system within a unified framework. Although achieving this objective is still beyond our capabilities, great advances will be possible by approaches that enable different parts of the whole to be individually analyzed.

1.7 Concluding remarks

We have outlined the reasons for which networks are set to become a major paradigm in biology. Indeed the advantages of describing the interactions between the molecular components of a cell in terms of networks are becoming increasingly recognized. This approach enables to illustrate in an efficient manner the hierarchical organization of a cell, which consists of intertwined networks on a wide range of scales, from the interactions between individual metabolites or proteins to the interplay between metabolic and signaling networks.

2

Transcriptional regulatory networks

SARATH CHANDRA JANGA AND M. MADAN BABU

2.1 Introduction

Complex systems that describe a wide range of interactions in nature and society can be represented as networks. In general terms, such networks are made of nodes, which represent the objects in a system, and connections that link the nodes, which represent interactions between the objects. In mathematical terms, a network is a graph which comprises of vertices and edges (undirected links) or arcs (directed links). Examples of complex networks include the World Wide Webs, social network of acquaintances between individuals, food webs, metabolic networks, transcriptional networks, signaling networks, neuronal networks and several others. Although the study of networks in the form of graphs is one of the fundamental areas of discrete mathematics, much of our understanding about the underlying organizational principles of complex real-world networks has come to light only recently. While traditionally most complex networks have been modeled as random graphs, it is becoming increasingly clear that the topology and evolution of real networks are not random but are governed by robust design principles.

A number of biological systems ranging from physical interaction between biomolecules to neuronal connections can be represented as networks. Perhaps the classic example of a biological network is the network of metabolic pathways, which is a representation of all the enzymatic inter-conversions between small molecules in a cell. In such a network, nodes represent small molecules, which are either substrates or products of an enzymatic reaction, and directed edges represent an enzymatic reaction that converts a substrate into a product. Yet another cellular network which has been the focus of intense study in the last decade is the network of physical interactions between proteins and is usually referred to as the "protein interaction network." In such a network, nodes represent proteins and two nodes are connected if the proteins are experimentally determined

Networks in Cell Biology, ed. M. Buchanan, G. Caldarelli, P. De Los Rios, F. Rao and M. Vendruscolo. Published by Cambridge University Press.

to physically interact with each other. The focus of this chapter is on another important class of biological networks, referred to as the "transcriptional regulatory network." The expression of genes in all living systems, i.e. the synthesis of transcripts, can be positively or negatively regulated by the presence of other proteins called transcription factors (TFs). These TFs physically bind to specific sequences i.e. cis-regulatory motifs, near target genes (TG) and affect their expression. It is through the transcriptional regulatory network that a cell co-ordinates its response to both external and internal stimuli by controlling the expression of thousands of genes in appropriate amounts under different conditions and time. At an abstract level, all the regulatory interactions linking TFs to the set of transcriptionally controlled target genes (TGs) in an organism can be viewed as a directed graph, in which the TFs and TGs represent the nodes while the regulatory interactions that connect them are represented as directed edges (see Fig. 2.1). The resulting network is typically a complex, hierarchical, multi-layered graph that can be studied at several levels of detail. Though simple transcriptional regulatory networks involving only a few nodes were one of the first dynamical systems for which modeling attempts were carried out [311–313], substantial advance in our understanding of these networks and their modeling on a genomic scale has

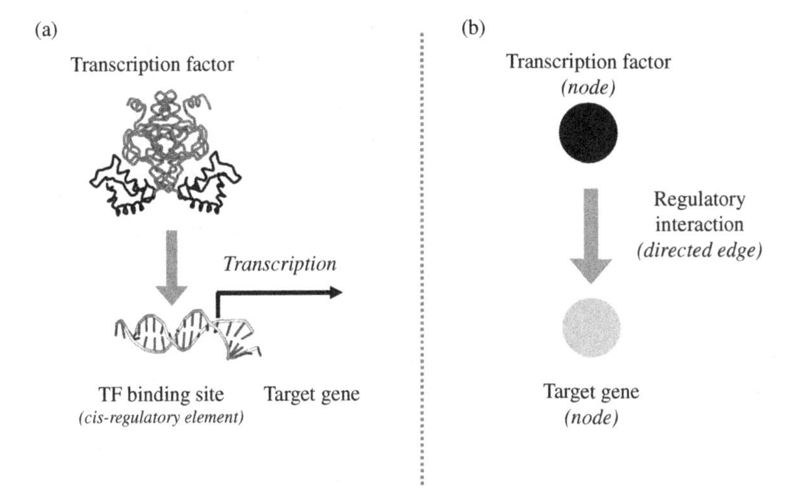

Fig. 2.1. Transcriptional regulatory interaction. (a) Standard view: regulation of gene expression is mediated by transcription factors that bind DNA through their DNA-binding domain. The specific DNA sequence to which they bind is called a TF binding site or cis-regulatory element. Binding to the cis-regulatory element may either activate or repress the transcription of a nearby target gene (black box). The arrow head represents the transcriptional start site. (b) Network view: transcription factors and target genes are represented as nodes and a directed link connecting a transcription factor to a target gene represents a regulatory interaction between the two.

Table 2.1. *Computer programs, databases and internet based platforms for investigating transcriptional regulatory networks*

Network visualization	**Website**
Pajek	http://vlado.fmf.uni-lj.si/pub/networks/pajek
Cytoscape	http://www.cytoscape.org/
Osprey	http://biodata.mshri.on.ca/osprey/index.html
GraphViz	http://www.graphviz.org/
H3Viewer	http://graphics.stanford.edu/munzner/h3/
Visant	http://visant.bu.edu/
Biolayout	http://cgg.ebi.ac.uk/services/biolayout/
Yed	http://www.yworks.com/
NetMiner	http://www.netminer.com/
Network analysis	**Website**
NEAT	http://rsat.ulb.ac.be/rsat/index_neat.html
Mfinder	http://www.weizmann.ac.il/mcb/UriAlon/groupNetworkMotifSW.html
FanMod	http://www.minet.uni-jena.de/wernicke/motifs/
Clique finder	http://topnet.gersteinlab.org/clique/
MCode	http://cbio.mskcc.org/ bader/software/mcode/index.html
Cytoscape	http://www.cytoscape.org/
Vanted	http://vanted.ipk-gatersleben.de/
Biotapestry	http://www.biotapestry.org/
TYNA / Topnet	http://tyna.gersteinlab.org/tyna/
NCT	http://chianti.ucsd.edu/nct/
Bioconductor	http://www.bioconductor.org/
Databases	**Website**
RegTransBase	http://regtransbase.lbl.gov/cgi-bin/regtransbase?page=main
RegulonDB	http://regulondb.ccg.unam.mx/
DBTBS	http://dbtbs.hgc.jp/
Coryneregnet	http://www.coryneregnet.de/
Prodoric	http://www.prodoric.de/
Yeast regulatory network	http://www.mrc-lmb.cam.ac.uk/genomes/madanm/tfcomb/tnet.txt

Table 2.2. *Genome-scale experimental methods to probe protein–DNA interactions*

Method	Description
Chromatin-Immunoprecipitation-Chip experiments (ChIP-chip) Chromatin-Immunoprecipitation-Sequencing experiments (Chlp-seq)	The DNA binding protein is tagged with an epitope and is expressed in a cell. The bound protein is covalently linked to DNA by using an in vivo cross-linking agent such as formaldehyde. After cross-linking, DNA is sheared and the protein–DNA complex is pulled down using an antibody against the tag. Reversal of the cross-link releases the bound DNA, allowing the sequence of the bound fragments to be determined by hybridization to a microarray (ChIP-chip) or by sequencing (ChIP-seq). In ChIP-chip experiments, intergenic regions are spotted on to a microarray chip. Following a chromatin immunoprecipitation step, the bound fragments are reverse cross-linked and hybridized onto the chip. Complementary sequences will bind to specific spots on the chip, thereby providing the exact intergenic region to which the protein was bound [90, 265, 355]. In ChIP-seq experiments, the bound fragments are directly sequenced using 454/Solexa/Illumina sequencing technology. The sequences are then computationally mapped back to the genome sequence. Fragments that were bound by the protein will be enriched and hence sequenced several times, providing a direct measure of its enrichment [194, 269, 296].
DNA adenine methyltransferase Identification (DamID)	To overcome any potential non-specific cross-linking of protein to DNA as could happen with ChIP-chip experiments, the DamID technique was introduced. The protein of interest is fused to an *E. coli* protein, DNA adenine methyltransferase (Dam). Dam methylates the N6 position of the adenine in the sequence GATC, which occurs at reasonably high frequency in any genome (1 site in 256 bases). Upon binding DNA, the Dam protein preferentially methylates adenine in the vicinity of binding. Subsequently, the genomic DNA is digested by the DpnI and DpnII restriction enzymes that cleave within the non-methylated GATC sequence, and remove fragments that are not methylated.

Table 2.2. *(cont.)*

Method	Description
	The remaining methylated fragments are amplified by selective PCR and quantified using a microarray [233].
Protein binding universal DNA microarrays (PBMs)	In contrast to the methods described above, this is an in vitro method to probe protein–DNA interaction. A DNA binding protein of interest is epitope tagged, purified and bound directly to a double-stranded DNA microarray spotted with a large number of potential binding sites. Labeling with fluorophore conjugated antibody for the tag allows detection of binding sites from the significantly bound spots [91].

come only recently. Despite the enormous interest in understanding transcriptional networks across organisms, our knowledge on transcriptional interaction graphs for a genome has been very limited and is mostly restricted to model organisms such as *Escherichia coli* and *Saccharomyces cerevisiae* (see Table 2.1). Transcriptional interactions between a TF and a TG are either identified from small-scale assays that are documented in regulatory network databases through extensive manual curation efforts (see Table 2.1 for databases) [5, 60, 409, 515] or are obtained from high-throughput screens (see Table 2.2 for high-throughput methods to probe protein–DNA interactions). These high-resolution high-throughput methods directly infer in vivo binding of a transcription factor to promoter regions of targets. This can be achieved by methods such as ChIP-chip or ChIP-seq, allowing the identification of regulatory interactions for a vast set of TFs in an organism on a genomic scale [6, 356]. Alternatively, lower-resolution high-throughput approaches that identify genes which are differentially expressed directly or indirectly can also be carried out. Such methods identify the set of genes which are differentially expressed using gene expression arrays upon over-expression or deletion of the transcription factor [112, 142]. The results of these experiments are best represented as the transcriptional regulatory network (TRN), where nodes represent TFs or TGs and links represent direct binding of the TF to the promoter region of a target or differential expression of the target upon deletion or over-expression of the transcription factor.

In this chapter, we will first provide a brief introduction to transcriptional regulation in prokaryotes and eukaryotes. We will then provide a detailed discussion of our current understanding of the structure of the transcriptional regulatory network. We will then discuss the evolution of regulatory networks both within and across different organisms. Finally, our understanding of the dynamics of transcriptional regulatory networks and future questions that remain to be addressed will be discussed.

2.2 Transcriptional regulation in prokaryotes and eukaryotes

Regulation of gene expression at the transcriptional level is a fundamental mechanism that is evolutionarily conserved in all cellular systems [482]. This form of regulation is typically mediated by transcription factors (TFs) that bind to DNA through their DNA-binding domain and either activate or repress the expression of nearby genes [87, 684](see Fig. 2.1). The specific DNA sequence to which a TF binds and through which it regulates expression of the TG is called a transcription factor binding site or a cis-regulatory element. TFs can be classified as activators, repressors or dual regulators depending on their mode of action on a particular promoter [87, 290, 383, 421, 471]. An activator induces the expression of its target gene by directly recruiting or enhancing the activity of the RNA polymerase at the promoter. Activation is known to occur typically by the binding of TFs upstream of the transcription start site and often upstream of the -35 promoter element [87, 383, 421]. For negative control of transcription, TFs act as repressors by binding to DNA to prevent RNA polymerase from initiating transcription. Repression normally occurs when TFs bind downstream of the transcription start site, thereby causing DNA looping or by binding in between the -35 and -10 elements of the promoter region, thereby sterically blocking the RNA polymerase from binding and initiating transcription [87, 383].

Although the process of transcription is conserved across all cellular life forms and is carried out by homologus RNA polymerases, the mechanisms of regulation involved and their related machineries vary considerably among the kingdoms of life, especially between bacteria and eukaryotes. In bacteria, the absence of a nucleus and the organization of functionally related genes into Operons (i.e. adjacent genes which are transcribed in the same messenger RNA but translated independently) facilitate transcription and translation to occur in the same cellular compartment. Since the genetic material in bacteria is primarily packaged in a single circular chromosome, the control of transcription in prokaryotes can be considered primarily to occur at the DNA sequence level through the use of cis-regulatory elements. However, the genome of eukaryotes is contained in the nucleus, condensed in a complex, hierarchical manner and is encoded in several different linear chromosomes. Therefore, the process of transcriptional regulation is highly complex and is known to be coordinated at least at three major levels. The first is at the DNA sequence level, i.e. the linear organization of transcription units and regulatory sequences. The second is at the level of chromatin, which allows switching between different functional states, i.e. between a state that suppresses transcription and one that is permissive for transcription. This level involves changes in the chromatin structure and is controlled by the interplay between histone modification, DNA methylation, and a variety of repressive and activating mechanisms (i.e. epigenetic regulation). This regulatory level is linked with the control mechanisms from the previous level that switch individual genes in the cluster to on and off, depending on the properties of the promoter. The third is at the level of whole nucleus, which includes the dynamic 3D spatial organization of the chromosomes inside the cell nucleus. The eukaryotic nucleus is structurally and functionally compartmentalized and epigenetic regulation of gene expression may involve repositioning of loci in the nucleus through changes in large-scale chromatin structure. All these differences add another layer of complexity and control to the inherent structure of transcriptional networks in eukaryotes. Even though there are fundamental differences in several basic principles, their organization, structure and evolution from a network perspective have been demonstrated to be similar in both the kingdoms of life [236, 356, 417, 558, 593, 680]. This will be discussed in the following sections.

2.3 Structure of transcriptional regulatory networks

One of the recent important developments in our understanding of gene regulation has been the application of graph theory to investigate transcriptional networks on a genomic scale [42, 55]. The accumulation of data on many factors that control the expression of genes or groups of genes together with the increased use of high-throughput techniques, such as DNA arrays and proteomics, has generated an overwhelming amount of data that has to be understood to infer relationships between genes, and between genes and signals. The reductionist approaches of molecular biology are not amenable to deal with large amounts of information giving rise to the increasing use of the notion of networks in biology. One major advantage of this approach is that it allows subjective description of complexity (number of interaction partners, weak or strong interaction, etc.) to be replaced by objectively quantifiable, numerical parameters (e.g. connectivity or strengths of interactions)[294, 508].

Most network analysis of transcriptional regulatory events in an organism involves representing genes and the proteins they encode as nodes. However, it should be noted that in contrast to protein interaction networks, transcriptional networks are directed. In such graphs, an edge has a starting node and a target node which captures the fact that a transcription factor regulates a target gene. Though more complex representations, which include the incorporation of other entities such as small molecules, RNA encoding genes, signal-transduction pathways or interacting proteins, are possible, most of our understanding of transcriptional networks has been limited to the holistic view of TFs and TGs as nodes and the regulatory interactions between them as edges. In this chapter, unless otherwise stated, we will be using "transcriptional regulatory networks" (TRNs) or "transcriptional networks" to refer to such a representation of the regulatory interactions in an organism.

2.3.1 Global level

At the global level, the overall structure or topology of the TRN can be described by parameters derived from graph theory. One of the important parameters that can be obtained is the connectivity, i.e. how many connections a node has, and the connectivity distribution, i.e. how many nodes have a particular number of connections. In the case of transcriptional networks we can calculate the incoming and outgoing connections separately. The incoming connectivity is the number of transcription factors regulating a target gene, which quantifies the combinatorial effect of gene regulation. For instance, the maximum number of incoming connections in the currently available transcriptional networks of *Escherichia coli* and *Saccharomyces cerevisiae* stand at 9 and 31. The fraction of target genes with a given incoming connectivity was observed to follow an exponential distribution in both *Escherichia coli* and *Saccharomyces cerevisiae* [236, 593]. The exponential behavior indicates that most target genes are regulated by similar number of factors. This could reflect the limits on the number of transcription factors that can affect a target gene owing to the constraints on the intergenic spacing available and the number of proteins that can simultaneously act in a promoter region. On the other hand the outgoing connectivity, which is the number of target genes regulated by each transcription factor, was found to be distributed according to a power law. This is indicative of a hub-containing network structure, in which a select set of transcription factors participate in the regulation of a disproportionately large number of target genes. These hubs can be viewed as "global regulators," as opposed to the remaining transcription factors that can be considered as "fine tuners." Global regulators, which were traditionally defined with ambiguous concepts [230], can now be defined based on the number of genes they regulate [41, 558], i.e. the number of target genes regulated by them. Others define global regulators in the model bacterium *E. coli* by taking into account additional factors such as the number of co-regulators and the number of conditions in which they are regulated [401]. Using this definition, seven global regulators have been identified that regulate more than 50% of all the genes in the entire network that is currently known for *E. coli*.

Since target genes encoding transcription factors can themselves be regulated by other transcription factors, the TRN is not a simple, flat network. Instead, it has been shown to possess a multi-layer hierarchical structure (see Fig. 2.2). Such a hierarchical structure has been determined using both top-down and bottom-up

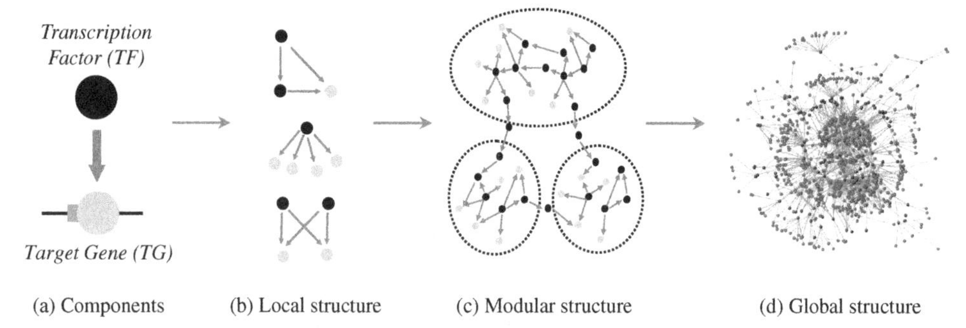

Fig. 2.2. Structure of the transcriptional regulatory network. Nodes represent transcription factors or target genes and directed edge indicates a regulatory interaction between the two. (a) Components of a regulatory interaction (b) Local structure of the network consists of patterns of inter-connections called network motifs. The three frequently occurring motifs are the feed-forward motif (top), single input motif (middle) and multiple input motif (bottom) (c) Motifs are interconnected to form groups of highly connected genes, referred to as regulatory modules (dashed circles) (d) the set of all regulatory interactions in a cell is referred to as the transcriptional network. (Adapted from [42].)

approaches [382, 680]. Analysis of the hierarchical structure has shown that the TRNs do not possess feedback regulation at the level of transcription. This means that TFs in the bottom of the hierarchy do not regulate TFs that are in the top, indicating the prevalence for alternative forms of feedback control of transcription. Typically such a feedback occurs through the usage of protein–protein interactions at the post-translational level or due to a complex interplay of cellular entities which control the activity of TFs by changing their conformation depending on the continuously varying intra- and extra-cellular conditions [402, 680]. It has also been observed that the TFs in the middle of this hierarchy (often from the levels 2 and 3 measured from the bottom) regulate more direct targets than those at the top suggesting that these middle level TFs act as managers and are indeed bottlenecks for cellular transcriptional response [680]. Based on our knowledge of protein interaction networks, one would expect that TFs at the higher levels of this hierarchy are more likely to be important for the survival of the cell [292]. However, surprisingly, it has been found that the TFs at the bottom level are more likely to be lethal and hence essential to the viability of the cell [680]. This has been attributed to the existence of indirect backup, where multiple TFs contribute to the functioning of a process, in regulatory networks [47]. The latter work also shows that the robustness in regulatory networks for mutations that disrupt regulatory hubs is mainly contributed because of this indirect backup and could be an inherent architecture of several transcriptional programs. Such a topology has been proposed to allow organisms to explore the adaptive landscape in changing environments. The multi-layer hierarchical transcriptional networks form the basis for investigating modules and motifs [51, 146, 277, 278, 382, 417, 500, 544, 558, 665], which are discussed in the next sections.

2.3.2 Modular level

The organization of the regulatory network can also be captured at an intermediate level by examining its modularity (Fig. 2.2). Although there is no general agreement on the definition of a regulatory module [659], a transcriptional regulatory module is typically defined as a set of genes that are regulated by a common set of TFs. Under this definition, it is intuitive to expect that various cellular processes will be regulated by discrete and separable modules that can co-ordinate the activities of many genes and carry out complex functions. Therefore, identifying transcriptional modules is useful for understanding cellular responses to internal and external signals under different cellular conditions. A first attempt to characterize the modular properties of regulatory networks was performed by measuring the extent of clustering in a network by calculating the clustering co-efficient. It was found that the clustering in the transcriptional network was five fold higher than what is expected by chance in a random network, suggesting an extensive inter-connection between the nodes in the network. This therefore provided indirect evidence for the existence of modularity in regulatory networks [236].

Datasets of genome-wide gene expression, location analysis (ChIP-chip), combination of motif-discovery methods and Bayesian network inference are all used to identify transcriptional modules controlling a variety of cellular processes [51, 278, 544, 584, 665]. Several of these studies have focused on yeast and other model organisms [51, 277, 278, 476, 544]. Despite several methods that have been developed to identify regulatory modules from expression data, the most frequently used implementations take into account that genes co-expressed in similar conditions are likely to belong to the same set of regulatory modules. More sophisticated approaches integrate additional data sources like TF binding data, motif information or functional annotation.

Although the different approaches to identify modules provide different results in terms of the number and size of the resulting modules, the general consensus has been that regulatory networks are highly interconnected and very few modules are entirely separable from the rest of the network. The major understanding has been that modules are frequently nested within each other in a hierarchical fashion at different levels. In fact, an analysis of the distribution of the commonly seen motifs (see Section 2.3.3 Local level), which are recurrent patterns of interconnections in the transcriptional networks [20], in the identified modules, suggests that network motifs themselves do not exist in isolation but rather integrate to form part of the modules by sharing some of their edges [146, 500]. Thus, many small, highly connected motifs group into a few larger modules, which in turn cluster into even larger ones. These nested modules are interconnected through local regulatory hubs. Such an organization not only explains the hierarchical organization, which is seen in other cellular networks [493] but also intuitively suggests the capacity for rapid regulatory changes through regulatory hubs with integration and fine tuning of the regulatory processes by downstream TFs, thereby linking the modules in a hierarchical manner.

2.3.3 Local level

At the local level, certain sub-graphs or patterns of interconnections appear more often than expected by chance and have been referred to as network motifs. Network motifs were originally described in *E. coli* TRN but the same patterns were subsequently found to occur in the TRN of yeast and other organisms [20, 558]. Three network motifs were found to predominantly occur in the transcriptional networks. (i) Feed-forward loop (FFL), in which a top-level transcription factor regulates both the intermediate-level TF and the target genes, and the intermediate-level TF regulates the target gene. FFL appears to be the most abundant motif among the best studied transcriptional networks. FFLs have been further classified into eight motif sub-types and two of them namely coherent type-1 and incoherent type-1 FFL appear to be much more predominant than others [20, 391]. The former was shown to act as a sign-sensitive delay element and a persistence detector while the latter was demonstrated to function as a pulse generator and response accelerator [392, 393]. (ii) Single-input module (SIM), in which a single transcription factor regulates several genes. For the genes regulated by a SIM, depending on the promoter strength of the regulated genes, it may respond to different concentration levels of the active TF. Therefore, if the concentration of the active TF changes with time, it has been shown that such a motif could set a temporal pattern in the expression of the individual targets [304, 685]. (iii) Dense overlapping regulons (DORs) in which several TFs regulate overlapping sets of genes. Since the individual TFs could potentially respond to different signals, such motifs could therefore integrate diverse signals and bring about differential expression of the relevant targets. Thus regulation of genes via network motifs provides distinct ways of regulating gene expression. Although motifs form over-represented sub-graphs in the entire network of transcriptional regulation, they do not appear independently but rather integrate to form superstructures that carry out a common biological function by sharing some of their edges [146, 500].

2.4 Evolution of transcriptional regulatory networks

With the availability of completely sequenced genomes and documented experimental evidence on transcriptional regulation for a number of the TFs in several model organisms [5, 409, 515], it has now become possible to address questions on the evolution of the structure and components of regulatory systems within and across organisms (see Table 2.3). Evolutionary relationships among genes either within a given genome or across a group of organisms are normally studied using sequence analysis tools such as BLAST [27] and domain assignments using Hidden Markov Models (HMMs) of known families [160]. While various implementations of BLAST are frequently used to study highly conserved sequences, remote homologs among evolutionarily distant genomes or within a given genome is typically analyzed using different implementations of HMMs.

2.4.1 Evolution of transcription factors and cis-regulatory elements

As such, TFs form one of the largest protein groups in most genomes with their fraction typically scaling as the square of the total gene number of a genome [33, 612]. Most TFs usually have at least one DNA-binding domain and one or more partner domains. The DNA-binding domains seen in transcription factors have been classified into a small group of families based on the structural folds [340, 654]. The assignment of the DNA-binding domains to different structural domain families allows one to assess the evolutionary relationships among transcription factors. In fact it has been shown in all the three domains of life that the repertoire of DNA-binding domains used by TFs comes from a relatively small, ancient conserved collection [34, 41, 354, 501]. Of these, the winged-helix domain and the zinc ribbon are encountered in all three principal superkingdoms of life whereas the ribbon-helix-helix domain is found only in the prokaryotes [34]. Most higher eukaryotes display a proliferation of several novel DNA-binding domains, such as the C2H2 zinc fingers, the AT hooks, the HMG1 domain and the MADS box [343]. DNA-binding regions of prokaryotic TFs based on structural domains comprise at least three folds, the helix-turn-helix, the winged helix and the beta ribbon [270] with the most abundant being the classical helix-turn-helix domain [33]. Analysis of the sequences of TFs from several model organisms has suggested a major role for recurrent, massive and lineage-specific expansions in the evolution of transcription factors in eukaryotes [129, 363]. In prokaryotes, several orthologous groups of transcription factors show a much wider spread across phylogenetically diverse organisms, suggesting a role for horizontal transfer, in addition to diversification through a lower level of lineage-specific duplication.

The availability of complete genome sequences of multiple strains of a single organism and those of phylogenetically closely related species has allowed investigation of a number of questions related to the evolution of cis-regulatory elements. Investigation of the inter-genic sequences and promoter regions of several genes in various organisms has revealed that cis-regulatory elements tend to be evolutionarily conserved upstream of orthologous genes in closely related species [414] and that upstream regulatory elements of duplicated genes tend to evolve faster. Genome sequences not only provide us with evolutionary insights on the conservation of cis-regulatory elements such as promoter regions and TF binding sites across organisms, but also enable us to predict them using a variety of comparative genome analysis techniques. Two general computational approaches have emerged for inferring TF binding sites in promoter regions: (i) analysis of co-regulated sets of genes and (ii) phylogenetic foot-printing of the upstream regions of orthologous genes in closely related genomes. Both methods aim to identify statistically significant patterns which are conserved in the background of the aligned inter-genic regions upstream of orthologous genes. The former method exploits the principle that genes which are regulated by the same transcription factor should have similar upstream binding sites. The latter method exploits the finding that selective pressure for gene regulation would conserve cis-regulatory elements in the inter-genic region over the background non-coding DNA among organisms at short evolutionary distances [116, 245, 315, 412, 480, 489]. Recent approaches exploit the fact that most prokaryotic transcription factors bind to DNA as spaced dimers [369, 426], while others use the observation that TFs often bind cooperatively to their targets, and hence statistically over-represented motif co-occurrence patterns were used to identify novel TF-TG associations [91].

2.4.2 Transcriptional network evolution within an organism

Recent studies have examined the effect of gene duplication on the evolution of transcriptional networks of an organism [404, 462, 591]. This has been assessed from three possible scenarios: duplication of the transcription factor, duplication of the target gene with its regulatory region and duplication of both the transcription factor and its target gene. Following duplication of a transcription factor, both copies of the factor will regulate the same target genes, until regulatory interactions are gradually gained or lost. Similarly, after duplication of a target gene, both copies will be regulated by the same set of transcription factors. Several groups compared the effect of gene duplication in the context of their transcriptional regulation and expression patterns. The common observation in such studies was that there is significant similarity in the transcriptional regulation and expression patterns of the duplicated genes. One group investigated paralogous genes in yeast and showed that the number of shared regulatory motifs in the duplicates decreases with evolutionary time, whereas the total number of regulatory motifs remains largely unchanged [462]. Another group found that upstream regulatory elements of duplicated genes evolve faster and demonstrated that, as the sequence identity of the duplicate genes decreases, the number of shared TFs between them also decreases [404]. In another study, paralogous transcription factors and target genes in *E. coli* and yeast were defined according to the domain assignments based on protein folds so that distant relationships can be identified [591]. This work showed that duplication of TFs and their TGs had given rise to a significant proportion of the currently known regulatory networks in both *E. coli* and *S. cerevisiae* [591]. The authors proposed that about 45% of the regulatory interactions in both organisms arose by duplication with inheritance of interaction. Surprisingly they did not find any strong evidence for the evolution of network motifs as a consequence of the duplication model proposed by them, suggesting that the network motifs might have arisen by convergent evolution [122, 591]. These observations suggested that, though gene duplication has had a key role in the evolution of the transcriptional network, the topology of the regulatory networks cannot be explained by a simple duplication–inheritance model. Apart from gene duplication and inheritance of regulatory interactions, significant contributions from other factors such as the continuous gain and loss of interactions after gene duplication have shaped network topology.

Rewired bacterial gene networks show extensive robustness A recent study that exploited the ability to synthesize genetic networks on a small scale [237] extended the strategy to understand artificially evolved networks in *E. coli* [282]. The study, which constructed 600 recombinations of promoters with different transcription and sigma factors, shows that most of the evolved networks are not lethal with majority of them not showing alterations in growth. The authors demonstrate experimentally that certain networks consistently survive over the wild-type, suggesting that addition of new links to the network is often not a barrier for evolution and that sometimes such events can offer a fitness advantage to the cell.

2.4.3 Transcriptional network evolution across organisms

From a cross-genome perspective one could ask how the networks evolve across genomes and if parts of the regulatory networks are conserved across organisms. This points to an important question: if there is a regulatory interaction between a transcription factor and target gene in one organism, does the orthologous transcription factor regulate the orthologous target gene in another organism? It has been shown in yeast, worm and fly that the regulatory interactions tend to be conserved if the sequences of the TFs are sufficiently similar. Regulatory interactions inferred by this way were called “regulogs” [681]. Therefore,

Table 2.3. *Evolution of transcriptional regulatory network*

	Transcriptional network evolution within an organism	Transcriptional network evolution across organisms
Components	Transcription factors and target genes have evolved as a consequence of duplication	Transcription factors are less conserved (i.e. evolve faster) than target genes
Local structure	Network motifs did not evolve by duplication of ancestral circuits, but via convergent evolution	Network motifs are not conserved as rigid units. However, organisms with similar lifestyle tend to conserve similar network motifs
Global structure	Transcriptional hubs did not evolve as a consequence of duplication of target gene followed by inheritance of regulatory interactions, but via independent gain of new interactions	Condition specific transcriptional hubs may be lost or replaced in evolution. However, organisms with similar lifestyle tend to conserve hubs and regulatory interactions
Key conclusions	Though gene duplication has played a key role in network evolution, the topology of the regulatory network cannot be explained by a simple duplication–inheritance model. Significant contributions from other factors such as the gain and loss of interactions after gene duplication have shaped network topology	Though transcriptional networks are extremely flexible in evolution, the lifestyle and the environment of the organism constrains network evolution. Despite these constraints, the presence of similar network topology in TRNs of different organisms suggests convergent evolution of network topology
Principal conclusion	Organisms evolve their transcriptional regulatory network rapidly and adapt to changing environments by tinkering individual regulatory interactions	

essentially a regulog corresponds to the transfer of annotation of a regulatory interaction between a TF and its target gene, from one organism to another. Owing to the predictive capacity of the regulog concept to identify putative regulatory interactions in an organism, it has been used by other recent studies to show that transcriptional regulatory networks evolve rapidly across genomes [43, 377]. These studies also demonstrated that TFs are less conserved across genomes than their target genes and are likely to evolve faster than their TGs [43, 377]. Interestingly, it was also found that global regulators which modulate the activity of a vast number of genes are not more conserved than general TFs suggesting a possible scenario for rapid evolution of gene regulatory mechanisms across bacteria. One of the above studies also observed that regulatory interactions within a network motif do not show any preference to evolve together and that organisms with a similar lifestyle are likely to preserve equivalent regulatory interactions and network motifs. These works demonstrate that regulatory networks are extremely flexible across organisms and evolve in a step-wise manner, with loss and gain of individual interactions probably happening more frequently than loss and gain of whole motifs or modules. This is interesting as well as important from an evolutionary perspective because it provides a contrasting view to the proposal that modular gain and loss of functions is easier than acquiring a particular function [349, 411, 566].

2.5 Dynamics of transcriptional regulatory networks

Cells continuously sense and respond to environmental changes and internal signals in a highly noisy environment and this is possible because of a complex cascade of interactions of bio-molecules with the transcription factors which ultimately controls the state of TFs themselves over a period of time. The interplay between the cis-regulatory elements and transcription factors provides a plethora of transcriptional programs which ultimately control the state of every gene in the cell tailored for different conditions. Despite several studies that focus on regulatory networks at a static level, it should be noted that the regulatory network of an organism is highly dynamic and different sections of the network are actively used under different conditions [183, 381]. Although informative, the topological properties of regulatory interactions themselves explain little of how a transcriptional regulatory network functions. To understand the principles governing the logic behind how various parts of the static network operate dynamically in a given condition, one must investigate the changes in the network structure of an organism across different conditions at a global level and at the level of individual regulatory interactions. Accordingly, recent studies to understand the dynamics of gene regulation can be split into two distinct categories, firstly those concentrated on understanding the large scale changes in gene regulatory interactions across different cellular conditions or over a period of time and secondly those focused on identifying the principles governing the functioning of local structures like motifs or circuits in the context of a larger system.

2.5.1 Temporal dynamics of transcriptional networks

Cells must have the ability to respond to most dynamical changes, whether simple (involving the changes in exogenous conditions such as variations in temperature or nutrient concentrations) or complex (involving the simultaneous change of many conditions). In order to properly process and respond to complex environmental changes, organisms use distinct transcriptional regulatory (TR) sub-networks, governed by sets of sensors which are specialized to detect complex environmental stimuli. By integrating gene expression data across five different conditions with the static transcriptional regulatory network, it was shown in yeast that the active sub-networks for the different conditions vary significantly both in terms of their local and global structure (Fig. 2.3) [381]. The five conditions were cell cycle and sporulation (both of which are developmental regulatory programs in a cell), as well as DNA damage, stress response and diauxic shift (all three of which are regulatory programs that are important for survival). The authors of this study identified the condition-specific transcriptional network by linking TFs present in a given condition to their differentially expressed target genes. TFs were classified as "present" or "absent" on the basis of their abundance in a given condition relative to their starting abundance. Although a TF might physically bind to its target site, the corresponding link was not considered active if the expression of the target gene did not change significantly, or if the TF abundance was low under the specific condition. Analyses of the resulting sub-networks revealed that the majority of regulatory interactions are condition-specific, and only a small subset is active in four or more conditions. Importantly, the authors found that most hubs are transient and their expression is not maintained between conditions. Only a small percentage of hubs were found to maintain a high out-degree in all conditions. However, even these hubs that maintained their high out-degree across conditions were found to switch their targets between conditions. As a result of extensive rewiring, the same TFs can be used in different conditions to regulate the expression of various sets of genes and to elicit a condition-dependent response, which implies that the response of the cell is commonly a result of combinatorial TF usage. Taken together, the authors suggest that the transcriptional networks are extremely dynamic and rapidly rewire their interactions to respond efficiently to changes in the external and internal environment.

In another genome-wide study in yeast [245], the authors determined the location of TF binding sites in the promoter regions for 203 TFs in rich medium. In addition, location data were obtained for 84 TFs in at least one of the 12 different environmental conditions. By investigating the resulting network, the authors addressed questions on the organization of regulatory elements in promoters in the context of their environment-dependent usage (Fig. 2.3). By compiling information from TF binding data, phylogenetically conserved sequences, and prior knowledge, they were able to infer a detailed transcriptional regulatory network for 102 TFs. In contrast to the previous approach, this approach experimentally identified all promoters that are bound by TFs in the different conditions. On the basis of the presence of TF binding sites, the authors classified the identified promoter regions

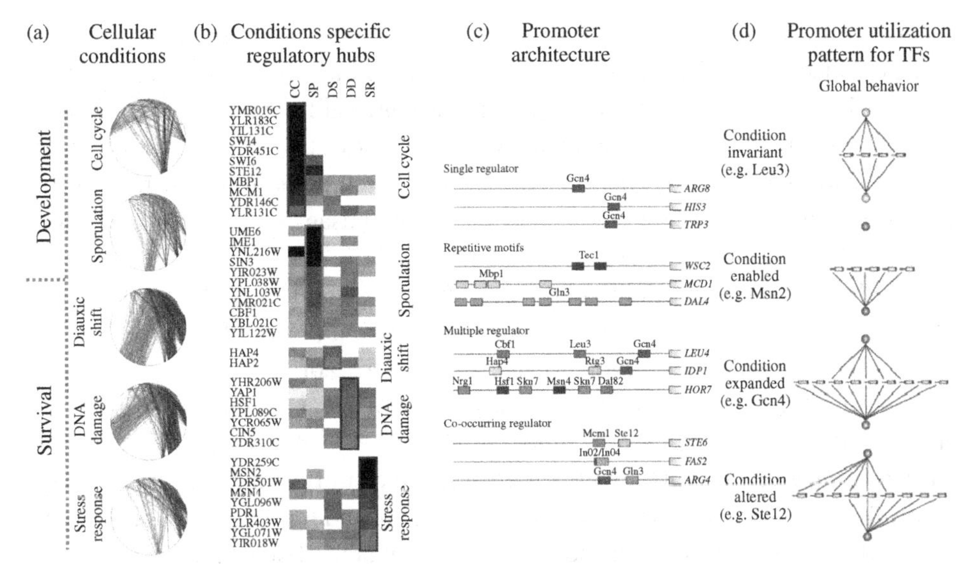

Fig. 2.3. Temporal dynamics of transcriptional regulatory network in yeast. (a) The active transcriptional regulatory network in a given cellular condition for the five major cellular processes is shown. These cellular conditions can be grouped into those that are involved in development and survival response. For each regulatory network that is active in a particular cellular condition, the transcription factors are shown in the top arc, the target genes are shown in the bottom arc and regulatory interactions are shown as a line connecting the two. (b) Condition-specific hubs under the five different conditions are shown. The gene name of the hubs is shown on the left. For each hub, the row to the right indicates the number of target genes regulated in each condition (column; CC: cell cycle; SP: sporulation; DS: diauxic shift; DD: DNA damage; SR: stress response). Darker cells represent high number of regulated target genes. For instance, YMR016C is a transcription factor that regulates a large number of genes during cell cycle (black box under CC) but regulates almost no genes in the other conditions (white boxes under SP, SR, DS, DD). (c) Classes of promoter architecture defined by analyzing the experimentally identified binding sites under different conditions. (d) Four patterns of genome-wide binding behavior are shown. Transcriptional regulators are represented by circles and are placed above (condition 1) and below (condition 2) a set of target genes/promoters. Note that the TFs can exhibit different behaviors when different pairs of conditions are compared. (Adapted from [381] and [245].)

into four different architectures – "single regulator," "repetitive motif," "multiple regulators" and "co-occurring regulator" types. The first two of these architectures are characterized by the presence of one or more binding sites for a single TF, whereas the other two contain binding sites for two or more TFs. The authors also classified promoter utilization patterns into four types: "condition invariant," in which the set of TF binding sites does not change across conditions; "condition enabled," in which TF binds in one growth condition but not in the other; "condition expanded," in which the set of binding sites in one condition includes those used in the other; and "condition altered," in which different sets of promoters are bound in the two conditions. In summary, the above two studies show that the model eukaryote, *S. cerevisiae* uses largely disjointed parts of its TR network in different environmental conditions via combinatorial usage of transcription factors in the promoter region of the relevant target genes.

Signal-specific sensors individually perceive the components of complex stimuli, whereas TR sub-networks reassemble the resulting pieces of processed information deeper inside the network. Thus, the existence of signal-processing units may make information processing simple and economical for the cell. If a simple environmental signal is sensed by only one TF, only the sub-network originating at the sensor may be dynamically affected while the rest of the network may remain relatively dormant. In contrast, complex signals may undergo parallel processing in semi-independent sub-networks before the development of an integrated response. In an attempt to extend this idea to bacterial systems, another study systematically identified topological units called origons under the assumption that different sub-networks from the completely known transcriptional regulatory network could be active under different environmental conditions depending on the signals sensed by the sensor TFs [48].

2.5.2 Dynamics of individual regulatory interactions

At the most basic level, the transcriptional network is composed of a binary interaction between TFs and their regulated genes. However, an even simpler system would be if the TF regulates its own expression, i.e. the target gene is the transcription factor itself. This mode of regulation is referred to as auto-regulation. In real regulatory networks, TFs are known to be both negatively and positively auto-regulated with abundance for negative auto-regulation in several experimentally known TFs. Negative auto-regulation occurs when a TF represses its own transcription. This simple circuit is known to have two important functions: speeding up the response time and reducing the cell-to-cell variation in protein levels [62, 100, 509]. Similarly, positive auto-regulation occurs when a TF activates its own transcription by up-regulating it. Positive auto-regulatory circuit has been shown to have effects which are opposite to what are seen for negative auto-regulatory feedback loops [384]. Indeed, strong positive auto-regulation has been shown to lead to extensive variations in protein levels among cells and is believed to lead to a differentiation-like partitioning of cells in a population [61]. Therefore, this circuit is proposed to help cell populations to maintain a mixed phenotype in order to survive when exposed to a mixture of external conditions.

In addition to auto-regulatory motifs, small patterns of interconnections called network motifs are known to occur frequently in transcriptional networks and have been shown to possess distinct dynamic and kinetic properties. The feed forward loop is one of the most characterized multi-gene motifs in terms of dynamics. As discussed in the section on local network structure, of the eight different subtypes of FFLs, coherent type 1 and incoherent type 1 have been shown to be the most abundant in real transcriptional networks. Each of the coherent and incoherent types of FFLs can have an AND or OR input function at the promoter of the target gene depending upon whether both or only one of the two TFs are needed to regulate the target. Coherent FFL with AND input has been shown to be responsible for an initial delay in the transcription of the target genes involved in the arabinose-utilization system in *E. coli* [393]. Since the input function is AND, any sporadic signals are filtered out by this motif and only persistent signals lead to the expression of the target gene. On the other hand, the same motif with OR input shows an initial quick response but exhibits a lag when the signal is shut down as has been demonstrated in the flagella system of *E. coli* [303]. In this study, the FFL regulating expression of the flagellar genes was found to prolong their expression even after the input signal stopped. However, no delay was detected when the input signal appeared. In an incoherent FFL, the effect of regulation from the two TFs involved is opposite. This motif gives rise to a pulse-like behavior because at the onset of the expression of the TF, directly controlling the target gene can lead to its sudden increase/decrease of expression. However, once the second TF regulated by the first reaches its repression/activation threshold with some time lag, it can lead to a complementary effect on the transcription of the target gene. This motif structure has been also shown to act as a response accelerator due to the sudden changes it can bring about in the expression of the target gene [392].

Single-input modules (SIMs), in which a single transcription factor regulates several genes form another class of regulatory network motifs whose dynamics have been well characterized in recent years. The main function of this motif has been shown to provide just-in-time transcription of the set of genes regulated by the TF. Typically in this motif, the TF which regulates different genes has different activation/repression thresholds for each gene. Therefore, when the concentration of the TF increases it crosses the thresholds in a defined order, following a temporal order of expression of the target genes. This motif has been studied in great depth in the case of the arginine biosynthesis pathway in *E. coli* where the arginine repressor is known to regulate several operons involved in the biosynthesis of arginine. It was found that when arginine is removed from the system, promoters of the enzymes responsible for its biosynthesis are activated in a temporal order just in the order needed for its biosynthesis [685].

Although other motifs like multi-input modules (MIMs) are known to be enriched in transcriptional networks, owing to their complex topology, our understanding from the experimental side has been rather limited. However, in future years with the availability of high-resolution mapping approaches it might be possible to elucidate the design principles of these complex circuits in the context of larger biological systems.

2.6 Conclusions

In this chapter, we have introduced how transcriptional regulation of target genes by transcription factors can be conceptualized as a network. Such a representation has allowed us to gain a better understanding of the complexity of gene regulatory systems on a genomic scale. In particular, research in the last decade has allowed us to gain insights into the structure, evolution and dynamics of transcriptional regulation in different model organisms.

At the level of global network structure, transcription networks have been shown to adopt a scale-free structure which is postulated to confer robustness to such systems, with hubs assuming importance. At the level of local network structure, detailed analyses of transcriptional networks have shown that they can be dissected to define network motifs, which are over-represented sub-graphs or patterns in the regulatory network. Each type of motif has been further demonstrated to encode a specific information processing task. From an evolutionary perspective, recent studies have clearly revealed that organisms evolve their transcriptional regulatory network rapidly and adapt to changing environments by tinkering individual regulatory interactions. Such plasticity in the network structure has also been observed between distinct cellular conditions, thereby enforcing the notion that regulatory networks are dynamically rewired between different growth states. These findings indicate that such rewiring permits organism to efficiently respond to changing conditions.

This understanding of the structure, evolution and dynamics of gene regulatory networks can be exploited in emerging areas such as synthetic biology and systems biology. Specifically, it can be used to design synthetic gene circuits with defined dynamic properties for the regulation of gene expression in a precise manner. Especially, with the advancements in genomic technologies, future years will foresee more extensive and detailed maps of transcriptional regulation in a number of organisms. The availability of such information is likely to transform our understanding of gene regulation and will permit us not only to generalize the principles governing gene regulatory systems but also to address fundamental questions linking different types of regulation in the near future.

Acknowledgments

The authors acknowledge the MRC Laboratory of Molecular Biology, Cambridge Commonwealth Trust, Darwin College and Schlumberger for generous support. We thank R. Janky, A. Wuster, G. Chalaucon and members of the TCB group for critically reading the manuscript and for providing helpful comments.

Some supplementary information is available at:

`http://www.mrc-lmb.cam.ac.uk/genomes/madanm/net_evol/`

`http://www.mrc-lmb.cam.ac.uk/genomes/madanm/evdy/`

`http://www.ccg.unam.mx/Computational_Genomics/TRNS/conservation/`

`http://www.ccg.unam.mx/Computational_Genomics/regulonbd/CellSensing/`

`http://www.mrc-lmb.cam.ac.uk/genomes/madanm/tfcomb/`

`http://www.mrc-lmb.cam.ac.uk/genomes/madanm/sarath/t_genorg/`

`http://sandy.topnet.gersteinlab.org/`

`http://www.mrc-lmb.cam.ac.uk/genomes/madanm/blang/methods/`

3

Transcription factors and gene regulatory networks

MATTEO BRILLI, ELISA CALISTRI AND PIETRO LIÓ

3.1 Introduction

Specific sensory and signaling systems allow living cells to gather and transmit information from the environment. All perceived signals are used in order to adjust cellular metabolism, growth, and development to environmental conditions. At the same time the cell is able to sense the intracellular milieu, e.g. energy and nutrient availability, redox state and so on and it accordingly adapts its physiological state. The importance of such changes in cellular processes is underlined by the presence of multiple regulatory systems (see Table 3.1), the most important of which controls the rate of transcription of a gene. The extremely different cell types in higher eukaryotes are a consequence of expression pattern differences, as well as cellular proliferation and differentiation, which are controlled by complex regulatory circuits originating space- and time-dependent transcriptional patterns. Thus, understanding the dynamic link between genotype and phenotype remains a central issue in biology [487, 653, 663].

Signals sensed by the cell are translated into changes in the rate of transcription of well-defined groups of genes through the activation of specific proteins (transcription factors, TF). TFs have high affinity for specific short sequences located upstream of genes and regulate transcription either positively or negatively. The binding of a TF to its target on the gene's promoter controls when expression occurs, at what level, under what conditions, and in which cells or tissues [662]. Interactions with other proteins, chromatin remodeling, modification complexes and the general transcription machinery affect the DNA-binding characteristics of a TF thereby influencing the rate of transcription. During transcription pattern changes, rates of RNA synthesis can fluctuate by orders of magnitude in a very short time, and differ in adjacent cells: most genes have spatially and temporally heterogeneous expression profiles and the presence of common binding sites is at the basis of co-expression. Given that eukaryotic genomes may contain tens of thousands of genes, a complex machinery is needed to control their expression.

Networks in Cell Biology, ed. M. Buchanan, G. Caldarelli, P. De Los Rios, F. Rao and M. Vendruscolo. Published by Cambridge University Press.

Table 3.1. *References on eukaryotic regulatory systems. For detailed reviews on mechanisms of eukaryotic transcriptional regulation see [103, 135, 346, 357, 368, 643, 650]*

Type of regulation	Reference
Chromatin condensation	[18, 96, 279, 362]
DNA methylation	[75, 92, 93, 191, 229, 263, 364, 424, 486, 496, 498, 521, 534, 553, 556, 602, 614, 631]
Alternative splicing of RNA	[31, 126, 170, 525, 541, 603, 661]
mRNA stability	[39, 66, 109, 207, 238, 261, 388, 429, 454, 560, 600]
Translational controls	[78, 256, 353, 407, 610]
Protein degradation	[16, 368]
Regulation by non-coding RNA	[29, 161, 347, 608]

This chapter is about transcriptional regulation in eukaryotes and it is organized in three parts: the first one describes eukaryotic promoters, focusing on human ones which are often remarkably complex; the second is an overview of the most important families of TFs, representing the physical link between a molecular signal and gene expression changes. The third part is dedicated to the bioinformatics for regulatory network characterization.

3.2 Promoters' complexity/eukaryotic gene promoters

Transcriptional promoters play a key role in determining gene transcription by integrating the influences of DNA sequence, combinatorial TF binding, epigenetic features and signal transduction events [253]. The presence of multiple target sequences for distinct TFs (which may be activators or repressors) generate combinatorial control of gene expression [135, 346, 375]. The inputs that a promoter integrates can take many forms: developmental genes have spatial and temporal inputs which, in combination, produces highly dynamic patterns of transcription in defined regions of the embryo [135, 653]; housekeeping genes are constitutively expressed, but their absolute level in the cell may change in specific conditions [477]; other promoters are activated by specific hormonal, physiological, or environmental signals [64, 563] and are otherwise off.

Two functional features are always present in eukaryotic promoters.

(i) The basal (or core) promoter: the site upon which the transcription complex assembles; it is not able to generate functionally significant mRNA levels by itself [341, 357, 360].
(ii) A collection of binding sites specific for one or more TFs, which collectively determine the expression pattern of a gene.

Protein coding genes are transcribed in eukaryotes by the RNA polymerase II holoenzyme complex, formed by several proteins [108, 330, 357, 453, 483, 595]. In yeast and mammalian cells this complex comprises several general transcription initiation factors and other proteins. The docking site for transcription complex assembly is a region of about 100 bp [357, 484, 499] often containing the critical binding site of the basal promoter, the TATA box, located 25–30 bp upstream of the transcription start site. The TATA box is not universal and many genes contain a distinct initiator element spanning the transcription start site. At the very beginning of transcription initiation, TATA-binding proteins (TBP) bind DNA [286, 341] and then several associated factors (TAFs) guide the RNA polymerase II holoenzyme onto the DNA, an event influenced by TFs eventually bound to other sites [346, 357, 360]. Basal promoters determine very low rates of transcription initiation [286, 341, 360]; moreover, most of the proteins binding them are too ubiquitously expressed to generate specific transcription patterns [103, 357, 360]: high rates of transcription and dynamic expression patterns require the intervention of specific TFs that bind DNA sequences located outside the basal promoter and generally thousands of nucleotides afar [103, 360, 643].

3.2.1 Human promoters

Considering the relatively low number of human genes (a few tens of thousands according to the most recent estimates), which are moreover conserved and shared by such different organisms as *Caenorhabditis elegans* and *Homo sapiens* the greater complexity in the evolutive time-scale may be produced, at the genetic level, by two mechanisms: the increased number of combination of binding motifs and increased number of proteins constituting the multiprotein transcription complexes.

It has been estimated that as much as one third of the human genome is composed by regulatory sequences, and the ENCODE project has provided evidence for dispersed regulation spread throughout the human genome. Regulatory sites, enhancers, silencers and insulators are scattered over distances of roughly 10 kb in fruitflies and 100 kb in mammals. A typical animal gene has several of these regulatory modules, that can be located in 5′ and 3′ regions, as well as within introns. They are composed of multiple binding sites for different regulatory proteins which provide the basis for specific expression patterns in distinct tissues or cell types.

A gene can gain new regulatory sequences through transposition: B2 SINEs in *Mus musculus* contain basal promoters [193] and some Alu elements in humans contain binding sites for zinc-finger TFs (e.g. Sp1, estrogen receptor alpha, and YY1) thereby influencing transcription [40].

Table 3.2. *Well-known promoter elements in humans (to go more in-depth see [671])*

Type	% in human	Description
TATA box	10–16%	Contrary to what happens in yeast or in *Drosophila* it is not a general promoter motif for human genes. Usually located 35 bp upstream of the TSS. Often associated with tissue specific genes.
TATA-less	84–90%	Genes with TATA-less basal promoters may generally be transcribed constitutively at relatively low levels [484].
Basal TATA and TATA-less		They can be associated with alternate start sites of the same gene [228].
CpG-associated	88%	Typical of mammals; they might play a role in gene expression more general than previously appreciated. They are found in the upstream region of transcriptionally active genes or immediately nearby. CpG islands are found in most of the least tissue-specific genes, which often code for proteins located in the nucleus or mitochondrion.
Sp1		Weak indicator of less-specific expression.
no CpG island or TATA box		The most common mid-specificity genes and commonly code for proteins located in a membrane.
YY1, Inr, DPE		Associated with the less-specific genes.

A large percentage of genes in humans (30–50%) have two or more active promoters; some of the regulatory sequences for one gene might be closer to another one and many regulatory elements may actually reside within the gene itself [104, 173, 179]. See Tables 3.2 and 3.3.

The basal human promoter is composed of binding sites whose fraction ranges from 10% to 20% of the nucleotides within the promoter, interspersed in larger regions whose function is still unclear. It has been observed that a marked increase in transcriptional complexity is correlated to the organization of the transcriptional units in the genome. There are GC-rich genomic regions that are highly enriched for transcripts (*transcriptional forests*, or *ridges*) and may be subject to shared epigenetic regulatory control; conversely, there are regions that are devoid of transcripts (*transcriptional deserts*, or *antiridges*) and generally contain tissue-specific genes [621].

The counterpart of this complex arrangement of regulatory regions representing an important source of complexity is the number and variety of TFs: 5–10% of the total coding capacity of metazoans is dedicated to this activity.

Table 3.3. *Functional complexity in human promoters according to [662]*

Promoters	Feature	Information
Housekeeping genes	simple	These promoters are often constitutively active, but at different levels in different cell types and they can be shut down in response to specific conditions, such as stress [477]. In principle, a promoter that is always and everywhere "on" needs to contain only one binding site for an ubiquitous TF. Additional binding sites might be present, however, to add robustness, to set levels of transcription precisely, or to modulate levels in response to extreme conditions.
Tissue specific	simple	Genes expressed exclusively in a single differentiated cell type should have simple promoters. Genes encoding the specialized products of terminally differentiated cells often have relatively simple promoters even though they produce spatially complex expression profiles [37]. They are typically activated by one or a few tissue specific TFs, and sometimes they lack binding sites for repressors [135]. The most tissue-specific genes typically have a TATA box, no CpG island, and often code for extracellular proteins.
Regulatory genes	complex	Regulatory genes expressed early in development possess some of the most complex expression profiles and they should have complex promoters. The promoters of genes that operate in early embryos drive temporally and spatially precise transcription, they often use cooperative protein binding to sharpen boundaries of transcription domains, and this requires additional binding sites. These promoters typically contain binding sites for several positive and negative regulators, as they must integrate multiple spatial and temporal inputs [36].
Multiphased expression	complex (modular)	Genes with several distinct expression patterns might have complex promoters with modular organization. Promoters that drive multiphased expression profiles should be more complex, on average, because they respond to many inputs through the interaction with a great variety of transcription factors. Multiphased expression is very common for genes encoding developmental regulatory proteins.
Genes with isoforms	complex	Loci that produce multiple isoforms of a protein may generally have more complex promoters. Alternate transcriptional start sites are often part of the way in which distinct isoforms are generated, adding complexity to such promoters.

Table 3.3. *(cont.)*

Promoters	Feature	Information
Genes responding to extracellular signals	very complex	Signal transduction systems communicate changing conditions in the cytoplasm or at the cell surface to the nucleus, often by phosphorylation or dephosphorylation of a specific TF already present in the nucleus. Contingent regulation of transcription should therefore require complex promoters with additional binding sites for these factors.
Co-expressed genes		Genealogically unrelated genes that are coordinately regulated should share some binding sites. The promoters of genes that are expressed in similar spatial and temporal patterns share similar functional requirements but they may sometimes contain binding sites that evolved independently and yet function in a similar manner.

3.3 Transcription factors

Most TFs are modular proteins consisting of a DNA-binding domain recognizing specific sequences and a domain for protein–protein interaction [662]. More in detail, TFs are formed by one or more of the following.

1. DNA-binding domain: several groups of structurally related proteins recognize and bind specific DNA sequences and slightly different DNA-binding domains may have distinct target sequences. The large TF repertoire of eukaryotes thus provides a wide range of combinatorial relationships for transcriptional regulation. The expansion and diversification of TF families relies on gene duplication of pre-existing TFs and divergence of DNA-binding domains. TFs are moreover often active as homo- and/or hetero-dimers, additionally increasing the possible combinations of functionally different TFs. The most common TF families are discussed below.
2. Protein–protein interaction domains, including leucine zippers [11, 141] and the pentapeptide motif of homeodomain proteins [346]. They are often present from one to several copies.
3. Localization signal: the activity of a TF may be modulated by controlling the ratio of cytoplasmic-to-nuclear localization (e.g. PEBP2/ CBF [634]).
4. Ligand-binding domain: TFs' activity can be affected by specific ligands, e.g. nuclear receptors bind steroid hormones [64].

Several gene families of TFs are known (Table 3.4, Figure 3.1 and [182, 246, 346, 375, 576]) whose size considerably differs within and among genomes. Existing

Table 3.4. *Classification of TFs according to Stegmaier et al. [576]*

Group	Class	#	Fam.	Subfam.
bZIP	bZIP	266	7	267
	bHLH	302	12	193
	bHSH	12	1	7
Zn coord.	Nuclear receptors	233	2	195
	C6 Zn clusters	50	1	88
	DM	2	1	2
	GCM	4	1	7
	WRKY	24	2	54
HTH	Homeo box	1007	10	$\geq$800
	Paired box	90	1	79
	Forkhead/winged helix	134	2	161
	HSF	27	3	35
	Tryptophan clusters	302	3	528
	TEA domain	9	2	9
β-Scaffold	RHR	28	2	35
	STAT	17	3	19
	p53-like	5	1	6
	MADS	274	3	241
	β-barrel α-helix	1	1	2
	TBPs	11	2	17
	HMG	88	5	110
	Histone fold	30	1	42
	Grainyhead	5	1	6
	Cold-shock domain	13	1	5
	Runt-like domain	26	1	19
	SMAD/NF-1	100	2	81
	T-Box domain	43	1	46
	Copper first	4	1	5
	Pocket domain	7	1	12
	AP2/EREBP-related	47	3	74
	SAND domain	3	1	3

paralogs are the result of duplications that occurred across a wide range of taxonomic lineages [69, 133, 156, 575] raising difficulties in classification. Here we will refer to the recent classification scheme of Stegmaier *et al.* [576] (Table 3.4) to briefly describe salient features of most common TF families.

3.3.1 bZIP transcription factors

bZIP domains are not structured when free in solution, and becomes α-helices when DNA interaction takes place (e.g. [646]). They are usually associated with a

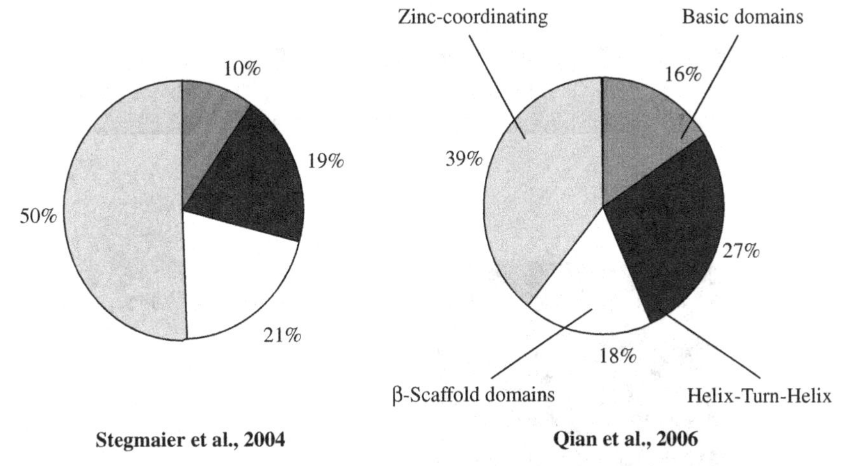

Fig. 3.1. TFs can be classified according to the structural fold of the DNA-binding domain. The pie charts show the percentage of TFs grouped by structural family in two different works. The classification of Stegmaier *et al.* [576] is based on a library of Hidden Markov Models exploiting available information from TRANSFAC, which contains data on transcription factors, their experimentally-proven binding sites, and regulated genes. Qian *et al.* [485] developed an automatic classifier where TFs are represented in terms of the functional domain composition and a data mining procedure to gather structural information on the domains composing a protein.

dimerization domain, a leucine zipper (ZIP), a helix-loop-helix (HLH) or a helix-span-helix (HSH) domain. They are involved in a wide range of processes, such as development, metabolism, circadian rhythms, learning, memory, and response to stress and radiation [141, 627]; their malfunctioning may lead to cancer and other diseases [514]. The highly conserved bZIP domain comprises a basic region (BR) and a leucine zipper (LZ) [274]: the BR is devoted to the interaction with specific DNA elements (generally hexameric sequences). In metazoans different bZIPs are able to recognize and bind six major classes of consensus [141]. The LZ domain is less conserved than the BR and it is composed by a short coiled-coil and amphipathic domain responsible for recognition and dimerization specificity [187]. There is a wealth of data on protein–DNA binding and also a very extensive data set of protein–protein interactions concerning bZIP proteins. The bZIPs from humans, fruit flies, and ascidians have been identified and classified [187, 622, 675].

c-Jun

The importance of the bZIP transcriptional activator c-Jun (Fig. 3.2) derives from its inherent oncogenic potential, which is strongly enhanced in v-Jun, a mutated retroviral variant. Levels of the protein are under transcriptional and translational

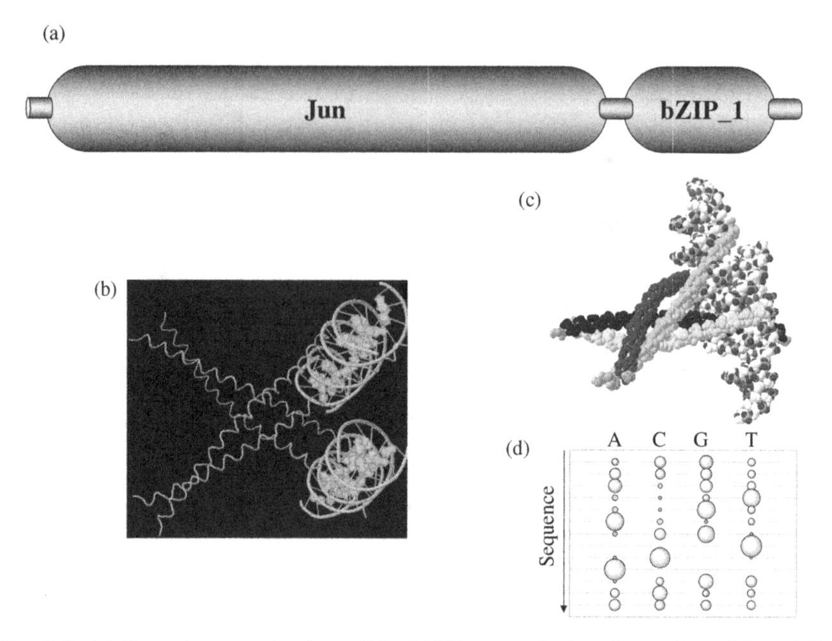

Fig. 3.2. (a) Domain organization of the bZIP transcription factor Jun; (b) and (c) structure of the heterodimer Jun:Fos complexed with CRE DNA from [224]; (d) bubble graph of the Position Weight Matrix describing the binding specificity of Jun, based on 47 binding sites retrieved from TRANSFAC. The uppermost row of bubbles corresponds to the first position of the binding sequence while the lowermost corresponds to the last position of the binding site; the four bubbles in each row are proportional to the frequency of each base in that position of the motif.

control, with a fine-tuning by phosphorylation, primarily through the activity of JNK (Jun N-terminal Kinase) and GSK-3 (Glycogen Synthase Kinase 3). GSK-3 phosphorylates c-Jun at threonine 239 (T239) inducing proteasomal degradation [640] but only if c-Jun has been previously phosphorylated at serine 243. One of the mutations of v-Jun determines an amino acid change at site 243, which can not be phosphorylated; this blocks the phosphorylation by GSK-3 and v-Jun degradation. This regulatory pathway connects c-Jun to two signaling pathways that stabilize c-Jun by negatively regulating GSK-3: the Wnt and the PI3K (phosphoinositide 3-kinase) pathways, both showing gain of function in numerous and diverse cancers.

JNK targets serines and threonines in the N-terminal region of c-Jun [644] and transforms c-Jun into a strong transcriptional activator [645]. v-Jun does not interact with JNK for a 27 residue N-terminal deletion destroying the interaction domain; nonetheless it acts as a strong activator because the very same deletion eliminates the binding to the inhibitory complex.

3.3.2 Helix-turn-helix domains

In 1982, Matthews and co-workers [408] showed that the three repressors cro, cI (from phage lambda) and LacI share a DNA-binding domain composed of three helices [448, 449, 523, 577]. The helix-turn-helix (HTH) motif, comprising the second helix and the third helix, appeared to be critical for DNA binding; the same motif was then found in eukaryotic TFs for development and differentiation (the homeo- and Myb domains, Figure 3.3) [202, 455]. The following works reinforced the importance of the HTH in DNA–protein interactions, across a wide phylogenetic spectrum [82, 148, 149, 455, 536]: it is present in several eukaryotic specific and basal TFs, in histone H1 and in other chromatin associated proteins [34, 81, 115, 208, 327, 490, 536, 586, 655]. Several classes of HTH transcription factors exist (e.g. [86, 209, 217, 333, 503, 641] where the HTH domain is often fused with additional globular domains.

c-Myb

The vertebrate myb gene family has three members:

(i) c-Myb, mainly expressed in hematopoietic cells;
(ii) a-Myb, mainly expressed in male germ cells and female breast ductal epithelium;
(iii) b-Myb, with a broader cell-type specificity, essential for inner cell mass formation in the early stages of development.

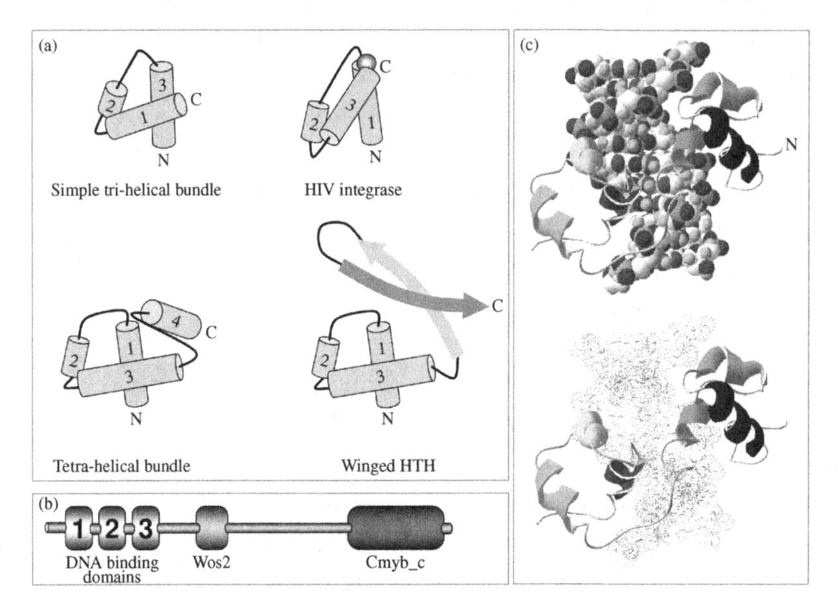

Fig. 3.3. (a) Some of the HTH domain architectures; b) domain architecture of c-Myb (according to pfam); (c) c-Myb DNA-binding domains 2 and 3. Proteins are in ribbon representation whereas DNA is represented with a space-filling model. Schemes have been modified from [33]; structures are from [447].

The c-Myb DNA-binding domain (Figure 3.3) consists of three imperfect tandem repeats of 51–53 amino acids with a HTH-like domain each. Myb controls hematopoietic cell-specific genes together with a group of TFs including C/EBPβ, AML1, Ets, and GATA; it recognizes the AACNG sequence (the AAC by repeat 3, and the G by repeat 2). c-Myb has three functional domains responsible for (i) DNA binding, (ii) transcriptional activation (Wos2 in Figure 3.3b), and (iii) negative regulation from the N-terminus. The negative regulatory domain is important to suppress the c-Myb activity, and the loss of functionality leads to oncogenic activation of c-Myb.

3.3.3 Zinc-coordinating domains

Five classes currently populate the superclass of zinc-coordinating domains according to the analysis performed by Stegmaier *et al.* [576] (Table 3.4). Zinc fingers are nucleic acid-binding domains widespread in most eukaryotes. The basic domain is composed of 25 to 30 amino acids including conserved key residues: two cysteines and two histidines in a $C-(X)_2-C-(X)_{12}-H-(X)_3-H$ motif. The twelve residues separating the second C and the first H are mainly basic, implicating this region in DNA binding. Each domain binds one Zn atom in a tetrahedral array formed by the C and H of the motif and that gives the domain's characteristic finger-like projection; the *zinc finger* interacts with nucleotides in the major groove of the double helix. Fingers have been found to bind to about five base pairs of nucleic acid containing short runs of guanine residues.

3.4 Bioinformatics of regulatory networks

We here describe some of the most used tools for DNA-binding protein identification and classification and a framework for the prediction of TF binding sites by integrating different information sources.

3.4.1 Transcription factors identification

The identification of the repertoire of TFs in a given organism is required to allow deciphering the regulatory network on a systems perspective. One of the features shared by all TFs is the presence of a DNA-binding domain which can be used to filter proteins from a given genome, retaining only those plausibly able to bind DNA. DNA-binding domains can be modeled using hidden Markov models (HMM) derived from known TFs and the models can then be used to scan the proteome of a given organism or entire databases. A similar approach has been

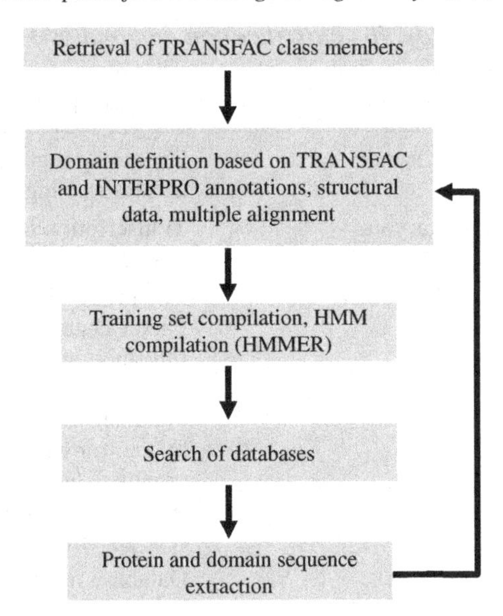

Fig. 3.4. A possible workflow for TF identification using hidden Markov models (modified from [576]).

followed by Stegmaier *et al.* [576] using HMMER (Fig. 3.4). Two types of hidden Markov models were developed: class-specific HMMs to annotate structural domains corresponding to TRANSFAC classes and subtype HMMs, to specifically assign domains to lower hierarchical levels, families and subfamilies. Popular databases and tools for TF analysis are reported in Table 3.5.

3.4.2 Motif finding

The computational identification of cis-regulatory elements is currently based on three main approaches: (i) identification of conserved motifs using interspecies sequence global/local alignments [470]; (ii) motif-finding algorithms that identify previously unknown motifs overrepresented in the promoters of co-expressed genes [46, 94, 178, 206, 271, 457, 585]; (iii) computational detection of previously known motifs in promoters of genes for which regulating TFs are unknown [314]. Limitations of the first approach are caused by the variability of promoter sequences, by rearrangements of binding sites within the non-coding regions or by changes in regulation of ortholog genes. The second approach requires a large number of sequences highly enriched for at least one motif or a system to recognize true regulatory sites and reduce false positives. The third approach seems promising but requires details on the motif models of each TF; if this information is available there is a good chance of good predictions for unknown target genes. An accurate analysis requires the use of an appropriate statistical background model

Table 3.5. *Web resources for TFs classification and analysis*

Database	Organism	Additional info
TransFac	Eukaryotic	Cis-acting regulatory DNA elements and transacting factors
Jaspar	Eukaryotic	Transcription factors modeled as matrices
Protein Lounge	Eukaryotic	The Protein Lounge Transcription Factor Database is a resource housing information relating to transcription factors of humans and other organisms. This database is organized by the transcription factor names, organism specification, binding sequences, binding elements and cell type specification
PLACE	Plants	Database of Plant Cis-acting Regulatory DNA Elements
TRRD	Eukaryotic	Transcription Regulatory Regions Database. Only experimentally confirmed information is included
TRED	Human, mouse, rat	Transcriptional Regulatory Element Database
SCPD	Yeast	The Promoter Database of *Saccharomyces cerevisiae*
EPD	Eukaryotic	The Eukaryotic Promoter Database. An annotated non-redundant collection of eukaryotic POL II promoters, for which the transcription start site has been determined experimentally
Noble TF database	*Medicago truncatula*	The prediction was based on transcription factor binding sites and hidden Markov models (HMM). The models were built mainly on documented transcription factors and their family information in *Arabidopsis* and a small number of known transcription factors from legumes (soybean, Alfalfa and *Medicago truncatula*)
Dragon TF database		TF names are collected from Swiss-Prot, LocusLink, GO and Entrez Gene databases. Users can browse the names of TFs, with links to the original sources

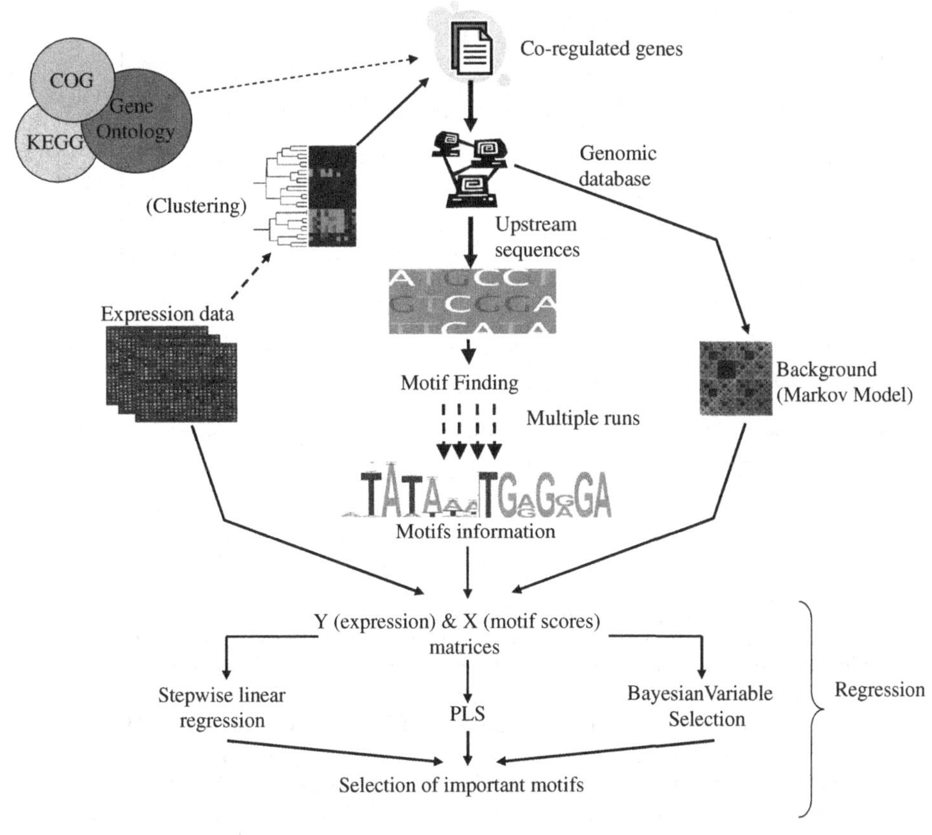

Fig. 3.5. A framework for regulatory motif identification integrating data. PLS = partial least squares regression. COG, Kegg and GO databases can be used for functional data mining. Multiple expression condition can be used both in a preliminary step to obtain groups of co-expressed genes, and in the regression step, allowing to filter out most false positives. Motif finding might be performed using different algorithms to overcome their specific limits. See Fig. 3.6 for an example output.

of DNA sequence and can be validated through the integration of several sources of data, e.g. sequence of gene promoters, ortholog genes, and gene expression data (Fig. 3.5). Background DNA can be modeled by using Markov Chains.

In Fig. 3.5 we schematize a framework that can be used to improve the specificity of the identification of DNA regulatory motifs; it has been used for the analysis described in Fig. 3.6 and by [28, 83, 123, 587].

Once motifs have been identified, we need some metric to score them and quantify how well a promoter matches each candidate motif; to do this we take into account the position weight matrix (PWM) describing the motif, the background model and the number of motif occurrences in a promoter. A commonly used

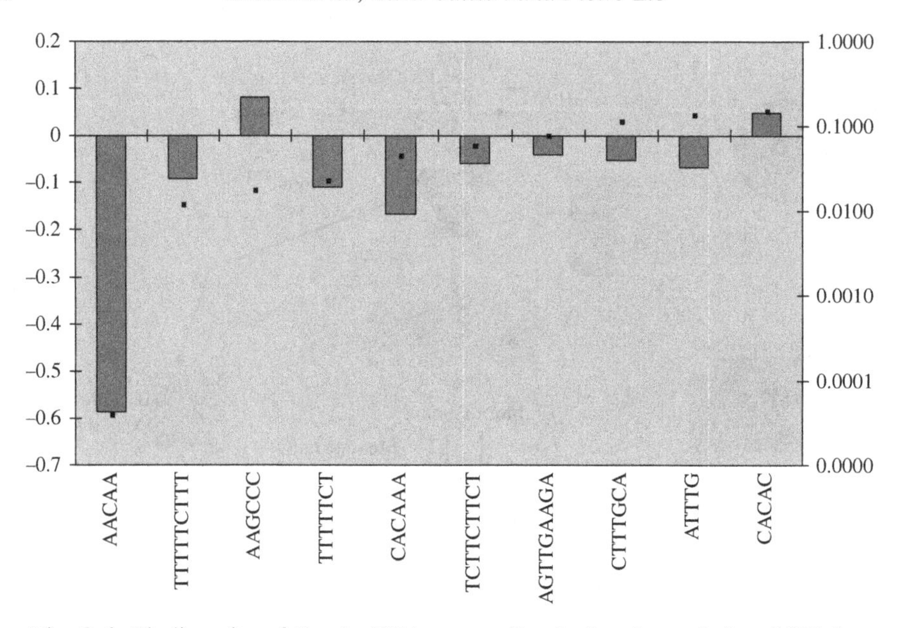

Fig. 3.6. Binding site of Rox1p TF in yeast: Rox1p is a heme-induced TF that recognizes YYNATTGTTY and represses genes normally expressed in hypoxic conditions. We integrated information from expression, *de novo* motif finding using Gibbs sampling and stepwise regression. We collected upstream sequences from yeast genes repressed in the experiment of Rox1p overexpression [213] and used them to search for motifs in the range 5–12 nucleotides with MDscan [374]. For each candidate motif we calculated scores of genes with MotifScorer [83] taking into account a sixth-order Markov model and then we performed a stepwise regression using the scores as predictors of the observed expression level in the experiment of [213]. We show the regression coefficients (histogram, left axis) and p-values (squares, right axis in log scale). The AACAA consensus is part of the known Rox1p binding site. Other motifs have a significant coefficient ($p - value < 0.05$): AAGCCC, TTTTTCTTT and TTTTTCT, both related to Azf1p binding site (TTTTTCTT), a TF preferentially synthesized in higher amounts under non-fermentable growth conditions, CACAAA, related to binding sites of Mig3p (cccCACAAAaat), a TF that when phosphorylated induces damage response genes, and Rme1p (gtacCACAAAa), a TF controlling cellular adaptation to the nutritional status of the environment.

scoring function is the following: given a motif μ of length w, and occurrences $\mu_i \in \mathbf{X_{w,g}}$ in gene g's promoter; given the corresponding PWM, $\mathbf{M}_\mu := (P_{w,n})_{w \times n}$, where $n \in \{A, C, G, T\}$, we calculate the motif probability on $\mathbf{M}_\mu$, $P(\mu_i|M_\mu)$ and on the background model $(P(\mu_i|M_{BKG}))$. Then, the score of the promoter of gene g is:

$$S_{\mu,g} = \log_2 \left[\sum_{\mu_i \in X_{w,g}} \frac{P(\mu_i|M_\mu)}{P(\mu_i|M_{BKG})} \right].$$

This formula has been implemented e.g. in MotifScorer [83] and in MotifRegressor [123] but there are others available, such as the one proposed by [533], which do not require a background model:

$$S_{\mu_i} = \frac{1}{w} \sum_j [2 + \log_2(P_{kj})],$$

where P_{kj} is the probability of base k at position j, accordingly to the motif's PWM. S_{μ_i} is in this case an information-based measure of potential binding for motif occurrence μ_i.

Future approaches will be based on extensive whole-genome analysis starting from a large number of previously identified groups of co-expressed genes, e.g. by clustering large collections of expression data recorded in many different conditions (i.e. a *compendium* of expression data), dedicated databases or bibliographical data mining. Upstream sequences of selected genes are then used as input for a *de novo* motif finding algorithm (e.g. a Gibbs sampler) to identify candidate motifs occurring at unexpected frequency. Each candidate motif can be *validated* by comparing its pattern of occurrence with the expression level of the corresponding genes by using regression procedures. The large number of candidate motifs returned by this approach requires regression techniques able to treat a number of predictors (i.e. motifs, stored in the **X** matrix) comparable to the number of observations (i.e. expression conditions, denoted here as **Y** matrix). In Fig. 3.5 we suggest partial least squares (PLS) regression [7, 658] and Bayesian variable selection [587].

PLS regression is a technique that generalizes and combines features from principal component (PC) analysis and multiple regression and it is particularly suited for large numbers of predictors and observations, and in the presence of multicollinearity between them. In PC regression the PCs are used as regressors on the observations. The orthogonality of the PC eliminates the multicollinearity problem but not the difficulty of choosing an optimum subset of predictors. A possible strategy is to keep only a few of the first components; however, they are chosen to explain the predictors and thus there is no guarantee that they are also relevant for the observations. By contrast, PLS regression finds components from the predictors that are also relevant for the observations. Specifically, PLS regression searches for a set of components that performs a simultaneous decomposition of **X** and **Y** with the constraint that these components explain as much as possible of the covariance between the two matrices. This step is followed by a regression step where the decomposition of **X** is used to predict observations. PLS has been widely used in microarray data analysis in terms of tumor classification and survival time prediction e.g. [83, 145].

Bayesian variable selection methods use a latent binary vector to index all possible sets of variables (i.e. motifs). This high-dimensional variable space is then explored by stochastic methods to identify those subsets of variables that best explain observations (i.e. gene expression level). The method provides joint posterior probabilities of groups of motifs and the marginal posterior probabilities for the inclusion of single motifs. Stepwise methods, on the other hand, perform greedy deterministic searches and can be stuck at local minima. Another limitation of the stepwise procedure is that it presumes the existence of a single "best" subset of variables and seeks to identify it. In practice, however, there may be several equally good models.

Another problem arises with mammals, because of the distances dividing a controlled gene and part of its regulatory sequences. A novel computational tool was recently released by [240] for the prediction of distal enhancer elements in mammalian genomes, based on both genomic sequence and conservation. This method tries to detect highly conserved sequences containing clusters of TFBSs by aligning large stretches (50 kb) of genomic DNA from two species.

4

Experimental methods for protein interaction identification

PETER UETZ, BJÖRN TITZ, SEESANDRA V. RAJAGOPALA
AND GERARD CAGNEY

4.1 Introduction

Protein interactions can be identified by a multitude of experimental methods. In fact, the IntAct database of molecular interactions currently lists about 170 different experimental methods and variations thereof that can be used to detect and characterize protein–protein interactions (the main classes are listed in Table 4.1). While we present the commonly used methods in this chapter we will focus on the few technologies which are used in high-throughput studies and thus generated the vast majority of interaction data available today: the yeast two-hybrid (Y2H) assay and protein complex purification and identification by mass spectrometry (MS) (Table 4.2). These two methods represent two fundamentally different sources of interaction data and thus it is important to understand how they work and what strengths and weaknesses each of them has. This is especially important for theoretical analyses which often draw conclusions from datasets which may not be adequate for certain studies. For example, membrane proteins are underrepresented in both yeast two-hybrid and complex purification studies.

4.1.1 Complex versus binary interactions

It is important to note that most methods detect either direct binary interactions or indirect interactions without knowing which proteins are interacting. The yeast two-hybrid system usually detects direct binary interactions while complex purification detects the components of complexes (Fig. 4.1). Complex data are often interpreted as if the proteins that co-purify are interacting in a particular manner, consistent with either a spoke or matrix model. The **spoke model** assumes that all proteins in a complex interact with the bait protein only while the **matrix model** assumes that all proteins interact with all others. Even a combination of Y2H and MS is usually not sufficient to establish the precise topology as some interactions

Networks in Cell Biology, ed. M. Buchanan, G. Caldarelli, P. De Los Rios, F. Rao and M. Vendruscolo.
Published by Cambridge University Press.

Table 4.1. *Methods to detect protein–protein interactions, based on the PSI MI classification. Listed are the top categories with important examples. The whole list contains more than 170 terms and can be found at http://www.ebi.ac.uk/ontology-lookup/browse.do?ontName=MI (go to interaction detection methods). Methods in italics are discussed or illustrated in this chapter*

protein complementation assay
- *cytoplasmic complementation assay*
- *ubiquitin reconstruction*
- *membrane bound complementation assay*
- *mammalian protein protein interaction trap*
- *transcriptional complementation assay*
- *two hybrid*
- *bimolecular fluorescence complementation*
- 3 hybrid method
- protein tri hybrid

biophysical
- nuclear magnetic resonance
- surface plasmon resonance
- *mass spectrometry studies of complexes*
- X-ray crystallography
- isothermal titration calorimetry
- fluorescence technology
- fluorescent resonance energy transfer (FRET)

biochemical
- cross-linking study
- *affinity technology*
- display technology
- far western blotting
- *affinity chromatography technology*
- *pull down*
- *tandem affinity purification*
- *coimmunoprecipitation*
- *array technology*
- *peptide array*
- *protein array*
- enzymatic study
- phosphotransfer assay

imaging techniques
- fluorescence microscopy

may be too weak to be detected individually. X-ray crystallography can provide a detailed model of the proteins in a complex. However, note that crystallized complexes often lack additional weakly associated proteins that do not co-crystallize and thus may not provide a complete picture either.

Table 4.2. *The contribution of various PPI methods to protein interactions in the IntAct database (as of May 4, 2008)*

Method	Number of interactions	Percentage
Two-hybrid	73 050	60.7%
Co-IP	11 719	9.7%
TAP purification	4530	3.8%
Pull-down	6340	5.3%
Protein array	1493	1.2%
Other	23 250	19.3%
Total	**120 382**	**100%**

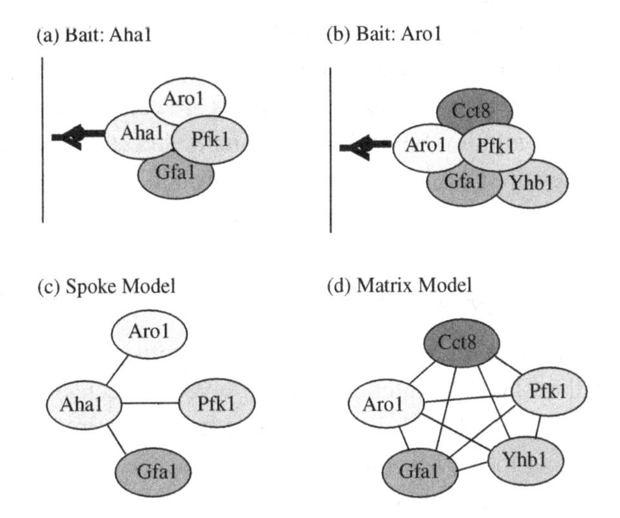

Fig. 4.1. Protein complexes vs. binary interactions. (A,B) When two proteins of a complex are tagged for purification and the components are identified, the two purifications rarely result in the same components. (C,D) Although proteins in a complex are associated, it remains usually unclear which proteins interact directly with each other. In order to predict direct interactions either the matrix (C) or spoke model (D) is applied to lists of purified proteins. To evaluate such interactions Gavin *et al.* have invented the socio-affinity index (SAI). Essentially, the SAI quantifies the tendency for a protein pair (e.g. Aro1 and Gfa1) to identify each other when tagged (B) and to co-purify when other proteins are tagged (A) relative to what would be expected from their frequency in the whole data set. High SAI values result when both proteins co-purify when either one of them is tagged (without purifying many other proteins) and when both are always seen together in purifications of other baits. Modified after [227].

4.1.2 The biological relevance of detected protein–protein interactions

We do not know whether most protein–protein interactions are biologically relevant and, if so, why. Given the fact that only about 20% of all yeast proteins and even smaller percentages of all proteins in bacteria are essential under laboratory conditions, it is clear that an even smaller number of all detectable interactions are essential for survival. However, recent studies show that 97% of all yeast gene deletions have a measurable phenotype when assayed under a large number of conditions [255]. Thus, a large fraction of all detected protein–protein interactions may only be required under specific biological conditions. Alternatively, they may not be relevant to a cell at all, for instance when two proteins that interact *in vitro* never interact *in vivo* because they are housed in different cellular compartments. Indeed, a major challenge for the future will be to distinguish "essential" from "non-essential" interactions and then to identify the non-essential interactions that have a biological role and thus provide a selective advantage. It is possible that a class of non-essential and "irrelevant" interactions are continuously generated and lost in the course of evolution but only occasionally selected. As long as they do not harm the cell they are simply subject to loss through random genetic drift.

4.1.3 Protein–protein interactions are incompletely studied

Information about protein–protein interactions is surprisingly incomplete. While we have tens of thousands of interactions for several species, we do not know much beyond that. A complete description of protein–protein interactions would require the structure of the proteins involved. Indeed, there are stucture models for about 50% of all microbial proteins now [682]. However, given the rapid evolution of microbial interactions in particular it remains unclear how many of these models are really useful for predicting interactions. Because proteins come together to carry out biochemical functions, ideally we also need to know their localization, precise concentration, and how the genes of their components are regulated, how stable the proteins are and thus how quickly they are turned around. This information is only available for a significant number of yeast and possible *E. coli* but not for other organisms although human proteins are currently intensely studied. Even more importantly, we would need to know the precise affinities and thus the dynamics and kinetics of complex assembly. Assembly of complexes often involves conformational changes about which we know very little. Neither do we fully understand the role of post-translational modifications and how they affect the assembly of protein complexes. We should keep in mind that we are still in the process of qualitatively cataloging protein–protein interactions without paying too much attention to quantitative and dynamic aspects. This will change as we approach complete catalogs of all protein–protein interactions for the major model systems. Some recent studies estimate that we have identified only 50% of all yeast interactions and only 10% of all human interactions [247]. We cannot make such estimates for other species for which there is still too little information.

4.2 Protein complementation techniques

The most popular protein complementation technique is the yeast two-hybrid system. All such complementation techniques are based on the reconstitution of split proteins that re-generate a functional protein from two fragments. After the yeast two-hybrid system was invented, researchers realized that they can apply its concept to many proteins. In fact, new complementation techniques continue to be invented. This chapter will focus on the classical yeast two-hybrid method as it is the only one that has been applied to a large number of protein interactions while the utility of the other methods is still being investigated.

4.2.1 The yeast two-hybrid system

The yeast two-hybrid (Y2H) system is a widely used genetic assay for the detection of protein–protein interactions. The original assay was developed by Fields and Song [195] and takes place in living yeast cells (Fig. 4.2a). It employs a transcription factor, e.g. the yeast transcription factor GAL4, which can activate a reporter gene when its DNA-binding domain (DBD) and its transcriptional activation domain (AD) are linked. When both domains are separated from each other, they do not have the capability to activate transcription of the reporter gene. To answer the question whether protein A interacts with protein B, each protein is fused to one of these transcription factor domains, AD and DBD, respectively. If protein A binds to protein B, an active transcription activator complex is re-established, the reporter gene is transcribed and its gene product can be used to detect the protein–protein interaction. The protein linked to the DNA-binding domain is called bait; the protein linked to the activation domain is called prey.

In the original Y2H system developed by Fields and Song [195] both fusion constructs are derived from the yeast GAL4 protein: baits contain the DNA-binding domain (amino acids (aa) 1-147) and preys the activation domain (aa 768-881) of this transcriptional regulator. However, any other transcription factor can be used as well. The lexA-based system is just one alternative that was developed by the laboratory of Roger Brent [226]. In this system, the DBD is provided by the prokaryotic LexA protein, which physiologically acts as a transcriptional repressor, when bound to LexA operators. The AD is formed by an 88-residue acidic *E. coli* peptide (B42) that acts as a transcriptional activation domain in yeast. In a

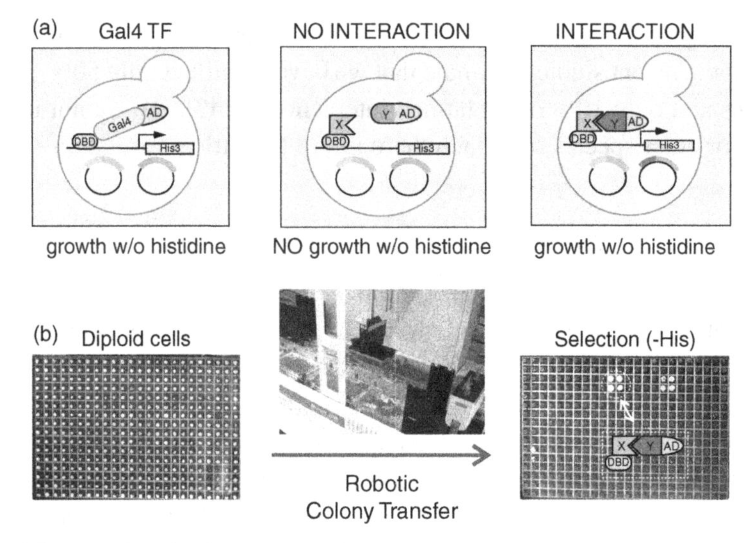

Fig. 4.2. (a) The classical Y2H system is based on a split transcription factor (Gal4). In the native Gal4 a DNA-binding domain (DBD) is linked to an activation domain (AD). The transcription factor activates the expression of a reporter gene (e.g. His3) in specially engineered yeast cells, which allows for growth under histidine-deficient conditions. For the Y2H assay, a protein X is fused to the DBD and a protein Y to the AD of Gal4. If X–Y do not interact, no growth without histidine is possible. However, when X binds to Y, an active transcription factor is reconstituted and the respective yeast cells can grow on histidine-deficient medium. (b) For the array-based Y2H system individual yeast colonies are arrayed onto agar plates in the 384-well format. Protein pairs that are to be tested for an interaction are combined at each position of the yeast array by a mating based approach (in diploid yeast cells). This yeast array is transferred to selective conditions (e.g. without histidine) employing a robotic procedure. Only at positions of the array which carry an interacting protein pair, can yeast colonies grow (note that the shown test was done in quadruplicates).

further modification of the lexA system, either the Gal4-AD or an acidic activation domain of the herpes simplex virus V16 protein was employed for prey protein construction [623].

Regardless of the Y2H system used, several different reporter genes are employed to indicate a protein interaction. Most reporter genes are under the control of artificial promoter constructs, which consist of an appropriate UAS (upstream activation sequence) and a TATA sequence. Most reporter genes are integrated into the yeast genome. Therefore, special yeast strains are used for Y2H tests. β-Galactosidase (lacZ) is a popular reporter, since several different assays for measuring its activity are available and quantitative values of the reporter activation can be obtained. However, lacZ is not well suited for library screening. Rather, reporters are chosen that allow the cell to grow if it harbors interacting bait and prey proteins. These reporters are often metabolic enzymes that are replaced by a modified version in the Y2H yeast strain that is only expressed when the tested proteins interact (Fig. 4.2). Common reporter enzymes are involved in histidine metabolism (His3), leucine metabolism (Leu2), adenine metabolism, and uracil metabolism (Ura3). This way, an interaction can be simply selected by growing the Y2H strain on media lacking, for example, histidine. Different Y2H systems use different sets of reporters. Their promoters vary slightly and testing for activation of different reporter genes is thought to reduce the number of false positives.

The Y2H system can be used for testing individual protein interactions. However, its most powerful application is the screening for protein interactions in large or genome-wide libraries. Traditionally, cDNA or (for bacteria) genomic DNA prey libraries are constructed. These prey libraries are tested against a specific bait construct – by either a co-transformation or a mating-based strategy. Under selective conditions, only yeast strains carrying an interacting bait/prey pair can grow. Plasmid DNA or PCR product purification and sequencing is required for the identification of the prey library clone. An alternative approach is the **array-based Y2H system** [97]. The first step of this system is the systematic cloning of all open-reading frames of a genome (or a subset thereof) into prey vectors. These vectors are transformed into haploid yeast cells individually and the resulting prey strains are systematically arranged on agar plates with 96 or 384 positions. A haploid bait strain is then spotted on top of the prey strains at each position of this array; bait and prey strains form diploid cells that express both bait and prey fusions and individual protein interactions can easily be identified by the reporter-gene activation at specific positions of the array (Fig. 4.2b).

Library screens do not require systematic cloning of all prey constructs, however, the prey library must be created. Therefore, the complete DNA sequence of the genome of interest is no prerequisite. Whereas the setup of the prey library is more time-consuming in the array-based Y2H system, the final workup of the interactions is faster: the position of the growing yeast colony on the array directly identifies the interacting protein pair. In addition, the array-based system can be controlled much better. Every bait is tested against the same set of prey proteins: on the contrary, the preys in the library screen are random and unknown. Thus, the signal-to-background ratio can be systematically evaluated in an array screen for each bait protein and the specificity of each interaction can be known, e.g. the number of baits a given prey is interacting with (Fig. 4.2b). Baits or preys that have many interactions (say, more than two dozens) are usually suspicious and most of these interactions should be considered as false positives (see below).

Pooled library screens combine both presented approaches. In this strategy, preys of known identity (systematically cloned or sequenced cDNA library clones) are combined and tested as pools against bait strains. The identification of the interacting protein pair commonly requires either sequencing or retesting of all members of the respective pool. Zhong *et al.* established a method which allows for pooling up to 96 preys [691]. It was estimated that this pooling scheme reduces the number of interaction tests required from 1/8 to 1/24 in the case of the yeast proteome. Two recent large-scale interaction mapping approaches for human proteins employed such a pooling strategy: Rual *et al.* tested baits against pools of 188 preys and identified individual interactions by sequencing [511]; Stelzl *et al.* tested pools of 8 baits against a systematic library of individual preys and identified interactions by a second interaction mating [579]. Recently, a smart-pool-array system was proposed, which allows the deconvolution of the interacting pairs through the definition of overlapping bait pools [295], and thus usually does not depend on sequencing or a second pair-wise mating procedure.

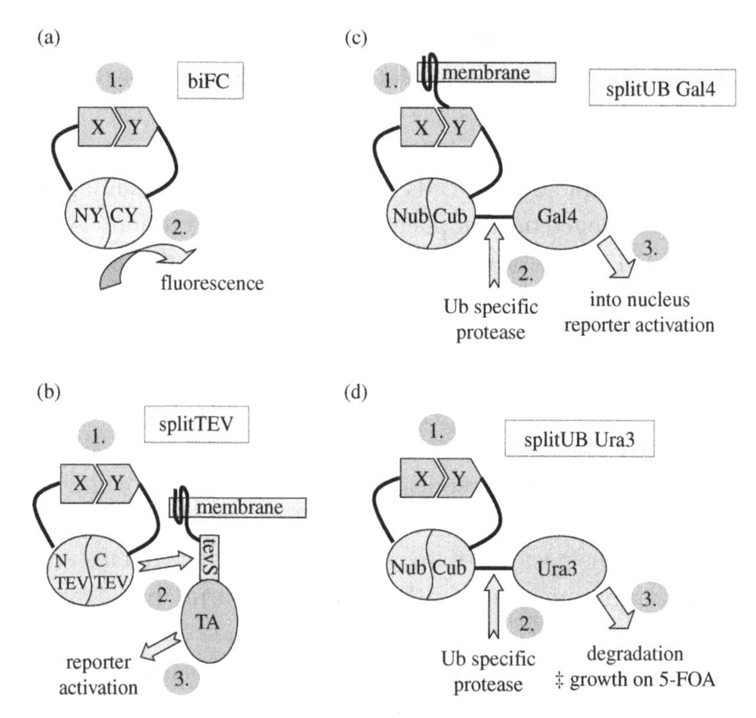

Fig. 4.3. Fragment complementation assays. (a) In bimolecular fluorescence complementation (biFC) methods the interaction of X and Y leads to the reunion of two non-fluorescent-protein fragments into a functional fluorophore. (b) In the splitTEV method the interaction results in the formation of an active TEV protease, which can, for example, release a membrane bound transcription factor (TA). (c) In split-ubiquitin methods reunited ubiquitin is recognized by ubiquitin-specific proteases. This can lead to release of membrane bound transcription factors or (d) to the degradation of an enzyme (Ura3), which mediates toxicity of 5-FOA.

Table 4.3. *Variations of the yeast two-hybrid system. Y2H = yeast two-hybrid, B2H = bacterial two-hybrid, M2H = mammalian two-hybrid, FC = fragment complementation, 3H = three-hybrid*

Class	Method	Principle	Reference
Y2H	classical Y2H	Reconstitution of active transcription factor, here based on Gal4 transcriptional regulator	[195]
Y2H	lexA-based Y2H	Reconstitution of active transcription factor, based on lexA (DBD) and VP16 or Gal4 (AD)	[226, 623]
Y2H	SOS recruitment system	Activation of Ras signaling pathway made dependent on interaction	[38]
B2H	split adenylate cyclase	Reconstitution of adenylate cyclase	[306]
B2H	RNA polymerase recruitment	Activation of reporter gene by RNA polymerase recruitment (similar to Y2H)	[300]
M2H	MAPPIT	Activation of cytokine signaling	[180]
M2H	mammalian two-hybrid system	Reconstitution of active transcription factor	[380]
FC	split-ubiquitin (splitUB)	Protein fragment complementation: analysis of membrane proteins	[297, 574]
FC	split TEV protease	Protein fragment complementation: flexible choice of reporter system	[639]
FC	biFC	Protein fragment complementation: fluorescent proteins (allows to localize an interaction)	[266]
3H	Three hybrid/kinase co-expression	Classical Y2H with kinase co-expression (detects phosphorylation dependent interactions)	[399]

4.2.2 Other fragment complementation techniques

Several other methods for the detection of protein interactions rely on the co-expression of two-hybrid fusion proteins (Fig. 4.3, Table 4.3). All these methods have been proven to work with a selected set of protein interactions. Unfortunately, hardly any systematic attempts have been undertaken to compare the quality and methodological biases of these approaches. Some approaches offer additional advantages such as the localization of protein interactions as in the case of "bimolecular fluorescence complementation" but also require additional equipment such as fluorescent microscopes; other differences are subtle and it remains to be seen how they compare in high-throughput screens. For the lack of comparative data we do not discuss these methods here. Readers are referred to the literature cited in Table 4.3 for more details.

4.3 Affinity purification methods

While protein complementation techniques are usually used *in vivo*, affinity purification requires that the interacting proteins be purified from a cell and then identified *in vitro* (even though the interaction takes place *in vivo*). Historically, GST pulldowns (see below) and co-immunoprecipitation (co-IP) have been the most popular methods, although they have been supplanted by refined high-throughput methods that use mass spectrometry for protein identification. However, all these methods are based on the principle that interactions involving affinity-tagged proteins formed *in vivo* are preserved during biochemical purification steps.

4.3.1 GST pulldown

A standard method for *in vitro* interaction assays uses glutathione-S-transferase (GST) as a tag (Fig. 4.4). Traditionally GST pulldowns have been used to verify interactions that were found in two-hybrid screens and other screening procedures. For re-testing purposes, the two proteins are expressed in a heterologous system, e.g. human proteins in *E. coli,E. coli*so that additional interacting proteins are not co-purified. While it is often desirable to co-purify all members of a complex (see below), in this case we want to have only two defined proteins present in the experiment. GST fusion proteins can be easily expressed and purified from *E. coli* by running a cell extract through a matrix of glutathione-coated beads, usually glutathione sepharose. Only GST fusion proteins and a few cellular glutathione-binding proteins bind to this matrix. Non-specifically bound proteins can be washed off with a salt solution such as PBS. Usually the fusion protein can be left on the matrix and incubated in a second protein solution, either a purified protein or an extract. Proteins from this solution will bind to the GST fusion protein. Often radio-labeled proteins are used (which can be generated by *in vitro* transcription/translation from a PCR fragment containing a promoter and the ORF of the protein in question). Commercial kits are available for such *in vitro*

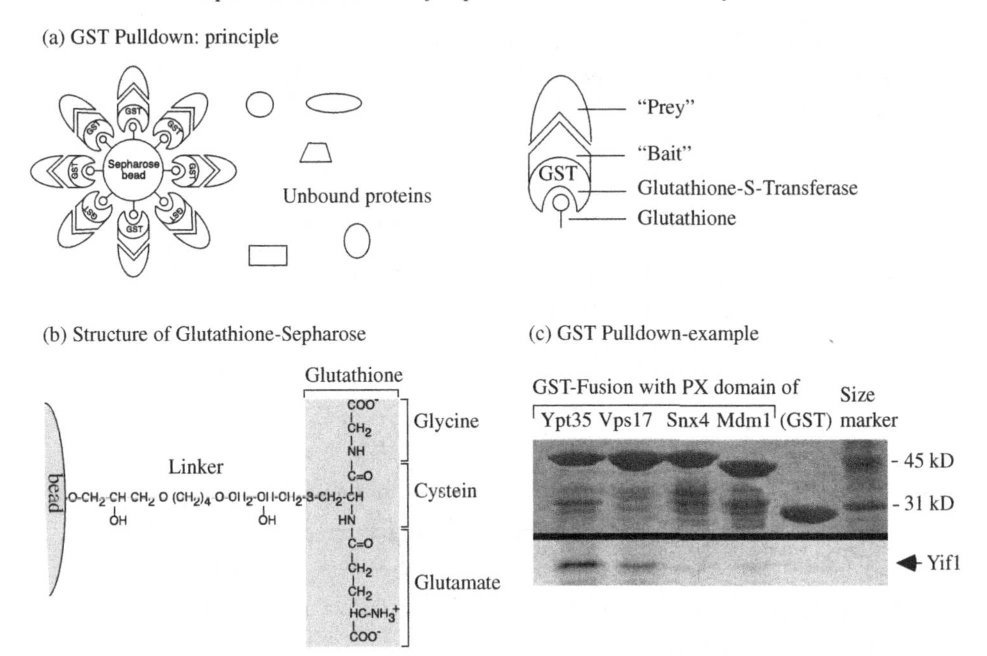

Fig. 4.4. GST pulldown. (a) General principle (see text for details). (b) Structure of glutathione-sepharose. Glutathione is a natural tripeptide, which is also stable as oxidated dimer when two molecules react through their SH groups. (c) Example: the yeast protein Yif1 only binds to certain PX domains of yeast. Four PX domain–GST fusion proteins as well as GST are visible in the top part of the gel (Coomassie stained). The lower panel shows an autoradiogram of radiolabeled Yif1 protein that binds to the PX domains of Ypt35 and weakly to Vps17. Yif1 was translated and radiolabeled *in vitro* and then incubated with bead-coupled GST fusion protein, washed, separated on a gel and then exposed to X-ray film. Modified from [624].

translation reactions to which only the PCR product and radio-labeled methionine has to be added. Alternatively, epitope-tagged proteins can be used that can be detected by Western blotting. In either case, the tagged or labeled protein is mixed with the matrix-bound GST fusion proteins and incubated. Subsequently the beads are washed so that only the GST fusion protein and the bound interacting protein are retained. Note that the concentration of salt in the washing buffer influences the experiment because it determines the stringency through progressive disruption of electrostatic interactions as the salt concentration increases. The next step of the experiment is to boil the glutathione sepharose in sample buffer (containing SDS = sodium dodecyl sulphate as a detergent) and to separate the protein solution on a poly-acrylamide gel (SDS-PAGE = SDS polyacrylamide gel electrophoresis). If the sample contains enough protein it can be stained (e.g. with Coomassie Blue dye) in the gel and thus its molecular mass determined. However, often the amount of protein is not sufficient for staining. In such cases the protein needs to be blotted onto a membrane and detected by Western blotting and autoradiography or by mass spectrometry.

4.3.2 Co-immunoprecipitation

Co-immunoprecipitation (co-IP) is very similar to GST pulldowns (Fig. 4.5). However, instead of glutathione sepharose co-IPs usually use a sepharose matrix coated

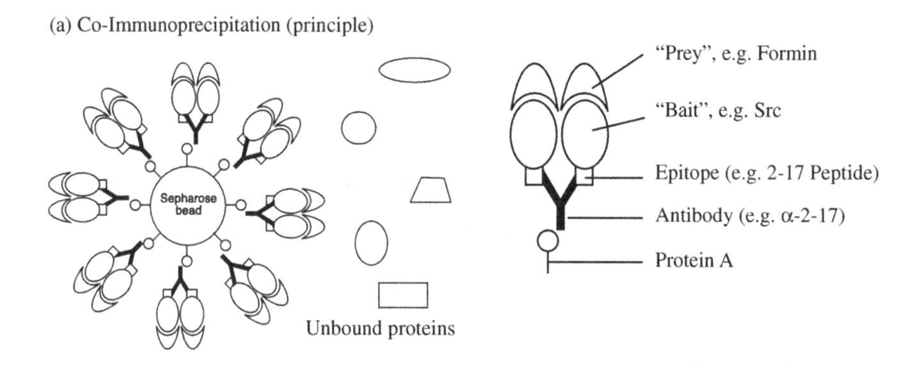

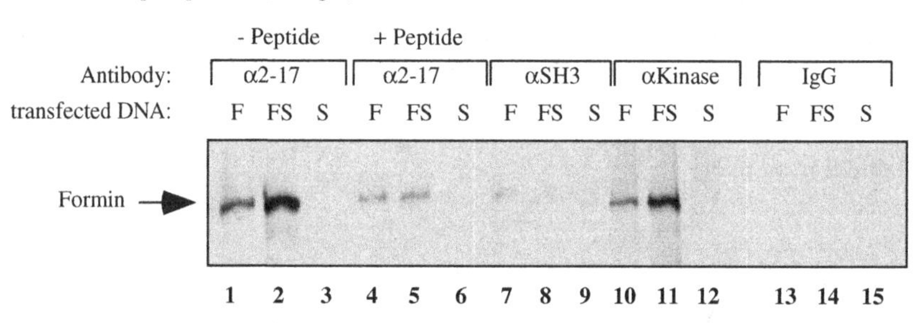

Fig. 4.5. Co-immunoprecipitation. (a) General principle (see text for details). (b) Example: a protein (here: formin) was co-precipitated with the oncoprotein Src. In this experiment four different anti-Src-antibodies have been used: one against Src peptides 2–17 (α-2-17), one against the SH3 domain (α-SH3), one against the kinase domain (α-kinase) and one control antibody mixture without binding specificity towards Src (IgG). The interacting proteins Formin and Src were expressed separately (F, S) or together (FS) in tissue culture. Cells were then lysed and incubated with the bead-bound antibodies. After washing the beads and elution in sample buffer the antibody-bound proteins were separated on a gel, Western-blotted, and detected using an anti-Formin antibody. Note that Formin cannot be co-precipitated with the anti-SH3-antibody because this antibody competes with Formin for a binding site on the SH3 domain. The peptide 2–17 also competes for the same binding site as addition of peptides (+peptide) can block binding of Formin to Src. Modified after [605].

Table 4.4. *Commonly used peptide affinity tags (see text for details)*

Affinity tag	Capture reagent	Sequence
FLAG	Monoclonal antibody	DYKDDDDK
c-myc	Monoclonal antibody	EQKLISEEDL
S-tag	S-fragment of RNaseA	KETAAAKFERQHMDS
Strep II	Streptavidin variant	WSHPQFEK
poly-His	Ni^{2+}-NTA	HHHHHHHH
poly-Arg	Cation exchange media	RRRRR
Calmodulin-binding domain	Calmodulin	KRRWKKNFIAVSAAN RFKKISSSGAL

with protein A, a protein originally isolated from *Staphylococcus aureus*. Protein A binds with high affinity to the constant chains of IgG antibodies and thus sepharose–protein A columns can be easily coated with antibodies of any specificity. Such a matrix can now be incubated with proteins, e.g. from a cell or organ extract. All proteins from this extract that are recognized by the antibody bind to the matrix. All other proteins can be removed by washing with buffer. Bound proteins can now be detected by boiling the matrix in sample buffer and subsequent separation on a protein gel and Western blotting. Alternatively, proteins can be identified by mass spectrometry. The requirement for specific antibodies is currently a major limitation of co-IPs although this may be relieved in the future by commercial production of antibodies against all proteins of a genome. In addition, new technologies have emerged that are not dependent on antibodies but rather use other proteins that can be engineered to have binding specificity for almost any given protein or small molecule (e.g. "affibodies" which are based on genetically engineered protein A). Co-IPs are often used to confirm yeast two-hybrid interactions. If antibodies are not available, proteins can be labelled by specific epitopes such as hemagglutinin (HA) or myc peptides for which commercial antibodies are available (Table 4.4). In fact, all yeast proteins have been epitope-tagged, purified and their interacting proteins identified by mass spectrometry.

4.4 Protein complex purification and mass spectrometry

The GST pulldown and co-affinity purification/MS (**coAP/MS**) approaches have been improved by using novel affinity tags and automated procedures for protein identification. These approaches are treated separately here, but in biochemical terms are similar in principle to the pulldown and co-IP protocols described above.

4.4.1 Purification of proteins using affinity tags

Purification of proteins can be carried out under conditions that preserve stable interactions with accompanying proteins. The proteins can later be identified using methods like Western blot or mass spectrometry (see below). The standard approach is to use an antibody that recognizes the protein of interest (bait protein) immobilized on solid phase media (e.g. sepharose beads) that are packed into a chromatography column, through which a cell lysate or protein mixture is passed. Alternatively, the media is suspended in the cell lysate. The bait is allowed to bind accompanying proteins for a period of time sufficient for equilibrium to be established, after which non-bound proteins are washed away. The washes may vary in stringency, and are sometimes applied in steps of increasing stringency and at other times as a continuous gradient. The eluted proteins are recovered and identified using approaches described below. For high-throughput projects involving dozens or hundreds of baits, more generic approaches are needed, that are independent of the requirement for production of an individual antibody or binding reagent for each bait. A wide range of affinity tags have been developed for this purpose.

Affinity tags are genetically encoded protein fragments that can readily be recovered and are expressed as a fusion with the bait protein. Desirable properties of an ideal tag are:

- compact (so that it does not disrupt the functions of the bait or its interactions)
- high affinity for a capture reagent, so that it can ideally be recovered in a single step
- compatible with economic recovery methods
- non-toxic
- couples to a capture reagent that is non-reactive with endogenous cellular proteins
- can be readily assayed during purification
- works for all proteins.

No single affinity tag satisfies all these desired properties, and a range of strategies is used to express individual proteins, and to recover expressed proteins as well as their bound partners. Affinity tags can be broadly classified into small peptides (e.g. FLAG, poly-His) or large peptides/proteins (e.g. glutathione S-transferase, calmodulin-binding domain) [592]. Small peptide tags are less likely to alter the tertiary structure, disrupt the function of the bait, or to be immunogenic. Larger peptides or proteins may increase the solubility of the bait but may need to be removed for applications such as antibody generation or crystallization. Some common tags are summarized in Tables 4.4 and 4.5.

The **poly-His tag** is often used for protein expression because it consists of a short tag (6+ residues) that can be recovered by using immobilized metal affinity chromatography (IMAC) systems that house nickel or other divalent metal ions. These form coordinate bonds with the histidine side chains [478], and the

Table 4.5. *Commonly used protein affinity tags (see text for details)*

Affinity tag		Capture reagent
Cellulose-binding domain	(CBD)	Cellulose
Chitin-binding domain	(CBD)	Chitin
Glutathione S-transferase	(GST)	Glutathione
Maltose-binding protein	(MBD)	Amylose
Green fluorescent protein	(GFP)	Monoclonal antibody

bait and interacting proteins can be recovered by competition with imidazole. The **FLAG tag** is an eight-residue hydrophobic peptide that is recognized by a number of antibodies with slightly different binding properties (M1, M2, M5). Tandem FLAG peptide units (e.g. 3×) are often employed for increased affinity. The bait proteins are eluted by competition with a synthetic FLAG peptide or using low pH. The **c-myc tag** is an 11-residue epitope from the c-myc protein that is also bound with high specificity by an antibody (named "9E10" after its affinity). The **S-tag** technology is based on an interaction between the 15-residue S-tag and a ~ 100-residue S-protein fragment, both derived from RNaseA, so that assays based on the activity of this enzyme can be used to monitor the purification. The interaction is very strong but is disrupted by strongly acidic conditions. The 26-residue **calmodulin-binding peptide** (CBP) binds calmodulin in the presence of calcium, and the interaction can be disrupted using the chelator EGTA. Several types of protein are able to bind **cellulose** with high affinity, some irreversibly. Severe conditions are generally required for elution of cellulose-binding protein, involving denaturing agents, so this tag is less suitable for detecting protein interactions.

Glutathione S-transferase (GST) is widely used for protein expression and protein interaction studies [507]. The GST protein is quite large (26 KDa) and dimerizes, but binds with high affinity to reduced glutathione. Binding is tight under non-denaturing conditions, so that bait–prey protein interactions may be maintained. The bacterial proteins **Protein A** and **Protein G** (from *Staphylococcus* and *Streptococcus*, respectively), bind with high affinity to the Fc portions of immunoglobulins. Many other epitopes have been used effectively as affinity tags, including **V5** from bacteriophage **T7** and the **HA** tag from hemaglutinin A.

Biotinylation is often used to label biological compounds for subsequent capture due to the extremely high affinity between biotin and streptavidin ($K_a \sim 10^{-15}$ M). Until recently, the introduction of the biotin group was carried out chemically, effectively precluding *in vivo* applications in protein interaction studies. The biotin ligase protein (BirA) from *E. coli* can be used to biotinylate a lysine side-chain within a 15-residue peptide (namely, biotin acceptor peptide). By expressing this tag as a fusion with the bait protein in a cell expressing BirA, the bait can be biotinylated *in vivo*, allowing effective capture of even poorly expressed proteins from complex cell lysates.

4.4.2 Tandem affinity tagging

Bertrand Séraphin and coworkers pioneered the use of tandem tags (Tandem Affinity Purification, TAP), separated by proteolytically cleavable regions [504]. After binding of the bait and associated proteins to chromatography media via one tag, the media is washed and a protease that recognizes and cleaves a sequence in the inter-tag region is introduced. This results in release of the bait which still retains the second tag (Fig. 4.6). A subsequent step introduces media with affinity for the second tag. The advantage of this tandem approach is that very stringent conditions can be used to ensure that minimal background binding (by non-specific proteins) takes place. Potential tags can be drawn from the list discussed above, but effectively the FLAG, His, HA, Protein A, myc, and calmodulin-binding domain tags have been used in many systems because they can bind under non-denaturing conditions where interactions of the bait with associated proteins

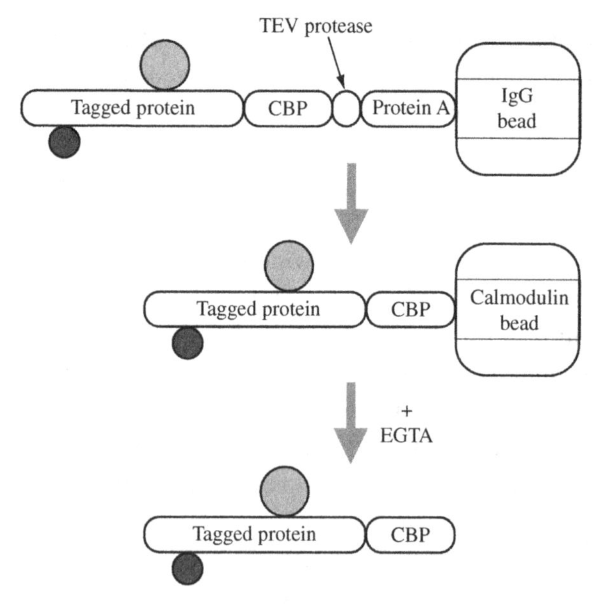

Fig. 4.6. Tandem affinity purification (TAP) coupled to mass spectrometry. Using the TAP approach, a protein of interest (tagged protein) is expressed as a fusion with two affinity tags separated by a protease cleavage site (here calmodulin-binding protein, tobacco etch virus protease, and Protein A). Associated proteins are represented by gray spheres.

can be maintained. The initial description of the TAP method used Protein A and calmodulin-binding domain tags separated by a tobacco etch virus (TEV) protease cleavage site. The TEV protease recognizes a seven-residue sequence (EXYXQS, cleavage C-terminal to serine) that is uncommon in the proteome. TEV cleavage is efficient at low temperature and can be improved by placing the recognition sequence between two domains (as in the case of the TAP tag method). Other protease cleavage sites may be used. **Enterokinase** recognizes the sequence DDDDKX and cleaves C-terminal to the lysine, although some non-specific cleavage occurs at alternative sites with low frequency. Note that the FLAG tag contains an enterokinase recognition site. **Factor Xa** can also function at low temperatures and cleaves C-terminal to the sequence IEGR. In recent years, inteins have also been successfully used for self-cleaving tagged protein release without the need for protease [673].

4.4.3 Genetics and cloning of affinity tagged proteins

Nearly all affinity tags currently used for high-throughput protein–protein interaction studies are introduced via expression cloning, the DNA encoding the tag being inserted genetically at some point in the gene encoding the bait. Both amino- and carboxy-terminal tags are commonly used. However, each protein is obviously unique, and alternative tagging sites may need to be examined. In some cases, tags may be inserted into non-terminal regions, or into mutant proteins or proteins with regions deleted. In many cases, structural or functional information may help. For example, in cases where post-translational cleavage of the amino terminus generates the mature protein, or where the carboxy-terminus contributes a structural fold essential to function, then these positions would be avoided as sites for introducing an affinity tag.

Large-scale studies are dependent on methods to introduce the DNA into specific genomic locations in a high-throughput manner, and so to date have been most common in model organisms for which such methods are available, notably yeast. Both *Saccharomyces cerevisiae* and *Schizosaccharomyces pombe* contain high efficiency recombination machinery that permits introduction of exogenous DNA in a sequence-specific manner and a number of very powerful genome-wide resources have been generated. Sets of *S. cerevisiae* strains, each encoding an individual gene tagged with GST, GFP, TAP (Protein A-TEV-calmodulin-binding peptide), or FLAG have been generated [44]. Similar sets comprise strains in which each single gene has been replaced with a marker. The *E. coli* genome has also been extensively tagged [35, 95], while a smaller yet significant number of human genes have also been tagged (e.g. [77]). Most of these strains are publicly available.

4.4.4 Isolation of protein complexes

Generally, a strain or cell line containing the bait fused to an affinity tag is grown and the cells are lyzed by using methods appropriate to the organism. (An exception is where the fusion proteins are generated by *in vitro* methods such as cell-free translation, where no lysis is necessary.) It is important to ensure that the methods are consistent with maintaining the protein–protein interactions. Furthermore, because lysis can lead to mixing of cellular compartments, care should be taken to avoid exposure to proteinase or other enzymes that might degrade proteins or disrupt their interactions. For this reason, proteinase inhibitors and low temperatures are routinely used, and early steps in the procedure (when proteolysis is most likely) should be carried out as rapidly as possible. Affinity-tagged proteins may bind either to media packed into chromatography columns through which lysate is passed, or to media suspended in the lysate. The choice of approach depends on issues such as protein abundance, binding affinity, and other factors like automation or cost. Ideally, the procedure will be optimized for each individual bait. However, in high-throughput projects, this is often impossible, so compromise conditions compatible with the other elements of the project are used.

4.4.5 Proteomics by mass spectrometry

Mass spectrometry (MS) is the study of gas phase ions as a means to characterize molecular structure [9]. Mass spectrometers separate the ions in space or time based on their mass-to-charge ratio (m/z). Currently, proteomics relies especially on two ionization techniques, electrospray ionization (ESI) and matrix-assisted laser desorption ionization (MALDI). ESI involves ionization of peptides at atmospheric pressure by nebulizing a stream of solvent under a potential difference of several thousand volts. The technique, which is usually coupled with triple quadrupole and ion trap or time of flight detectors, can determine protein masses in excess of 150 000 to an accuracy of 0.005%. Nanoelectrospray is a refinement of ESI in which miniaturization of the electrospray source increases the sensitivity of the analysis to the low femtomolar range.

For identification purposes, protein mixtures are typically digested with trypsin before analysis by MS. Peptides resulting from trypsin treatment, which primarily cleaves at lysine and arginine residues that occur approximately every 10–15 residues in proteins, often have size and charge properties that render them effective candidates for ionization. In MALDI, the peptides are embedded in a matrix that absorbs laser light, allowing desorption of ions and analysis by the mass spectrometer. MALDI is most often used in conjunction with time-of-flight (TOF) mass spectrometers, which use transit time differences through a drift region of the instrument to separate ions of different m/z. MALDI-TOF MS permits very sensitive and accurate measurement of peptides up to about 500 kDa. These peptides are generally identified using peptide mass fingerprinting, a technique that compares the peptide masses observed by MS to a set of masses predicted from an *in silico* digest of proteins encoded by the DNA sequences from genomic databases. Computer programs search for matches between actual and theoretical fragment masses, with strong matches leading to identification of the protein under investigation.

Tandem MS (or MS/MS) is also frequently used in proteomics, particularly with ion trap and quadrupole instruments. In this approach, two or more stages of mass analysis are conducted sequentially. Introduction of an inert gas at the position of the second and/or subsequent mass analyzers causes fragmentation of the initial ionized peptides to produce daughter ions. The product ion spectra can be interpreted to deduce the amino acid sequence of a protein by comparison with predicted patterns obtained from translated protein databases (as with peptide mass fingerprinting). The development of these algorithms was a major advance because it removed the need to manually interpret each mass spectrum and so opened the door to truly high-throughput proteomics.

4.4.6 Identifying interacting proteins using mass spectrometry

Both MALDI and MS/MS are commonly used to identify proteins purified using affinity tagging (or other) strategies. Such purifications may vary widely in sample complexity and dynamic range, so protein mixtures may be directly analyzed by MS, or may require some fractionation before or during MS analysis to reduce the complexity of the mixtures entering the mass spectrometers. For example, purified protein preparations may undergo electrophoresis on an SDS polyacrylamide gel (PAGE), stained with silver or Coomassie blue dye, and the visible bands removed and identified by peptide mass fingerprinting using MALDI-TOF MS (Fig. 4.7). Alternatively, an aliquot of purified protein preparation can be digested directly following the purification experiment and the peptides separated by online HPLC (liquid chromatography–tandem mass spectrometry, LC/MS/MS). Advantages of the SDS-PAGE MALDI approach include the fact that the identification is linked to a specific band on a gel (that can be compared to a theoretical protein mass), while advantages of the LC/MS/MS method include less manual effort and in some cases greater sensitivity.

4.4.7 Quantitative proteomics

Although gels can be used to estimate relative protein abundance differences between samples (using staining intensity), this information is lost upon analysis

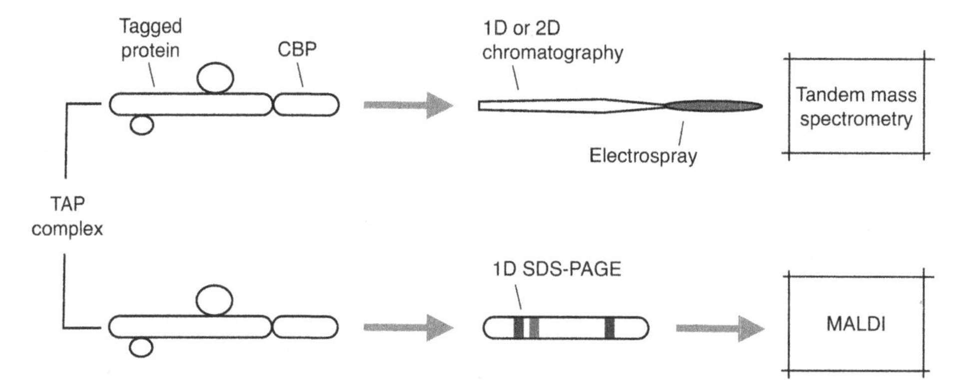

Fig. 4.7. Protein identification by mass spectrometry (MS). The purified TAP complex of proteins can be digested using trypsin and the resulting peptides introduced directly to a mass spectrometer by electrospray ionization following separation by one- or two-dimensional chromatography (top). Alternatively, the intact proteins may be separated by gel electrophoresis and the individual bands digested using trypsin before introduction to a mass spectrometer using the MALDI process. Both approaches result in mass spectra that can be used to identify the protein components of the purified mixtures.

by LC/MS. Quantitative proteomics data are useful for protein interaction studies because they can help to distinguish "true" interacting proteins from the background of non-specific proteins identified in control experiments. For example, a purified complex should contain all proteins in stoichiometric amounts; quantitivative measurements can determine the stoichiometry and thus tell "true" components of a complex from contaminants. Several approaches have been used to obtain quantitative data, mostly involving the use of stable isotopes such as ^{2}H, ^{13}C or ^{15}N. Proteins or peptides obtained from two or more sources, each labeled so that the equivalent peptides have different masses, can be tracked by mass spectrometry using these mass differences. Quantitative proteomics strategies can be classed as either "pre-experiment" labeling, or "post-experiment" labeling.

In pre-experiment strategies, the isotope is introduced during culture of the organism or cell line. For example, one sample may be grown in normal media while another is grown using media containing deuterated leucine [450]. Peptides containing deuterated leucine will contain mass shifts relative to those containing natural leucine and the relative signal emanating from each ion in the mass spectrometer can be used to estimate the relative abundance of the parent peptides. In post-experiment labeling, the label is introduced chemically or enzymatically after the proteins have been purified (Fig. 4.8). Several variants have been described. The isotope-coded affinity tag (ICAT) method described by Gygi and coworkers [239] uses a biotinylated thiol-active reagent containing a linker that is produced in alternative heavy (containing eight deuterium atoms) and light (containing eight

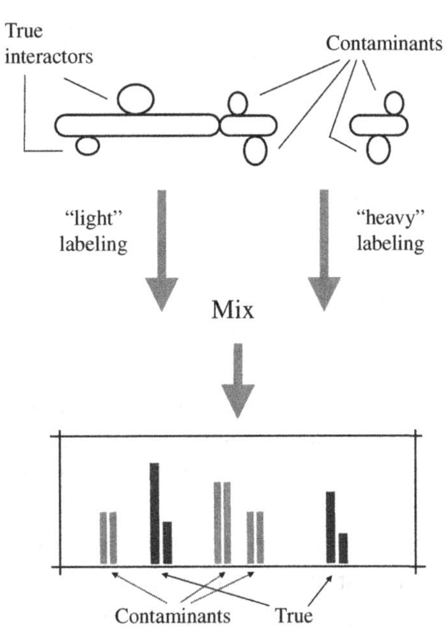

Fig. 4.8. Quantitative proteomics to assess specificity in pulldown experiments. If peptides derived from a control pulldown are labeled using a "heavy" stable isotope-containing reagent, while those from the sample pulldown are labeled using a "light" reagent, true interactors can be distinguished from contaminant proteins. This is because the heavy and light contaminant peaks are derived from both samples in approximately equal amounts, while light peaks will be significantly more intense than light peaks for peptides that are only present in the true sample fraction.

hydrogen atoms) forms. The thiol-active component of the reagent binds cysteine-containing peptides in the lysate. When comparing two protein samples, both forms are used, so cysteinylated peptides in one sample are heavy labeled while the corresponding peptides in the other sample are light labeled. The two fractions are pooled, purified by solid phase extraction, and analyzed by LCMS. The heavy–light pairs for each peptide can be recognized by their mass difference during total ion scans, and the relative peak area for these peaks is used to determine the relative abundance of the corresponding peptides in the original sample.

4.5 Protein and peptide chips

Proteins can also be printed onto glass such as microscope slides and these protein chips can be then screened with labeled proteins. Zhu *et al.* (2001) were the first to make a proteome-wide chip with almost all GST-His6-tagged yeast proteins attached on a single slide that was coated with nickel [694]. They screened these slides with biotinylated calmodulin and detected six previously known calmodulin

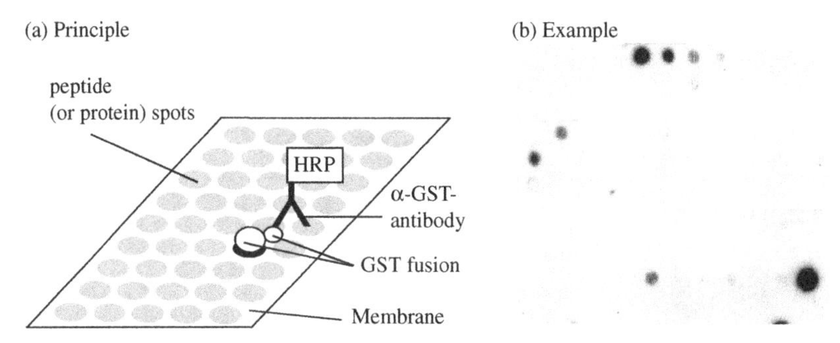

Fig. 4.9. Peptide and protein arrays. (A) Proteins or peptides are spotted on membranes or glass slides and detected by interacting proteins and pertinent antibodies. (B) Peptides can also be synthesized directly on the membrane using a technology called SPOT synthesis as shown here. Peptides on the membrane were probed with GST-fusions of an FF-domain from yeast. Courtesy of Claudia Ester.

interactors and 33 new ones. So far, not many additional screens have been published and thus it is difficult to judge how protein chips do in comparison with other methods. In addition to full-length proteins, glass slides or membranes can also be coated with short peptides, usually of 10–30 amino acids in length. Such peptide arrays can be screened for interacting proteins when labeled proteins are incubated with them, similar to protein chips. Their main application is the mapping of interaction epitopes (Fig. 4.9). This is based on the fact that most proteins bind to relatively short linear peptides. Since the membrane-bound peptides are not sterically constrained as when they are in a folded protein, they can be induced to fit into a protein that is used to screen the library. However, this flexibility also may lead to false positives.

4.6 Other methods for interaction detection and functional analysis

There is a far wider range of methods to investigate protein–protein interactions than can be discussed here. We can only mention a few of them briefly but they will play an increasing role with more data being generated.

Genetic interactions

It has been known for a long time that genes can "interact" genetically, that is, they are involved in the same or related processes. Usually such interactions are investigated through mutagenesis screens in which one mutant does not have a strong or lethal phenotype. However, when this mutant is combined with another mutant that also does not have a strong or lethal phenotype, their combination may be lethal or much more severe than the individual genes. In fact, many mutations do not have a phenotype at all (under standard laboratory conditions) but when combined they can be lethal. In yeast such "synthetic lethal" screens have been done on a large scale and revealed thousands of genetic interactions [598]. More recently, many genetic interactions have been related to physical interactions [67] although the majority of phenotypes can still not be explained in molecular terms. In any case, it will be only a matter of time until such "functional" interactions will be described in molecular terms, either involving protein–protein or protein–small molecule interactions.

Functional interactions such as post-translational modifications

The protein chips mentioned above can also be used to detect "functional" interactions such as phosphorylation events. "Functional" means that an interaction involves a certain consequence beyond the interaction. Proteins that modify other proteins have to interact with them and kinases, proteases, or chaperones are just a few examples for such dynamic interactions. In fact, most methods such as coAP/MS or Y2H appear to miss such interactions because of their transient nature. Thus, the most efficient way to study such cases is to use the actual modification reaction. For example, Ptacek *et al.* [481] have analyzed protein phosphorylation in yeast by incubating protein arrays with individual kinases and radiolabeled ATP and discovered more than 4000 phosphorylation events involving 1325 proteins. This and other studies showed that an estimated one third of all proteins in yeast, flies, or humans are phosphorylated. Note that there are hundreds of possible modifications in proteins and only a few of them have been studied systematically [210].

4.7 Quality of large-scale interaction data

Although several methods for the detection of protein–protein interactions exist, no method is able to identify all protein–protein interactions – that is, each experimental strategy generates a significant number of **false negatives**. The reasons for this systematic error are only partly understood. Two-hybrid false negatives might be caused by insufficient expression or nuclear localization of the tested proteins, by sterical effects due to the usage of two fusion proteins ("two-hybrid"), or involve weak interactions within complexes that require cooperative effects to be stabilized [23]. Mass spectrometry analysis, on the other hand, often has problems with low abundance proteins and proteins that are only weakly associated with protein complexes (or transient interactions) and hence tend to get lost during purification.

False negatives lead to gaps in our picture of the internal structure of the cell. However, equally serious are **false positive** interactions, which result in erroneous data and thus misleading conclusions. Two types of false positives need to be distinguished: technical false positives and biological false positives. In yeast two-hybrid studies, technical false positives can arise by bait constructs, which activate the reporter gene without interacting with a prey ("self-activating baits"). In addition, mutations in the reporter genes or incorrect folding in the unnatural environment are sources for technical false positives. On the contrary, "biological false positives" represent true interactions that take place in the Y2H system but have no biological relevance [284]. Examples are proteins that are interacting in the Y2H system but are expressed in different cell types or different organelles *in vivo*.

A number of studies tried to estimate the number of true positives for high-throughput interaction studies. The critical point of any attempt to estimate the number of true and false positives in a HTS interaction study is the choice of the "true positive" data set against which the new interactions are evaluated.

To estimate the overall interaction reliability, Deane *et al.* compared the co-expression profiles of known interacting proteins with protein pairs from high-throughput screens [138]. Based on this comparison, they estimate a false-positive rate of 50–70% for yeast two-hybrid experiments. Sprinzak *et al.* tried to estimate the interaction quality based on the observed degree of co-localization and shared functional role of the interacting proteins [573]. This estimation yielded a false-positive rate of ~50% for large-scale yeast two-hybrid studies. Patil and Nakamura used a combination of three genomic features (known interacting domains, gene annotations, and sequence homology) with a Bayesian network approach [467] and estimate that 56% of the high-throughput interactions for yeast have high reliability.

Edwards *et al.* [162] selected known interactions from 3D-structures (RNA polymerase II, proteasome and the Arp2/3 complex), and additionally, complexes from the literature. The crystal structures of complexes approximate the "absolute truth" about stable protein interactions because they reveal all interactions in atomic detail, at least for the proteins that have been co-crystallized. Based on crystal structures, Edwards *et al.* found a false negative rate of 51–96% for yeast two-hybrid and of 15–50% for *in vivo* pulldown experiments, respectively. In this context it is remarkable that conventional "low-throughput" methods also produce a large fraction of false positives – for example 61% in a pulldown study of RNA polymerase II [162]. Another method proposed by D'Haeseleer and Church [144] does not rely on a gold standard and involves comparing two interaction datasets to each other and to a reference dataset. In this study, a false positive rate between ~50–90% (depending on the dataset) was calculated both for Y2H and for coAP/MS experiments.

Overall, several approaches to estimate the reliability of two-hybrid interactions conclude that 50% or more are true positive interactions. This is underlined by the finding in recent large-scale yeast two-hybrid studies that between 50% and 70%

of the identified interactions can be reproduced by an independent method (which also has a certain false-negative rate) [511, 604].

However, several approaches were devised to minimize the number of false positives further. These approaches rely either on the identification of intrinsic properties, which lead to unspecific interactions, or on the integration of several datasets (and data types). Uetz *et al.* [607] systematically evaluated the signal/background ratio and discarded yeast two-hybrid interactions, which could not be reproduced. Ito *et al.* [283] defined interacting protein pairs found three or more times as the (supposedly reliable) "core" dataset. Rain *et al.* screened bait proteins against a genomic fragment prey library and considered overlapping prey fragments as the most reliable. This approach combines reproducibility and identifies the interacting domain at the same time [488]. Giot *et al.* employed a statistical technique (logistic regression) together with a set of known "gold standard" interactions to identify properties of true positive interactions. The authors estimated their filtered high confidence network to retain 40% interactions of biological significance [221].

All these approaches are based on the evaluation of intrinsic properties of the respective system. Further evidence for reliable interactions can be obtained by integration of several data sources and data types. Several studies showed that interacting proteins tend to be co-expressed at the mRNA level under various experimental conditions [214, 234, 288]. However, while co-expression of the two partners increases the confidence in a protein–protein interaction, it is only an indirect measure of its reliability. Proteins in a complex may need to be expressed at similar levels in order to maintain their stoichiometric ratios, but this is certainly not true for all complexes and even less so for transient interactions that are often found in Y2H screens. Reliability can be also gained by looking at the interactions of homologous proteins. Interactions reproduced with paralogous proteins were labeled as highly reliable by Deane *et al.* [138]. However, many proteins do not have paralogs and paralogs with diverging functions do not need to retain the same (or even similar) interactions.

4.8 Comparison of methods

4.8.1 Y2H vs. co-AP/MS

Detection of protein interactions by Y2H and coAP/MS differ in a number of important aspects (Fig. 4.10). First, Y2H assays mainly detect direct binary interactions, whereas coAP/MS detects one-to-many relationships, which usually also include indirectly interacting proteins. Second, Y2H analysis is often conducted under non-physiological expression conditions, whereas for coAP/MS, detection

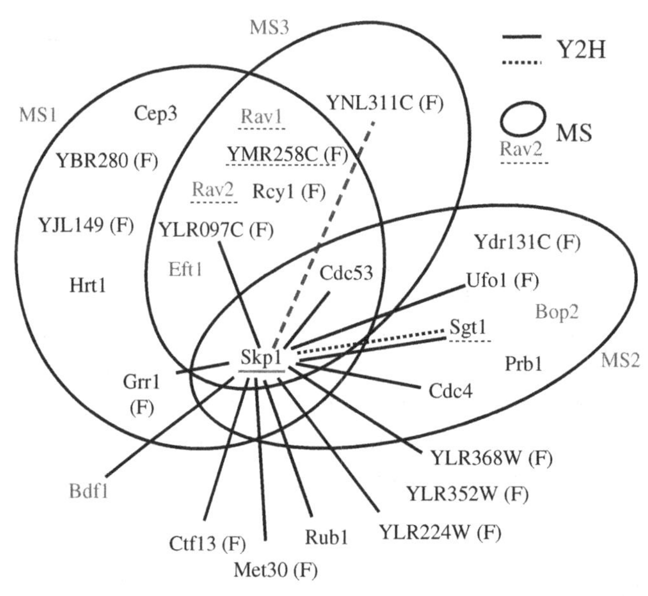

Fig. 4.10. Comparison of interaction data gained by Y2H and MS using Skp1 as a bait in both Y2H screens and coAP/MS analysis. The purified complexes of Skp1 from three independent MS studies (circles, MS1-3) and the binary interactions from two Y2H studies (straight and dashed lines) are shown. Despite the differences in the data sets, most of the discovered interactions seem to be plausible: most proteins are known to be involved in protein degradation. Skp1 is directed to its target proteins via so-called F-box proteins, which contain a short peptide motif, the F-box (F). Note that neither Gavin *et al.* (2006) nor Krogan *et al.* (2006) found any proteins with Skp1 as bait. However, Gavin *et al.* found Skp1 in purifications with Rav2, Ymr258c, and Ypt52 as bait and Krogan *et al.* found Skp1 with 12 bait proteins, namely Cdc34, Csn9, Gcd11, Pcl8, Pol30, Rav1, Ri1, Sgt1, Vip1, Vma2, and Ymr258c. Proteins found in these purifcations are indicated by dashed underlining. Modified after [596].

can be done at endogenous expression levels. Third, because of the previous reason, Y2H interactions are hypothetical interactions which may not take place *in vivo* or only under conditions that are not yet known. Purified complexes are forming under the condition under which the complex was purified. However, different complexes may form under different conditions and laboratory conditions may not even reflect the normal circumstances in the life of a cell.

The published experimental approaches for high-throughput interaction analyses employing these methods have taught us already one important lesson: Y2H and MS datasets are strikingly different but also complementary. This difference between datasets – even between datasets derived by a similar method for the same species – is exemplified by two recent coAP/MS studies for yeast [214, 336]. Goll and Uetz found that only 28% of the core complexes from Gavin *et al.* are completely contained in the complexes identified by Krogan *et al.* Only six out of about 500 complexes were identical [227].

The complementary nature of Y2H and coAP/MS methods seems to depend on molecular properties of the interactions: transient interactions are often found by yeast two-hybrid analysis but not AP, whereas stable interactions (such as those in protein complexes) are more reliably identified by *in vivo* pulldown techniques [23]. This finding is not surprising, given the highly cooperative forces that stabilize a protein complex: weak interactions in a complex will not be detected by Y2H analysis as long as only pairs of proteins are tested that are not stabilized by the other subunits of a complex. In addition, coAP/MS methods are biased towards highly abundant proteins [625]. Y2H assays are less biased in this respect as they use proteins expressed at sufficient levels from a heterologous promoter.

4.8.2 coAP/MS vs. protein chips

In the first large-scale ErbB-interactome study, Schulze *et al.* (2005) used a quantitative proteomic approach for identifying proteins associated with phosphotyrosine motifs of the ErbB receptor members [537]. To this end, they generated 89 different "bait" peptides in phosphorylated and unphosphorylated forms covering all intracellular tyrosine residues of the ErbB-family members. In the next step, these peptides bearing phosphorylated and unphosphorylated tyrosine residues were incubated with HeLa cell lysates and pulldown experiments were performed to enrich specific binding partners for the phosphorylated bait peptides. Lastly, these proteins were identified and quantified by mass spectrometry. Surprisingly, only 40 out of the 89 examined tyrosine residues did have an interacting protein in their phosphorylated form. Most of the tyrosine residues that interacted with specific partners accumulated at the C-terminal regions of the receptors. This study also indicated that the distribution of interacting partners of the different ErbB members shows clear differences between individual receptors, but also a significant overlap. For example, both EGFR and ErbB4 have multiple binding sites for the adaptor proteins Shc and Grb2, but only EGFR binds the ubiquitin ligase Cbl, whereas ErbB4 is unique in binding Nck. Altogether, Schulze *et al.* found 10 ErbB interactors, all of which have either an SH2 or PTB domain, thus emphasizing the specificity of these signal transduction modules [468].

In a related project, Jones *et al.* used protein microarrays to identify ErbB interactors [298]. Since the phosphotyrosines in ErbB receptors are primarily bound by SH2 and PTB domains, Jones *et al.* successfully expressed and purified 106 (out of 109) SH2 domains and 41 (out of 44) PTB domains encoded in the human genome. These domains were then spotted onto glass slides and subsequently probed with ErbB peptides that contained phosphorylated and nonphosphorylated tyrosines, respectively. While Schulze *et al.* used all possible 89 tyrosine-containing ErbB peptides for their pulldowns, Jones *et al.* concentrated on 29 peptides that were known to be phosphorylated in EGFR, ErbB2 and ErbB3, and four peptides that were predicted to be phosphorylated in ErbB4. These experiments revealed that each phosphotyrosine on EGFR binds seven different proteins on average, whereas those on ErbB2, ErbB3 and ErbB4 bind 17, nine and two proteins on average, respectively. This adds up to many interactions, namely 54 with EGFR, 59 with ErbB2, 37 with ErbB3 and 8 with ErbB4, most of which were new. The sobering fact is that the mass spec study by Schulze *et al.* found quite different, and, in fact, much smaller numbers, namely nine, four, four, and eight different interacting proteins for the four receptors, respectively. While at least the number of identified ErbB4 interactors by both approaches appears to be identical, only one protein (Shc) is actually common to both ErbB4 data sets [606].

Where do these dramatic differences come from? Well, simply from the fact that the two groups looked at very different things: while Schulze *et al.* pulled down proteins that most likely bind to ErbB receptors in HeLa cells, Jones *et al.* looked at the whole SH2/PTB interaction space of ErbB receptors *in vitro*. That is, Jones *et al.* told us which SH2 and PTB domains may bind to which receptor if both are present in a cell. In contrast, Schulze *et al.* told us which SH2 and PTB proteins bind to ErbB receptors specifically in HeLa cells. Unfortunately, we have to wait until further studies reveal which proteins can be pulled down in the other 200+ human cell types or, alternatively, which of the ErbB receptors and SH2/PTB proteins are expressed together in those cells. At least we know that the proteins detected by Schulze are coexpressed in HeLa; some differences (e.g., Grb2 interacts with all four ErbB in Schulze but only with ErbB2 in Jones; STAT5 interaction is missed altogether in Jones) might also be the result of possible methodological differences.

4.9 Conclusions

It should have become clear from this chapter that none of the technologies available today is perfect. No technology can identify all interactions, and all of them have a certain fraction of false positives and false negatives. Furthermore, quantitative technologies are only being developed and none of them has been applied on a large scale. Computational methods are required to assess the quality of interaction data and thus to prioritize them for further study. One way to do this is to integrate various datasets including non-interaction data such as structural information. The other chapters in this book will detail some of these strategies and tools.

Glossary of abbreviations in Chapter 4

5-FOA	fluoro-orotic acid, a chemical that inhibits the HIS3 enzyme often used in the yeast two-hybrid system
AD	activation domain, i.e. transcriptional activation domain
B2H	bacterial two-hybrid
B42	artificial bacterial transcriptional activator; used as an alternative to the yeast GAL4-AD
CBD	cellulose-binding domain or chitin-binding domain
CBP	calmodulin-binding peptide (CBP) binds calmodulin in the presence of calcium, and the interaction can be disrupted by EGTA.
Co-AP	co-affinity purification
Co-IP	co-immunoprecipitation (see text for details)
DBD	DNA-binding domain
EGFR	epidermal growth factor receptor
EGTA	ethylene glycol tetraacetic acid, a chelating agent that is related to the better known EDTA, but with a much higher affinity for calcium than for magnesium ions.
ESI	electrospray ionization
FC	fragment complementation
FRET	fluorescent resonance energy transfer
GAL4	a yeast transcription factor that served as basis for the first Y2H assay which used the DBD and AD of this protein
GFP	green fluorescent protein (GFP)
GST	glutathione-S-transferase
HA	hemagglutinin, a protein and antigen of Influenza virus
HIS3	imidazoleglycerolphosphate (IGP) dehydratase, an enzyme that catalyzes the seventh step in the histidine biosynthesis pathway
HRP	horse-radish peroxidase
HTS	high-throughput screening
ICAT	isotope-coded affinity tag, a chemical tag used to label proteins for subsequent MS analysis
LacZ	beta-galactosidase. This enzyme from *E. coli* is often used as reporter gene in Y2H assays
LCMS	liquid chromatography mass spectrometry
LEU2	3-isopropylmalate dehydrogenase, an enzyme that catalyzes the third step in the leucine biosynthesis pathway
LexA	bacterial repressor and DNA-binding protein that is used as an alternative to GAL4
M2H	mammalian two-hybrid

MALDI	matrix-assisted laser desorption ionization
MBD	maltose-binding protein
MS	mass spectrometry
PBS	phosphate-buffered saline, a pH-stabilized buffer solution
PTB	phospho-tyrosine-binding domain
SAI	socio-affinity index, a measure that describes the tendency of a protein to interact specifically with others
SDS-PAGE	sodium dodecyl sulphate poly-acrylamide gel electrophoresis
SH2	Src homology domain 2
SH3	Src homology domain 3
TAP	tandem affinity purification
TEV	tobacco etch virus
TOF	time of flight, usually used in combination with MALDI ("MALDI-TOF"), a special kind of mass spectrometer
URA3	orotidine-5′-phosphate decarboxylase, an enzyme that catalyzes the sixth step of pyrimidine biosynthesis
Y2H	yeast two-hybrid system

5

Modeling protein interaction networks

FRANCESCO RAO

5.1 Introduction

Most of biology is interpreted via macromolecular interactions. Protein interaction networks represent the first genome-wide drafts of those interactions and have been explored as models for understanding cellular processes [45, 305]. Given the constant flow of new experimental data on the correlation of genes and proteins within an organism or even between different species, there is the need to rationalize in a solid framework the thousands of possible protein interactions inferred by experiments. Computational models can help in this task investigating the microscopic mechanisms responsible for the behaviors observed in experiments as different as yeast two-hybrid, mass spectroscopy, gene co-expression, synthetic lethality, just to mention the most popular [25, 512, 517].

In protein interaction networks, nodes and links represent the proteins and the interactions between them, respectively (Fig. 5.1). However, depending on the approach that has been used to generate the map, a link does not always indicate a direct physical interaction [625]. It can also represent correlated expression in the cell, performance of successive steps in a metabolic pathway, similar genomic context and so on. Observation of direct binding is a good indication that two interacting proteins cooperate in the same biological pathway and whenever possible this chapter will restrict to this type of interactions.

Modeling can focus on different network properties. Some approaches use evolutionary arguments [570, 616]. Others take into account genomic information as well as the physico-chemical properties of proteins in a statistical way [139, 397]. Finally, there are models that consider the full complexity of the problem treating proteins as three-dimensional polypeptide chains with a given structure and dynamics [25, 512, 517]. In other words, protein interaction models span different resolutions: from shapeless objects to the full atomistic detail. This diversification is very important because while low-resolution models aim to investigate

Networks in Cell Biology, ed. M. Buchanan, G. Caldarelli, P. De Los Rios, F. Rao and M. Vendruscolo.
Published by Cambridge University Press.

Table 5.1. *Summary of the predictions that the different methods described in this chapter are able to provide (○ = in some cases, • = always)*

Model	Large scale properties	Predicting interactions	Structural properties	Section
Evolutionary	•			5.2.1
Binding affinity	•			5.2.2
Gene sequences		○		5.3.1
Bayesian analysis		○		5.3.1
Protein docking			•	5.4.1
Homology			•	5.4.2

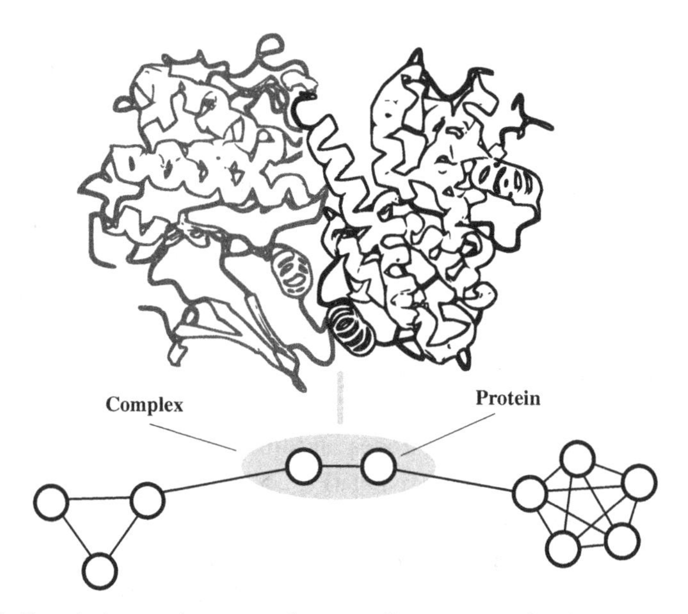

Fig. 5.1. Protein interaction networks naturally represent the thousands of physical bindings observed experimentally. In these maps, nodes and links are proteins and binary complexes, respectively.

the general principles for protein organization, high-resolution models provide a full mechanistic explanation of the interactions. A short summary is provided in Table 5.1.

In this chapter a perspective on some of the strategies so far proposed is presented. Starting with low-resolution models, the text moves towards approaches providing more and more structural detail. A full review of the argument goes far beyond the scope of this chapter and the interested reader is invited to refer to the original research papers cited throughout the text.

5.2 Scaling laws and network topology

From high-throughput experiments, such as the two-hybrid approach [607], it has been observed that protein interaction networks from different organisms share several universal features. These include scale-free topology and small world behavior [292]. The former indicates that the number of connections per protein is not distributed randomly [53]. They follow a power-law distribution such that most nodes have only few connections, and only a small fraction of them are highly connected, the so-called "hubs." The latter indicates that a connection can be established between any two nodes of the network by following only a small number of links [637]. This property is partially due to the presence of hubs which have been recently recognized to play an important biological role in the cell [242, 292]. Models which reproduce the observed scaling properties of experimentally resolved protein interaction networks provide an important step towards the understanding of the biological meaning of those properties.

5.2.1 Evolution and duplication of proteins

Proteins are divided in families according to their sequences and functional similarities [254, 589]. The existence of these families can be explained by using the evolutive hypothesis that all proteins in a family evolved from a common ancestor [688]. This evolution is thought to take place through gene or entire genome duplications, resulting in redundant genes. After the duplication, redundant genes diverge and evolve to perform different biological functions.

Using this fundamental property of genomes, models for the protein interaction network of yeast from two-hybrid experiments have been built [570, 616]. Those models do not take into account the information on the function or structure of the proteins involved and aim to reproduce the scaling behavior observed in protein interaction networks [292]. Starting from an arbitrary graph of a few nodes, a network of any size is built using the rule of duplication and mutation. The first step of the procedure is to randomly select a node of the network and to duplicate it. The new node is characterized by the same connections as the parent. This stage models the process of gene duplication in which the original and the duplicated protein share identical properties, i.e. the same links. To implement the possible mutations occurring at the gene level, the new created links can be removed and rewired with probability p and q, respectively. The two parameters of the model can be estimated from empirical data [570]. Model networks obtained by a duplication mechanism reproduce the behavior of experimentally resolved protein interaction maps [292, 570, 616] showing that their scale-free topology and small-world property can be interpreted and reproduced in terms of gene duplication and diversification.

On the other hand, other studies indicate that the topology of protein interaction networks cannot be attributed solely to this mechanism [139, 168]. Using phylogenetic trees the "age" of any protein in the yeast organism has been estimated [168]. This analysis indicates that the age correlates very well with the number of connections a protein makes, suggesting a mechanism for network growth compatible with the classical preferential attachment model [53].

5.2.2 Protein binding physical models

The scaling properties observed in the protein interaction maps from yeast two-hybrid experiments [292] have also been explained in terms of simple physical models that explicitly account for the physics of protein binding [139]. This approach starts with the assumption that a large fraction of the binding free-energy between two proteins is caused by the burial and desolvation of hydrophobic groups at the binding interface. This hypothesis is not strong and takes into account in a reasonable way some of the most fundamental aspects of protein binding and folding [167].

Given two proteins i and j, the desolvation of K_{ij} hydrophobic residues at the interface is related to the binding affinity A_{ij}, i.e. the strength, of the interaction between the two proteins [139, 167]:

$$A_{ij} \sim \exp(K_{ij}). \tag{5.1}$$

Once a cutoff A_C is set, a link between two proteins is made if the binding affinity between them is larger than the cutoff. It is important to note that this model is mathematically related to a large class of scale-free networks static models [99]. Within certain ranges of A_C, network topologies created by this model show very good agreement with respect to yeast protein interaction networks from experiments [283, 607], hence the power-law behavior of the node degree (i.e. the scale-free behavior) as well as higher-order node correlations like the clustering coefficient and the average neighbor connectivity. Moreover, this approach predicts a correlation between the hydrophobicity of the protein surface and the number of partners (i.e. links), that is confirmed by yeast two-hybrid experiments [139].

These findings indicate that the scaling laws observed in protein interaction experiments may arise as well from a set of largely non-specific interactions that contain no evolutionary information; a result that goes in neat contrast with respect to the previously presented evolutionary models. At the same time, this approach puts the unexpected low correlations between different realizations of yeast two-hybrid experiments on some physical grounds [26, 138].

Taking together the results from the evolutionary and the physically based models, no final interpretation is given to the biological meaning of the scaling laws observed in protein interaction maps. Those models represent a substantial improving of our understanding of the role of scale-free networks in biology but, at the same time, they reflect the limitations in interpreting the output of high-throughput experiments in terms of simple models only.

5.3 Predicting protein interactions

The universal scaling laws observed in high-throughput experiments represent a first glimpse of protein organization. However, the first drafts of whole-organism protein interactions are still far from complete [26, 625] and can therefore be complemented by computational predictions. Interestingly, protein–protein interactions predicted computationally have accuracies that are comparable to those of large-scale experiments [625].

Nowadays there is an emerging number of experimental data that can be used to infer whether two proteins are correlated, i.e. involved in the same biological pathway (but not necessarily physically interacting), or not. The final goal of this type of analysis is to predict, given a set of protein sequences, which pairs of proteins are likely to have interactions (see [565, 609] for a review).

5.3.1 Genome analysis and expression

The continuous flow of information coming from genome sequencing projects has boosted the development of methods to infer protein interactions from this data [565, 609]. A large computational effort has been devoted to the comparison of complete genome sequences with the aim of predicting genome-wide protein interaction networks. The analysis of *phylogenetic profiles* from several organisms has shown that interacting proteins tend to be either present or absent together from the genome [275, 469]. Another approach is based on the observation that interacting proteins are likely to have similar *phylogenetic trees* [218, 225, 491]. *Gene fusion events*, i.e. proteins that are separated in one genome but appear "fused" together in other organisms, often represent signatures for protein interactions as well [174, 396]. Monitoring the levels of *expression of mRNA* in varying cellular conditions is another (indirect) source of information for protein interactions. Genes that react in a similar way for different conditions suggest interactions between the encoded proteins within those groups [216]. An *in vivo* technique, called *synthetic lethality*, monitors the lethality when two non-essential genes are mutated at the same time. Encoded proteins from this type of genes may also interact physically [597].

Other methods propose integrating evidence from many different sources [289, 397]. A Bayesian approach for integrating interaction information that allows for the probabilistic combination of multiple data sets has been applied to yeast [289]. The approach is able to predict protein interactions *de novo* combining several interaction data sets as different as co-expression temporal profiles of mRNA along the cell-cycle, co-expression of mRNA in different cellular conditions, information on biological function, and information about whether proteins are essential for survival. Interestingly, none of these sources are interaction data per se, containing information weakly associated with interaction. The basic idea is to assess each source of evidence for interactions by comparing it against samples of known positives and negatives yielding a statistical reliability. Then, by extrapolating at the genome-wide level, predictions of possible pairwise protein interactions are carried out by combining each independent data set according to its reliability. The result is an overall confidence score for each interaction which makes it difficult to estimate whether the interaction is a physical one or not, because two proteins can cooperate in the same biological task without being in direct contact. Comparison of the predicted interactions against existing interaction data, as well as new affinity purification experiments, has strongly indicated that higher scores are more likely to indicate direct physical contacts [289]. The great challenge for the future would be to correlate the association scores obtained from this method with more biophysical measures like protein–protein dissociation constants [214].

5.4 Towards models at an atomic level of resolution

The last step towards modeling protein interactions maps is to investigate *how* proteins are interacting together at an atomic level of detail. The aim of this type of description is a mechanistic understanding of biochemical, cellular and higher order biological processes. X-ray crystallography provides atomic-resolution models for proteins and complexes [49, 687]. Despite the incredible success of this approach it is often difficult to grow crystals of large complexes. NMR is limited to smaller complexes (usually less than 300 residues), but it plays an important role in defining interaction interfaces between proteins for which X-ray structures are not available [390]. There are currently fewer than 2000 known different structures (using homology considerations) of assemblies from a variety of organisms, involving two or more protein chains. Theoretical estimations have predicted that the number of different interactions occurring in nature might be of the order of 10 000 [24]. Considering the current rate for structure determination it will take decades to obtain a representative set of them. Therefore, there may be thousands of biologically relevant macromolecular complexes whose structures are yet to be characterized, calling for computational approaches which help to fill the gap.

5.4.1 Protein–protein docking

Classic *docking* approaches attempt to find the best docked complex on the basis of shape and/or electrostatic complementarity between protein surfaces [232, 567,

657]. To be accurate, the docking method generally needs high-resolution structures of the interacting proteins, but clearly not of the docked complex, which is the output of the approach. Structures for the docking procedures can come either from X-ray or NMR experiments as well as high-resolution models.

Docking strategies usually rely on a two-stage approach: they first generate a set of possible orientations of the two docked proteins and then score them in the hope that the *real* conformation of the complex will be ranked highly according to some quality function. This means that, given two protein structures, docking algorithms generate an ensemble of possible conformations which are compatible with the two input proteins. From this ensemble the algorithm should be able to extract the "best" solution according to the scoring function.

Several methods have been developed in the past few years [657]. They usually differ in the protein representation, in the ranking procedure of the different configurations and/or in the search for the best solutions. Innovative new methods have been developed to combine docking with NMR experiments (for example, HADDOCK [150]), which have been shown to enhance remarkably the quality of the results. In fact, any information on where the interacting surface of the two proteins is located usually improves the results substantially [30, 147]. Introduction of protein flexibility in the docking procedure can account for some of the most relevant structural rearrangements taking place during binding [361].

Docking algorithms are systematically assessed through blind trials in the Critical Assessment of PRedicted Interactions (CAPRI) [232, 361, 657]. CAPRI meetings are sort of competitions where structure predictions of a number of protein complexes are made just before the high-resolution structures are solved experimentally, allowing for the assessment of the proposed models.

For the greatest applicability in the future, docking techniques will need to work with coarse protein structures. In fact, despite protein structures for many single globular proteins or domains having been determined by high-resolution techniques, it will still take decades for a full set of experimental structures to become available. Finally, docking methods provide the best configuration of the protein complex according to a given scoring function, but they are still unable to indicate whether or not a pair of proteins are really interacting in an organism.

5.4.2 Modeling by homology

An alternative approach for modeling the protein interaction network at a structural level is to use information for experimentally known complexes. In this way, the unknown structure of a protein complex is built on the basis of those that have been seen previously. This approach is known as *homology modeling* and has been shown to be crucial for obtaining protein interaction networks at the atomic level [21].

Homology modeling has been used for several years in the prediction of the three-dimensional structure of single proteins [220]. The approach takes a known three-dimensional structure as a modeling template for a homolog that has been selected on the basis of sequence similarity. In protein interaction studies, algorithms are based on empirical pair potentials which assess how well a homologous pair of sequences fit onto a previously determined structure of a complex. For example, this approach has been implemented in InterPReTS [26]. A similar algorithm, which is implemented in MULTIPROSPECTOR [379], is applied to more distantly related protein sequences by a threading approach followed by an analysis on how well the individual sequences fit into the proposed folds. The latter has even been used to predict interactions on a genome scale by applying the technique to all of the possible interacting proteins present in the yeast organism [378]. Structure-based interaction predictions (including protein complex prediction and the prediction of cross-talk between complexes) has recently culminated in the delineation of the first network of modeled protein complexes in yeast [21].

Once a template has been selected for a given interaction it is possible to model the unknown complex by using standard comparative modeling techniques [400]. The problem here is that there might be multiple templates for the same interaction, as well as the same template modeling multiple target complexes. Therefore, it is important to assess the likelihood of these potential interactions, particularly in the absence of experimental validation [334].

The accuracy of homology-based methods depends greatly on the degree of sequence identity between the target and the template onto which it is being modeled. It has been observed that when sequence similarity is larger than 30% proteins are likely to interact in the same way [22]. This result is similar to what is known for single protein structure prediction [220] and it is just an empirical rule. In fact, there might be cases in which interactions are structurally similar despite of low sequence similarity or vice versa.

There are two great limitations of homology methods: they are limited in their applicability by the number of experimentally resolved structures and that they cannot account for any conformational change which might be involved in the modeled interaction.

5.5 Concluding remarks

Protein interaction maps, whether identified experimentally, or predicted computationally, are static and do not take into account either spatial or time relations. However, proteins are expressed at different times during the cell cycle, under different cell conditions or cellular compartments and every interaction needs to be considered together within its protein context. This implies that it is possible to see interactions, experimentally or computationally, that would never occur in nature. Moreover, the lack of an accepted benchmark, i.e. a large set of experimentally verified interactions, makes it difficult to derive error rates.

In the future, models for protein interaction networks should try to focus more and more on biologically relevant issues observed in experiments. Notwithstanding

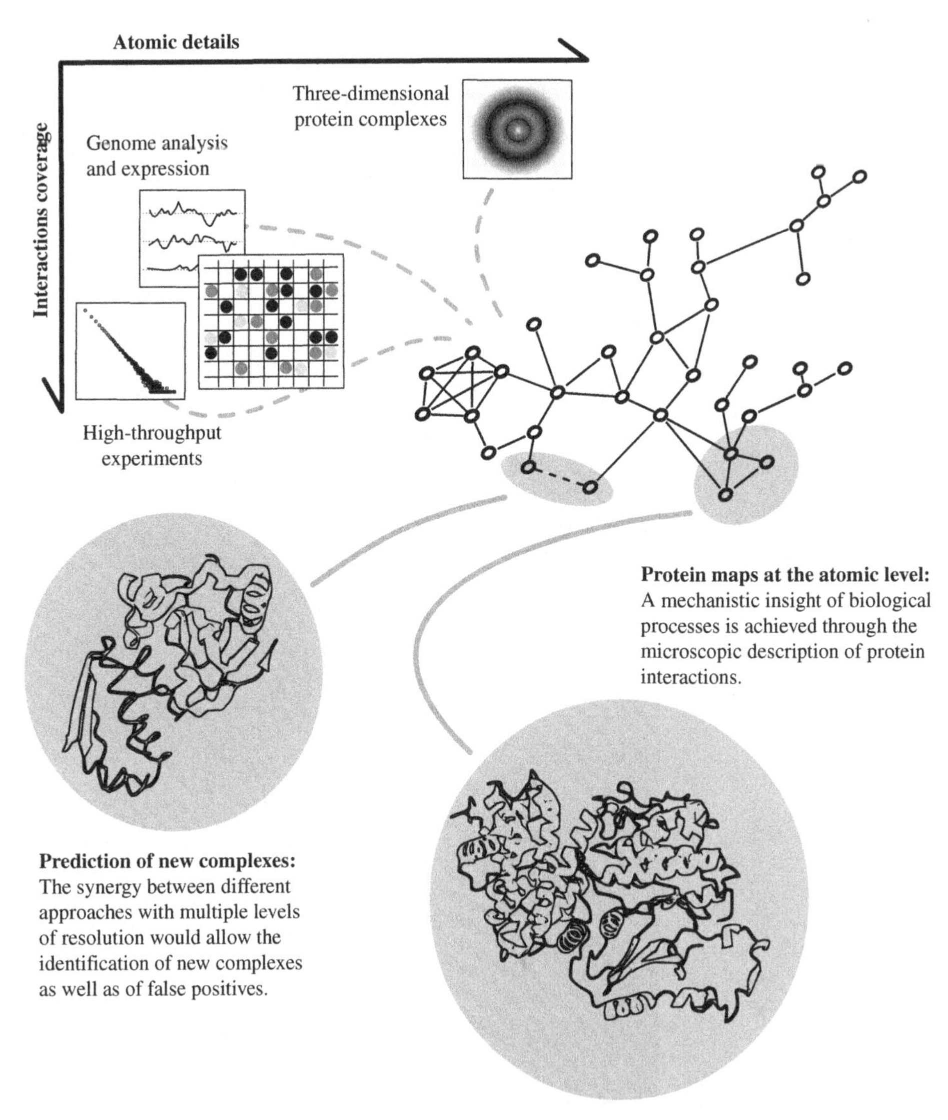

Fig. 5.2. The combination of different experimental and computational approaches is providing more detailed and reliable maps.

the value and importance of the single techniques, a combination of approaches is likely to be more powerful than any single method alone (as shown pictorially in Fig. 5.2). Integrating the information gathered at multiple levels of resolution will provide a better description of the multitude of interactions at molecular level, allowing the general structural principles that underlie cellular processes to emerge.

6

Dynamics and evolution of metabolic networks

DANIEL SEGRÈ

6.1 Introduction

This chapter deals with the complex network of biochemical reactions known as cellular metabolism. Understanding how the different components of this network coordinate their action towards generating coherent pipelines of chemical transformations, how the pipelines themselves are promptly assembled, disassembled and controlled as a function of changing environmental conditions, and how evolutionary adaptation shapes this whole system, constitute fundamental ongoing challenges. These questions are not only intellectually fascinating, but also practically important for many biomedical, engineering, and environmental problems. Because of the complexity of these networks, mathematical models and computer simulations are an essential component of this challenge. This chapter aims at providing a concise and elementary introduction to some basic concepts on mathematical modeling of metabolic networks, with a few examples of recent research applications. Those interested in serious background should refer to classical biochemistry textbooks [65, 430] and recent books on metabolic engineering [582, 599] and computational models of biological networks [19, 252, 458].

6.2 Cellular metabolism and its regulation

In the busy economy of a cell, the balance of resources is essential for survival and reproduction. The main currencies, free energy stored in chemical bonds, and molecular building blocks, can be used for a variety of purposes, from the synthesis of new molecules, to the maintenance of gradients across the membrane; from the capacity to move and find more food, to the production of all components necessary for self-reproduction. This economy involves several hundred to thousands of types of small molecules and biochemical reactions (Fig. 6.1) [2, 3]. Metabolic reactions

Networks in Cell Biology, ed. M. Buchanan, G. Caldarelli, P. De Los Rios, F. Rao and M. Vendruscolo.
Published by Cambridge University Press.

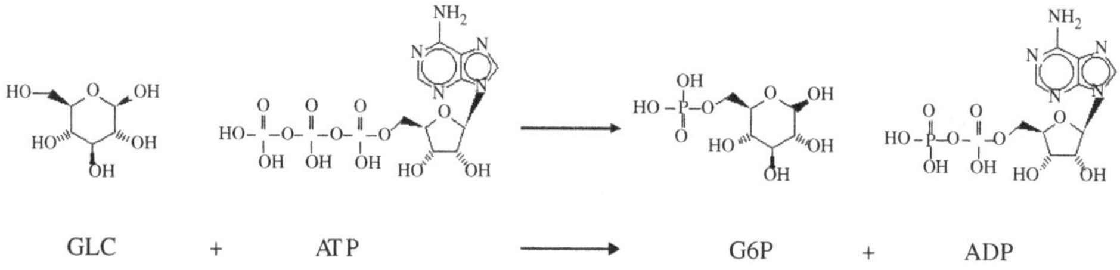

GLC + ATP ⟶ G6P + ADP

Fig. 6.1. An example of a metabolic reaction, transferring a phosphate group from ATP to glucose, to produce glucose-6-phosphate. The reaction (EC 2.7.1.2) is catalyzed by an enzyme called glucokinase, and is the first step of glycolysis.

work around the clock, constantly breaking chemical bonds and making new ones, or transferring ions and electrons in a highly organized fashion. These molecules and reactions constitute the network of cellular metabolism. Importantly, almost all metabolic reactions would happen extremely slowly if it were not for specialized macromolecular machines, the enzymes, that act as catalysts, enhancing reaction rates by several orders of magnitude. The fact that the cell relies on these complex macromolecules to perform almost any chemical transformation at a reasonable speed provides the opportunity for several levels of control and regulation. If an enzyme is absent, or present but temporarily inhibited, the corresponding reaction is shut off. In general, to appropriately control its metabolic processes, the cell can tune the amount of messenger RNA (mRNA) transcribed, the amount of mRNA translated into protein, the rate of degradation of the protein, or the degree to which the enzyme can freely perform its function (e.g. owing to covalent or noncovalent modifications involving other molecules). Furthermore, the different timescales at which these control mechanisms operate, offer the opportunity for a broad spectrum of temporal organization patterns. It is important to remember (see also Section 6.4) that – in analogy to a widely open faucet connected to an empty reservoir – abundant enzyme is a necessary but not sufficient condition for a reaction to occur: the molecular substrates for the reaction have to be available upstream in an amount large enough to guarantee a free energy fall. In a cell-scale metabolic network, several reactions typically share multiple substrates and products, and cycles of reactions are quite abundant. This makes metabolic homeostasis and regulation a highly sophisticated – and often seemingly unintelligible – product of evolution.

A successful cell has to be able to manage energetic and molecular resources in a smart way, regulating the flow through the different chemical reactions in response to intracellular self-assessment (e.g. excess presence of a given metabolite) and

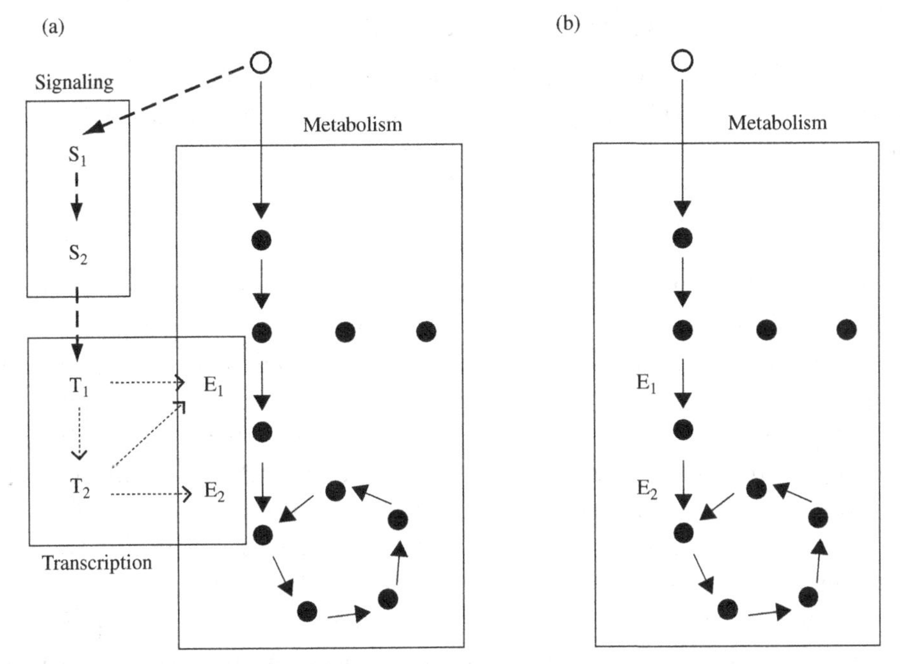

Fig. 6.2. An extremely simplified diagram of how different networks in the cell contribute to metabolic dynamics. (A) The metabolic network (continuous line arrows) is controlled at multiple levels. A major component of metabolic regulation is the transcriptional network (dotted arrows), which determines the level of expression of each enzyme gene (E_1, E_2). The transcription factors (T_1, T_2) that control gene expression can often respond to environmental changes, mediated by signaling cascades (dashed arrows) involving sensing of external molecules and phosphorylation chains. While modeling the complete metabolic, regulatory and signaling system of a cell is currently beyond reach, predictive models that take into account only the metabolic subsystem (B) can provide useful predictions and insight.

external "news" (e.g. sudden availability of a better nutrient). The genetic regulatory (or transcriptional) network, that largely controls how much of a given enzyme is present in the cell, often integrates multiple sources of information through complex interactions between transcription factors, akin to logical gates. On top of the transcriptional network, a signal transduction (or signaling) network provides the interface with the external world and with internal self-assessment.

Since the metabolic, regulatory and signaling networks are highly intertwined, developing predictive cell-scale models of metabolism that can take into account all levels is currently beyond reach (Fig. 6.2). However, as illustrated later, significant biological insight can be gained by properly modeling the network of metabolic reactions as a standalone system.

6.3 Metabolism across disciplines

Metabolism plays a key role in several biological processes. Below is a partial list of research areas in which a focus on metabolic networks is particularly exciting and relevant for addressing important open questions.

Origin and evolution of life

As evident from the high degree of conservation of central metabolic pathways among different organisms, cells must have refined their chemical transformation skills very long ago. Whether an intricate and organized metabolic network is a prerequisite for the emergence of life or a consequence of it is an open and debated question [205]. Some views of early evolution ("replication first") suggest that simple molecular self-replication (e.g. autocatalytic RNA molecules) had to be the first step in the emergence of life, followed by the recruitment of increasingly complex supporting systems, including a pipeline of energy and building block supply [105, 165, 452]. Other views ("metabolism first") propose that the current metabolic infrastructure must have gradually emerged from a primitive, less-organized network of chemical reactions, abiotically feasible in a primordial Earth environment [157, 313, 422, 451, 547, 549]. According to the latter view, the first replicating system may have been a self-sustaining, globally autocatalytic network of reactions, namely a rudimentary form of metabolism. A more in-depth analysis of how one can use mathematical models to explore a "metabolism first" scenario is presented in Section 6.9. Obtaining conclusive evidence in favor of any detailed scenario for the emergence of life is an ambitious goal. At the same time, proposing scenarios, developing computational models, and testing hypotheses experimentally can have important implications for understanding how living systems function and evolve, for engineering novel minimal forms of life, and for planning space observations and explorations in search for signs of extraterrestrial life.

In addition to asking how life started and what role metabolism played in this process, other questions about the ancient history of biochemical networks may benefit from computational models. For example, one may wonder why, among the uncountable chemical reactions possible, the specific subset we observe today was selected to serve as the hardware of living systems. The answer might be that this is the outcome of a sequence of historical accidents. On the other hand, it is also conceivable that natural selection might have been able to search the combinatorially large space of possible networks for a topology that is in some sense optimal. Evidence for this has been reported, for example, in computational studies of ATP-producing pathways ([159]). The question of whether metabolic optimality principles may help explain some fundamental properties of the biochemistry of living systems has also been addressed from the perspective of nonequilibrium thermodynamics [321].

Metabolic engineering

While optimality criteria may be useful for understanding how metabolic networks evolve, a more straightforward use of optimization in metabolism is in the design of modified metabolic systems for increased production of valuable molecules [473, 580, 581, 599]. Several metabolites naturally produced by certain microorganisms can be highly valuable for medical (e.g. antibiotics), farming (e.g. food supplements) or energy technology (e.g. biofuels) applications. Often, even if a culturable microorganism cannot naturally produce a given molecule, it is possible to "copy and paste" genes from other organisms and artificially modify the metabolic network to enable a new biochemical pipeline. The existence of a pathway for the production of a given molecule does not necessarily imply that the organism will actually produce that compound, nor that it will produce it with a rate (amount per unit time) or yield (amount per unit of nutrient) high enough to make it commercially useful. Hence, metabolic engineers have developed and refined system-level analyses of biochemical networks to try and optimally redirect metabolic fluxes towards the desired product. In addition to predicting which reaction rates should increase or decrease to approach optimal production, this process involves understanding how to modify the natural regulatory network of the cell to achieve the desired enzyme activities.

Biomedicine and metabolic diseases

Biomedicine is another area that can benefit from understanding how metabolic networks are utilized and controlled in the cell. Several hundred genetic diseases associated with defects of metabolic enzymes are known, with detailed information available at public databases such as the Online Mendelian Inheritance in Man (OMIM) [4]. In addition, genome-wide association studies are producing an increasing number of predictions of genetic variants associated with clinically relevant phenotypes and complex diseases. Some metabolic diseases may be related to mutations of a single enzyme (e.g. fructose intolerance [386]), either in the coding region of the gene, or in the upstream region that determines the transcriptional regulation rules. Other metabolic diseases, such as diabetes, are deeply associated with metabolic malfunctions, but at the same time involve several other non-metabolic processes and are associated with multiple loci [308, 543, 686].

Microbial ecology

Cells usually share their environment with other cells, be it eukaryotic cells forming a tissue, or microbes interacting with other species in the oceans or in our guts. In any case, communication between cells (in particular between bacteria [58, 635]) often involves the transfer of metabolites, e.g. as chemical signals, or as symbiotically exchanged food sources. An exciting and difficult challenge for computational biologists in the next few years will be to make sense of the large amount of data that will become available on the genomic and physiological properties of microbial communities [619]. In addition to being relevant for addressing global environmental challenges (e.g. related to carbon cycling), the study of complex cell–cell interactions will be very important for understanding how microbial diversity is generated and maintained, and what role the microbes that populate our body (the human microbiome) play in preventing or causing disease [101, 219].

6.4 Dynamics of a metabolic system

The complete dynamics of a metabolic system (isolated from its regulatory layer, see Fig. 6.2) is ideally represented as the collection of time-dependent concentrations of all the metabolites participating in the reaction network. In a network comprised of n reactions and m metabolites, the different metabolites can be labeled X_1, X_2, ..., X_m. The same notation can be unambiguously used to indicate the metabolite concentrations, which form a vector $\mathbf{X} = [X_1(t), X_2(t), ..., X_m(t)]$. These time-dependent concentrations obey a set of m differential equations, dictated by the rules of chemical kinetics. For example, for a single irreversible monomolecular reaction that has X_1 as a substrate and X_2 as a product

$$X_1 \xrightarrow{k} X_2 \tag{6.1}$$

one can write, in the simplest formulation, a differential equation based on the law of mass action

$$\frac{dX_1}{dt} = -kX_1, \tag{6.2}$$

where k is the rate constant.

Most metabolic reactions in the cell, however, occur at a meaningful rate only when catalyzed by an enzyme. The elementary reaction (6.1) should therefore be rewritten to explicitly take into account that an enzyme (E) binds the substrate to form an intermediate complex (C), before liberating the product:

$$X_1 + E \underset{k_{-1}}{\overset{k_1}{\rightleftharpoons}} C \xrightarrow{k_2} X_2 + E. \tag{6.3}$$

A standard approach to write the expected rate of production of the product (X_2) involves, in addition to mass action kinetics, a quasi-steady state approximation (assuming fast equilibration of formation of the complex, [545]). The final expression for the rate (or flux) of this reaction, known as the Michaelis–Menten equation, can be written as

$$v = \frac{k_2[E_T][X_1]}{[X_1] + K_M} \tag{6.4}$$

where $K_M = (k_{-1} + k_2)/k_1$ is the Michaelis constant, and $[E_T]$ is the total concentration of the enzyme ($[E_T] = [E] + [C]$).

There are various important properties of this equation that are worth emphasizing. First, it is nonlinear in the substrate concentration X_1. In particular, the rate of the reaction as a function of X_1 saturates to a maximal value. Second, the rate of the reaction is proportional to the amount of enzyme present (E). This means that, in principle, in order to model the metabolic dynamics in the cell in detail, we need to know the amount of all enzymes present in the cell at any given time. However, since the typical timescale for metabolites to equilibrate in the cell is much faster than the timescale for change in enzyme levels, the enzyme concentrations can be often assumed to be constant. Third, the equation includes two parameters (k_2 and K_M) whose value is required in order to numerically solve the kinetic equations. In reality, several biochemical reactions have multiple substrates and/or products, requiring kinetic terms that contain multiple metabolite concentrations and many more kinetic constants. These kinetic constants are usually difficult to measure, especially *in vivo*, and their values are often unknown. The equations increase in complexity if one takes into account reaction reversibility, allosteric effects and cooperativity.

In a network, each metabolite can participate in more than one reaction, leading to the presence of multiple terms analogous to expression (6.4) in the corresponding differential equation. In general, if S_{ij} indicates the number of molecules (or moles) of metabolite i that participate in reaction j (with the convention that S_{ij} is negative if the metabolite is consumed, and positive if it is produced), one can write the differential equation for metabolite X_i as:

$$\frac{dX_i}{dt} = \sum_{j=1}^{n} S_{ij} v_j(\mathbf{X}(t), \mathbf{P}), \tag{6.5}$$

where $v_j(\mathbf{X}, \mathbf{P})$ is the kinetic term representing the rate (or flux) of reaction j, dependent both on the metabolite concentrations and on a vector $\mathbf{P}$ of kinetic parameters. As in equation (6.4) the fluxes v_j typically depend nonlinearly on the metabolite concentrations, which are time-dependent. However, the overall rate of change of the concentration of a given metabolite is a linear combination (with fixed coefficients) of these nonlinear kinetic terms from different reactions. As discussed in detail in the next section, the fact that the kinetic terms combine linearly can be exploited for pursuing an entirely different approach to metabolic network modeling.

The overall nonlinearity of the kinetic equations, together with the challenge of knowing the enzyme concentrations and the kinetic parameters in vivo, make the use of the above differential equations usually limited to small systems (e.g. [403]). The most comprehensive metabolic network of *E. coli*, for example, counts approximately 2000 metabolic reactions and 1000 metabolites [188]. The corresponding nonlinear system of coupled equations (6.5) would hence contain several thousand parameters, largely unknown. Even if these practical problems were solvable, one may wonder to what extent the numerical solution of differential equations is truly useful for understanding how biological systems function and evolve. The large variety of external conditions encountered and the presence of intrinsic and extrinsic cellular noise [171], suggest that a large number of solutions should be computed in order to obtain a biologically plausible description. An approach that simplifies the study of kinetic equations by focusing on sensitivity of fluxes relative to parameters and concentrations is metabolic control analysis (MCA), which is described in detail elsewhere [252, 582].

An alternative category of mathematical approaches to model large metabolic networks, described next, is based on the analysis of network stoichiometry and steady states.

6.5 Stoichiometric analysis

The n by m matrix S used in equation (6.5) is called the stoichiometric matrix. It embodies a concise mathematical representation of the topology of the metabolic network and its molecular balance constraints. To illustrate the correspondence between a network and its stoichiometric matrix, consider, for example, the following simple reaction network composed of four metabolites ($m = 4$) and three reactions ($n = 3$):

$$X_1 \longrightarrow X_2 \tag{6.6}$$

$$2X_2 \longrightarrow X_3 + X_4 \tag{6.7}$$

$$X_3 \longrightarrow X_4. \tag{6.8}$$

The stoichiometric matrix S corresponding to this reaction network is:

$$S = \begin{pmatrix} -1 & 0 & 0 \\ 1 & -2 & 0 \\ 0 & 1 & -1 \\ 0 & 1 & 1 \end{pmatrix}. \tag{6.9}$$

Each column corresponds to a reaction, and each row to a metabolite. For example, from the second column, we can read that two molecules of X_2 (second row) are being consumed, producing one molecule of X_3 (third row), and one molecule of X_4 (fourth row).

The stoichiometric matrix serves as a standard and mathematically tractable description of the whole set of metabolic reactions of which the cell is capable. One could scan the genome of a fully sequenced organism and generate a list of all metabolic enzyme genes identified. To each enzyme one can then associate one or more metabolic reactions. The complete set of these reactions can finally be written in the form of a stoichiometric matrix. This process can be seen as a mapping from genome to stoichiometry (hence the name "genome scale" models). In theory, this process should be applicable to any of the hundreds of genomes sequenced. In practice things are more complicated, largely because quite a few of the genes found in an organism may have no known function. This is especially true for organisms that have not been the subject of detailed biochemical studies (unlike *E. coli*, for example). In general, building biologically relevant stoichiometric matrices has required so far a lot of manual curation and literature integration. Stoichiometric matrices (sometimes called "network reconstructions") are now available for many organisms, including several bacteria [163, 164, 188, 446, 557, 594], the yeast *S. cerevisiae* [201] and a generic human cell [155], either through published papers [479] or online databases [1].

In addition to serving as a repository of knowledge, the stoichiometric matrix can be used to characterize the capabilities of the metabolic network in a variety of ways. Even without using any additional information, one can ascribe potential biological significance to the left and right null spaces of the S matrix, defined respectively as $\{\mathbf{x} \in R^m | \mathbf{x}S = 0\}$ and $\{\mathbf{y} \in R^n | S\mathbf{y} = 0\}$ [186, 251].

The left null space of S defines possible conservation relationships in the metabolic network [251]. A conservation law can be expressed as the constancy of a given combination of metabolites (with coefficients forming a vector $\mathbf{g}$):

$$\mathbf{g} \cdot \mathbf{X} = const. \tag{6.10}$$

Taking the derivative of both sides with respect to time, and using equation (6.5), we obtain

$$\mathbf{g} \cdot \frac{d\mathbf{X}}{dt} = \mathbf{g} \cdot S\mathbf{v} = 0. \tag{6.11}$$

For this to be true for any flux value, it has to be

$$\mathbf{g}S = 0. \tag{6.12}$$

In general, one could find $m - \text{rank}(S)$ independent conservation vectors $\mathbf{g}$, forming a matrix G of conservation relationships. For these laws to be interpretable as rules of conservation of molecular groups (e.g. numbers of atoms of a certain type, or count of a specific moiety) all elements of the $\mathbf{g}$ vector have to be non-negative. The combination of equality constraints (equation (6.12)) and inequality constraints ($g_i \geq 0$) can be shown to define sets called convex polyhedral cones, whose properties are studied in the field of convex analysis. More information and references on these aspects of the study of stoichiometric matrices can be found in [251, 280].

The study of the right null space of S arises from imposing that the system of equations (6.5) is at steady state, namely $Sv(X(t), P) = 0$. While this may seem a strong and unrealistic assumption for a biochemical network, it is possible to envisage specific settings in which it would be approximately satisfied, e.g. a cell for which (as is typically the case) metabolite equilibration is faster than enzyme level changes, or a large asynchronous population of cells grown in a continuous culture (or chemostat). In the latter case, even if each individual cell may be undergoing cell-cycle related oscillations, the average behavior of the population can be thought of as a steady state.

It should be emphasized that, even under the assumption of steady state, the nonlinearity and the poor knowledge of the parameters would make it very difficult to solve or to interpret these algebraic equations for a genome-scale model. However, the system is much more manageable if we treat the reaction rates (v_i) as "black boxes," i.e. we interpret them as the main variables of the system, and forget about their dependencies on the metabolite concentrations and the kinetic parameters. At steady state, we would expect such variables to assume a certain constant value. Our problem now becomes finding a set of fluxes $\mathbf{v}$ that satisfies

$$S\mathbf{v} = 0. \tag{6.13}$$

Equation 6.13 represents a system of m linear equations (one for each metabolite) in n variables (the fluxes), with the elements of S as coefficients. The meaning of these equations may be easier to understand if we write the equation for a specific metabolite i:

$$\sum_{j=1}^{n} S_{ij} v_j = 0. \tag{6.14}$$

This equation tells us that, at steady state, the net stoichiometrically weighted sum of all the fluxes that produce or consume each metabolite i has to be zero. This is essentially a mass conservation (or flux balance) equation. Flux vectors that satisfy the flux balance constraints belong therefore to the right null space of S. From the computational perspective, there are two major advantages in this steady state formulation. First, we are now dealing with a set of linear algebraic equations rather than a set of nonlinear differential equations. Second, these new equations do not contain the kinetic parameters that are so difficult to determine.

Some stoichiometric analysis approaches have focused on defining mathematically the space of feasible steady state fluxes under additional constraints, e.g. non-negativity of the reaction fluxes v_j, giving rise to convex analysis problems analogous to the ones mentioned for the conservation relations (note that under these non-negativity constraints, reversible reactions must be represented by two opposite fluxes). As described in detail elsewhere [251] the ensuing decomposition into "extreme pathways" [461, 651, 652], similarly to the concept of "elementary flux modes" [539, 540, 578], can be useful for describing a metabolic network in terms of elementary units of metabolic organization [319, 320].

6.6 Constraint-based modeling: feasible states and optimality

The convenience of dealing with steady-state fluxes, instead of time dependent concentrations, together with the fact that the flux through a specific reaction (e.g. the outflow of ethanol) is what often counts in practical applications, has made stoichiometric models flourish in recent years.

In addition to the mass balance constraints (equation (6.13)), one can impose additional constraints on fluxes, based on experimental measurements (e.g. maximal capacity of a reaction), reaction irreversibility, or externally imposed nutrient uptake limitations. These constraints amount to inequalities on the fluxes and can be expressed in general as:

$$\alpha_i \leq v_i \leq \beta_i \quad (i = 1, ..., n). \tag{6.15}$$

Note that, as opposed to the stoichiometric decompositions described above, each reaction can be characterized here by a single net flux, which can assume positive or negative values. For example, if flux v_i corresponds to a reaction known to be irreversible, but otherwise not subject to other constraints, one would set $\alpha_i = 0$ and $\beta_i = \infty$. Alternatively, if v_i is the flux of uptake of a molecule (e.g. acetate) through the boundary of the cell, and the maximal availability in the medium per unit time (h) and per unit of total biomass dry weight (DW) is known (e.g. 10 mmol/grDW·h), whereas there is no reason to assume *a priori* a limit to the rate of secretion of the same molecule, one would set $\alpha_i = -\infty$ and $\beta_i = 10$. Inequalities (6.15) can also be used to set the value for a special flux that takes into account the energy that the cell needs to spend for some maintenance processes that are not directly metabolic. This maintenance amounts to an energy "leakage" (i.e. discharging of ATP) that must be quantified experimentally [164].

To achieve a reasonable description of cellular metabolism, it is fundamental to take into account the fact that the eventual fate of a large proportion of metabolites is to be incorporated into a new cell upon self-reproduction. This growth process can be added to the stoichiometric model of metabolism by introducing a sink flux

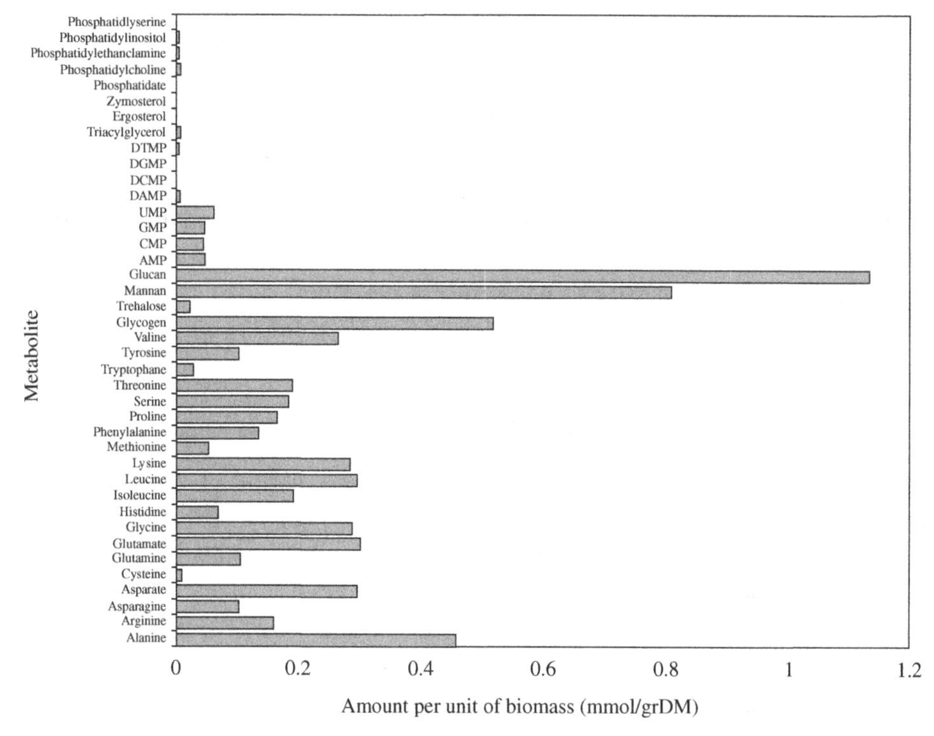

Fig. 6.3. Biomass composition for the yeast *S. cerevisiae* (from [201]).

that consumes, in appropriate proportions, all the molecular components necessary to build new cells. Materials needed in order to build a new cell (Fig. 6.3) include all amino acids (for proteins), nucleotides (for RNA and DNA), and sugars and lipids (largely for energy storage or structural purposes). The proportions of these different building blocks (or biomass components) need to be measured experimentally, and can vary significantly between organisms and conditions (although this last aspect is rarely taken into account). The growth process is hence encoded by an additional fictitious chemical reaction (with flux v_{growth}) that can be written as follows:

$$c_1 X_1 + c_2 X_2 + \cdots + c_3 X_3 \xrightarrow{v_{growth}} 1 \text{ Biomass}. \tag{6.16}$$

Here, c_i is the stoichiometric coefficient describing how many moles of metabolite X_i are needed to produce one unit of biomass. In this description of the growth flux, all components must be produced exactly in the proportions dictated by the vector **c**. If even just one of the biomass components cannot be produced by the cell under a certain condition, then v_{growth} must be equal to zero. Usually, the reaction for growth is incorporated into the stoichiometric model and treated similarly to any other reaction.

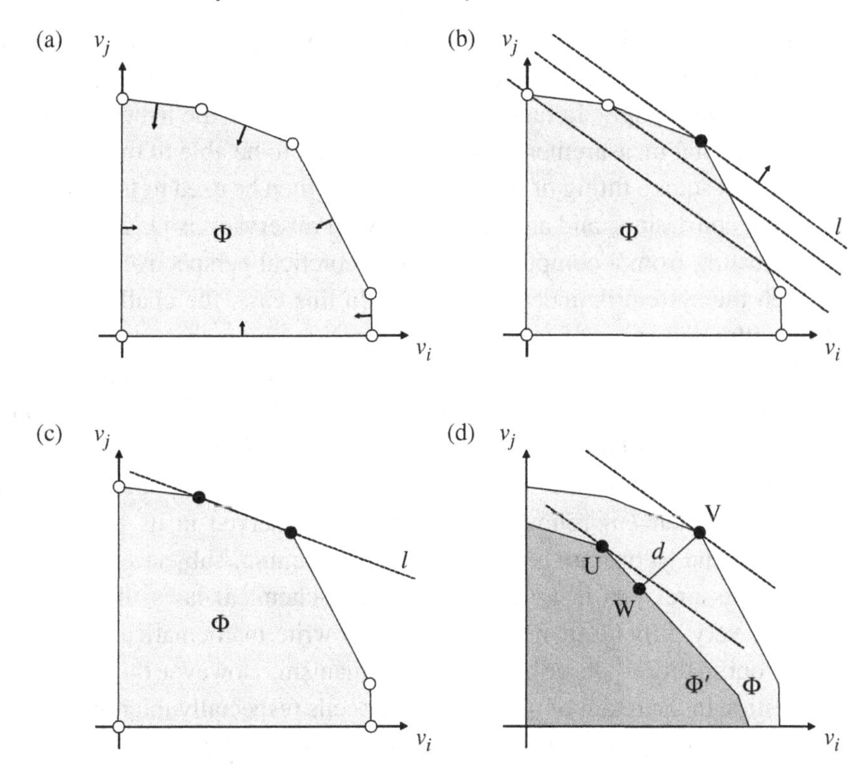

Fig. 6.4. Flux balance analysis and minimization of metabolic adjustment. (a) Equality and inequality constraints on n fluxes can be thought of respectively as hyperplanes and half-spaces in an n-dimensional space. Their intersection determines the convex polyhedron known as "feasible space" (Φ). (b) Among the feasible points in Φ, with linear programming (LP), one can search for a state that maximizes (or minimizes) a linear combination of fluxes (the objective function). This process can be represented as the search for the outermost point of Φ along a specified direction (the vector l). (c) In general, the solution to an LP problem can be an infinite set of optimal points (alternative optima). (d) Deleting a gene in a flux balance model corresponds to setting to zero the corresponding flux. The resulting feasible space (Φ') is contained in Φ. While one possibility for identifying the set of fluxes of a deletion strain is to repeat the same optimization process done in (b) on the reduced space (point U), there is reason to believe that this approach may be misleading. The mutated cell may not be optimal for the original objective. Alternatively, the method of minimization of metabolic adjustment allows one to compute the mutant state (W) as the point that has minimal distance from the original unperturbed point (V) (see text and [552] for additional discussion).

The constraints on the fluxes embodied in equalities (6.13) and inequalities (6.15) determine what is known as the feasible space of the metabolic network (Φ, see Fig. 6.4a). If we trust our knowledge of the network and the flux limitations, and believe that the steady state approximation is reasonable for the system under study, then we can say that any flux state of the cell must belong to the convex polyhedron identified by such constraints [251]. It is not obvious *a priori* that the feasible space is non-empty. In fact, one can try to include in the list of constraints enough experimental measurements of specific fluxes, to be able to over-constrain the system. Least square fitting or other methods can then be used to find the fluxes that best fit the constraints, and are compatible with observations [172, 524, 582].

More interesting from a computational and theoretical perspective is the situation in which the system is under-determined. In this case, the challenge is how to identify, within the space of feasible states, specific points that are biologically significant. Can we say where we would expect to find the cell, if we were able to measure all its metabolic fluxes? One possible way to address this question comes from thinking about evolutionary adaptation as an optimization process. Specifically one may reason that since evolution tends to select for organisms that are maximally fit for their environment, any organism observed in its natural environment might tend to operate close to a certain optimum, subject to constraints deriving from its previous history and to physico-chemical laws. In general, it is going to be very difficult, if not impossible, to write mathematically a similar criterion for optimality to describe any given organism. However this becomes a realistic question in the realm of metabolism, for cells (especially microbes) grown under certain controlled conditions. For example, bacteria that evolved under an abundant supply of nutrients might be close to optimal for growth rate, since any single bacterium growing slower than that would have been quickly outperformed by the faster ones. Hence, if we study the feasible space (Fig. 6.4) of a bacterium (such as *E. coli*) it makes sense to try and find the set of fluxes, within this space, that allow the cell to grow as fast as possible.

In general, it has been possible to formulate efficiently optimization problems in which the function to be optimized (objective function, Z) is a linear combination of the fluxes, with coefficients d_i. Since all the constraints on the fluxes are linear as well, linear programming (LP) algorithms can be used to efficiently compute possible solutions (Fig. 6.4b). Typical computation of a flux balance analysis (FBA) solution for a genome-scale model takes a fraction of a second. The general optimization problem, broadly known as FBA [310, 615], can be formulated as follows:

$$\begin{gathered}\text{Maximize } Z = \mathbf{d} \cdot \mathbf{v} \\ \text{Subject to:} \\ S\mathbf{v} = 0 \\ \alpha_i < v_i < \beta_i .\end{gathered} \tag{6.17}$$

If the objective to be maximized is the growth flux, then all coefficients of the $\mathbf{d}$ vector will be set to zero, except the one for the growth flux v_{growth}, that is set to one. The formulation of FBA allows one to maximize or minimize any single flux or combination of fluxes. For example, the biomass yield relative to the carbon source uptake (e.g. glucose) can be maximized by fixing the growth rate to a certain value, and minimizing the uptake of glucose. Alternatively, one can for example minimize or maximize flux through certain pathways if there is reason to believe that the cell is close to optimal for such task.

From the metabolic engineering perspective (e.g. trying to make the best possible prediction of growth under different conditions), having a good guess of the objective function is a crucial prerequisite for obtaining reliable results. Conversely, one can see the objective function as a metric for trying to understand what is the evolutionary pressure under which a given organism has adapted. If flux measurements are available, one can ask whether the observed system is close to optimal relative to one or more objective functions [535]. This can help us learn about the evolutionary history and natural habitat of the organism explored [552].

A natural question to be asked is whether, given all the approximations and assumptions made by FBA, the results obtained have any resemblance to experimentally observed growth rates and metabolic fluxes. Some comparisons between model-predicted and experimentally measured fluxes indicate a good agreement for specific organisms under certain conditions [552]. In addition, evolutionary experiments have shown that growth rate and uptake rates that are suboptimal prior to adaptation tend to converge to the values predicted by FBA upon several rounds of selection [276]. Unfortunately, the number of available flux datasets against which models can be compared is quite small, largely due to technical challenges involved in measuring fluxes (usually utilizing ^{13}C labeled compounds and mass spectrometry). A vast majority of the comparisons that have been made are at the level of gene deletions, asking whether genes predicted to be essential for a cell to grow are really essential when tested experimentally. This topic is developed in detail in the next section. One should also stress that metabolite concentrations cannot be predicted by FBA models. Conversely, one cannot rely, for example, on mass spectrometry measurements of metabolite amounts for validating FBA models. This is a consequence of the steady-state assumption and the transformation from concentration to flux variables. We can predict the rates of flow between the pools, but not the size of the pools themselves. This is perhaps the heavier price that one has to pay in exchange for the FBA speed and parameter independence.

A final remark is due here on the role of regulation in FBA models. While regulation is explicitly absent from the standard formulation of FBA, it is somehow implicitly taken into account in the optimization step. By asking what state is optimal for a given objective we are bypassing the question on how exactly the cell could achieve that state. Yet, we are assuming that among all the feasible states, the cell has a way to regulate itself to converge to the optimal one. Models that explicitly take into account transcriptional regulation in FBA are increasingly used to provide more accurate predictions [130, 562]. Broadly speaking, however, the issue of how to best integrate biochemical and regulatory networks (both transcriptional and kinetic) is an extremely important and open problem.

6.7 Predicting genetic perturbations

The most common way of experimentally studying the biological function of a gene in the context of an organism is to remove the gene and determine how its deletion (or knockout) affects the organism's phenotype [352]. For microorganisms such as yeast and *E. coli*, one of the most studied phenotypes is the capacity to grow. While some recent technologies allow for quantitative screening of growth rates for multiple gene knockout strains, several studies have just reported binary values (grows, or does not grow). If an organism is not able to grow upon deletion of a given gene in a certain condition, the gene is said to be essential under that condition.

The speed at which FBA can be implemented makes it possible to plan large scale *in silico* deletion experiments. To predict whether a certain gene knockout strain is able to grow or not, based on FBA, one can just modify the optimization problem solved for the wild type (the unperturbed organism) by adding an extra constraint that sets to zero the flux (or fluxes) catalyzed by the enzyme coded by the deleted gene. One of the first studies to perform such an analysis was focused on deletions of metabolic enzyme genes in the central carbon metabolism of *E. coli* [164]. The model was overall able to predict correctly 86% of the deletion cases analyzed, but failed to predict essentiality of a number of genes known to be essential. Similarly, various studies have used FBA, largely with biomass growth maximization as an objective, to predict the effects of gene deletions.

But is the hypothesis of optimal growth as reasonable for a mutant as it was for the wild type? When we search for the flux state with maximal growth rate upon performing a gene deletion, we effectively hypothesize that the cell, despite having lost a gene, is capable of rerouting its metabolic fluxes so as to achieve the best growth rate possible under the deletion constraint. But the loss of a gene, as opposed to an environmental change, is not something cells could have learned to cope with throughout evolution (if the gene was lost and they survived, then we would not see that gene in the cell today). Hence we probably should not expect a cell to be able to redirect fluxes optimally upon gene deletion, at least not immediately after the perturbation. The method of minimization of metabolic adjustment (MOMA, Fig. 6.4D) [552] addresses this issue by testing the hypothesis that a gene deletion causes a minimal flux redistribution with respect to the wild type metabolism, compatibly with the absence of the removed reaction (index k). If $\mathbf{v}^{(\mathrm{wt})}$ indicates the vector of fluxes for the wild type cell (obtained for example through a previous FBA calculation, or from experimental data), and $D(\mathbf{v}, \mathbf{v}^{(\mathrm{wt})})$ is a distance between the vectors $\mathbf{v}$ and $\mathbf{v}^{(\mathrm{wt})}$ in flux space, this problem can be formulated mathematically as follows:

$$\begin{gathered}\text{Minimze } D(\mathbf{v}, \mathbf{v}^{(\mathrm{wt})}) \\ \text{Subject to:} \\ S\mathbf{v} = 0 \\ \alpha_i < v_i < \beta_i \\ v_k = 0.\end{gathered} \tag{6.18}$$

If D is defined as a Euclidean distance, one can use quadratic programming (QP) to solve the minimization problem in flux space [552]. Alternative metrics have been explored as well [561].

By comparing FBA and MOMA predictions with experimental data for *E. coli*, it was confirmed that *E. coli* wild type may have evolved towards maximal growth, but mutants are more compatible with the suboptimal performance prediction of MOMA, than with FBA [552]. In general, the real flux distribution may lie between the FBA and MOMA solutions, as well as in other unexplored regions of the feasible space. While the FBA solution is likely to overestimate the growth rate, the MOMA solution might be too conservative when considering the capability of the cell to make adjustments in response to perturbations. In general, determining which of the two solutions is more accurate depends on how well the regulatory system of the cell can respond to the perturbation, and on how efficiently alternative pathways in the cell can sustain an unusually high flux.

It is important to emphasize that the MOMA hypothesis should be interpreted as a "regulatory inertia" rather than as an active mechanism by which the cell tries to stay close to the wild type state. Again, there should be no reason for the cell to have learned how to deal with a deletion. Rather, we can say that, upon the perturbation, the cell still does its best at doing what it evolved to do, i.e. behave as a wild type.

6.8 Double perturbations and epistatic interactions

If single gene deletions can tell us what reactions are essential for growth under a given environment, double gene deletions can serve as a more sophisticated method to probe the relationships between reactions. The way two perturbations combine can help group genes into modules associated with cellular functions (Fig. 6.5).

The collection of genes in an organism, its genotype, largely determines its observable traits, or phenotype. Two genotypes differing by a single gene may

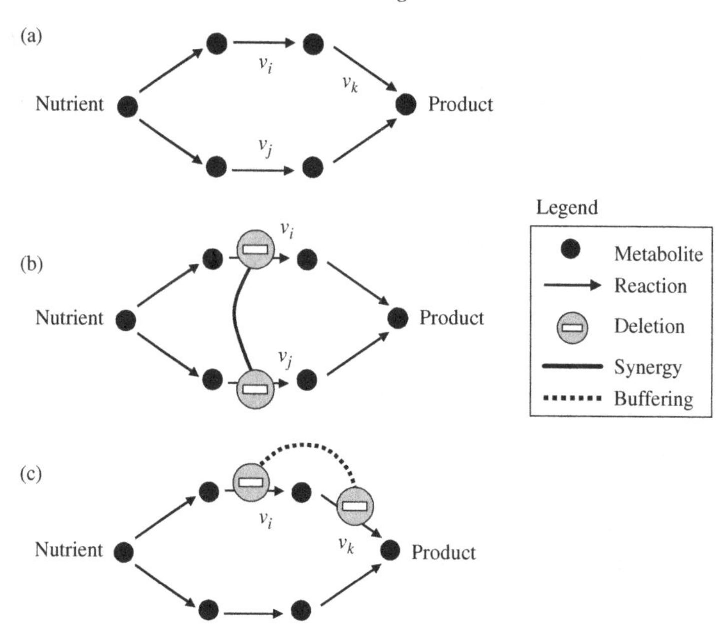

Fig. 6.5. A simplified view of epistatic interactions. (a) The network illustrates two alternative pathways from a nutrient molecule to a metabolic product essential for cellular growth. (b) If flux v_i is blocked (i.e. set to zero) the cell can still grow by using the bottom pathway. It is the same if only v_j is blocked. If both reactions are simultaneously knocked out, however, the cell will not be able to grow. This is a case of synergistic epistatic interactions between the two mutations. Each of them has little effect, but their combination is dramatically more pronounced. (c) Example of an antagonistic (or buffering) epistatic effect: upon blocking flux v_i, the removal of v_k has no additional effect. Hence one perturbation is masked by the other.

produce considerably different phenotypes, revealing important information about the biological function of the missing gene. In the case of genotypes differing by *two* distinct genes or loci, one may ask whether the phenotypic effect of this double difference can be predicted in a simple manner from knowing the phenotypic consequences of each genotypic difference. Sometimes this is not the case at all, and the phenotypic effect of complex gene–gene interactions, also known as epistasis, is considered a fundamental aspect of biological complexity, highly relevant for understanding evolutionary adaptation and functional roles of genes and gene sets (see [127] or [420]). In particular, recent experimental screening of double gene deletions in yeast ([598]), revealed that gene pairs with an extreme epistatic effect ("synthetic lethality," i.e. such that their simultaneous removal is lethal, whereas single mutants display little or no deleterious effects) are usually involved in similar functional roles in the cell (Fig. 6.5b). The systematic analysis of epistatic interactions constitutes therefore a useful functional genomics approach, that can be used to classify genes into functional categories or modules, and potentially ascribe functions to unannotated genes [548]. While synthetic lethality is probably the most striking example of a synergistic epistasis, other types of epistatic interactions may be expected in a network of genes [169], for example cases in which two single gene deletions each have a visible effect, but the simultaneous knockout of both genes displays only one of the two effects, masking the other (buffering or antagonistic effect, Fig. 6.5c).

In general, experiments for measuring epistatic effects at a genomic scale are laborious and delicate, because significant effects may be rare and hard to detect. Thus, one can employ computational models to perform systematic *in silico* experiments of single and double gene deletions, which will result in global maps of epistatic interactions, and provide the basis for studying the general principles governing gene–gene interactions in a complex biological network. Flux balance models are especially well suited for addressing these kinds of questions, because the prediction of the growth rate and genome-scale fluxes for each strain can be implemented very easily, and its computation is very fast.

A genome-scale study of epistatic interactions between all enzymes in a metabolic network was recently performed in yeast [548]. Using the flux balance model it is possible to compute the fitness w_i of a yeast strain lacking gene i, defined as $w_i = v^{(i)}_{growth}/v^{(wt)}_{growth}$, where $v^{(i)}_{growth}$ is the growth rate of mutant i and $v^{(wt)}_{growth}$ is the growth rate of the wild type strain. Similarly, one can compute the fitness for mutant of gene j, as well as the fitness of the double knockout $w_{ij} = v^{(i,j)}_{growth}/v^{(wt)}_{growth}$. The deviation from a non-epistatic, multiplicative behavior can be measured by introducing the following parameter:

$$\epsilon_{ij} = w_{ij} - w_i w_j. \tag{6.19}$$

An epistatic interaction corresponds to a significant deviation from $\epsilon_{ij} = 0$. Such a definition of epistasis has been used before in the analysis of experimental data [169]. For a more precise definition of "signifcant deviation," a rescaling operation on ϵ_{ij} can be used [548].

Genome-wide epistatic interactions can be represented as the distribution of the values of ϵ_{ij} across all gene pairs. More interesting however is to classify them based on whether they are less than multiplicative (synergistic) or more than multiplicative (buffering or antagonistic) (Fig. 6.5). Hence, the map of epistatic interactions can be also represented as a network, where links of different colors between genes reflect the type of epistatic interaction. Based on these bi-color interactions, it is possible to cluster genes into modules that have only one type of interaction (synergistic or antagonistic) with other modules. This property, named monochromaticity of the epistatic interaction network, disappears upon randomizations of the network, suggesting that it is a special consequence of the organization of metabolic pathways [548]. The modules identified using this approach are related to biological functions in a hierarchical way. For example, various modules associated with specific metabolic pathways group together hierarchically into two large modules associated respectively with fermentative and respiratory metabolism. The analysis of epistatic effects and the use of monochromatic clustering have been recently shown to be relevant for the study of interactions between the effects of drugs on cellular physiology [359, 676, 677].

6.9 The ancient history of metabolism: from cell-scale to biosphere-scale

While a vast majority of the efforts in metabolic network modeling focuses on understanding, perturbing, or optimizing metabolism in existing cells, some works have addressed the ancient history of biochemical networks. If a lot of uncertainty (e.g. about kinetic parameters and gene functions) affects our capacity to develop detailed models for existing organisms, we have to deal with even greater uncertainty when trying to model evolutionary transitions that happened millions or billions of years ago. In this section we discuss some examples of how to deal with uncertainty in this context.

Network expansion and the effect of oxygen on biochemical networks

The adaptation to new environments is often accompanied by the evolution of new metabolic pathways, especially when some component of that environment imposes a selection pressure. In particular, significant adaptive changes are thought to have occurred in response to major geological transitions, such as the appearance of oxygen (O_2) in the atmosphere of the Earth around 2.2 billion years ago [184]. The oxidation of the Earth's atmosphere, a result of the "invention" of oxygenic (oxygen-producing) photosynthesis, would have been catastrophic to a biosphere composed entirely of anaerobes. While many of these primitive organisms retreated to anoxic environments in the deep ocean and sediments, others adapted to utilize oxygen as a high-potential redox couple while simultaneously mitigating its toxicity. Understanding the changes in biochemistry and enzymology that accompanied O_2 adaptation is of considerable interest but has been restricted to analysis of enzymes that directly consume or produce oxygen. Recent research extended beyond these limitations by integrating metabolic network simulation and analysis with information on enzyme evolution to infer the network-level reorganization imposed by oxygen availability. Modeling this transition mathematically requires additional simplifications relative to the cell-scale flux-based approaches presented above. The recently developed metabolic network expansion algorithm [158, 244] can provide useful information on the interdependency between different metabolites in a reaction network merely from knowing its topology. The approach is substantially different from the stoichiometric analyses summarized in Section 6.5, as illustrated in Fig. 6.6. The network expansion algorithm determines, starting from a metabolic set seed, what is the set of all reactions and metabolites that can be potentially reached by iteratively "walking" through the available reactions. By differentially examining the reachable metabolic universes in absence and presence of a given molecule (e.g. molecular oxygen) one can infer sets of putative pathways whose evolutionary appearance on Earth may have been contingent on the availability of that specific molecule [495].

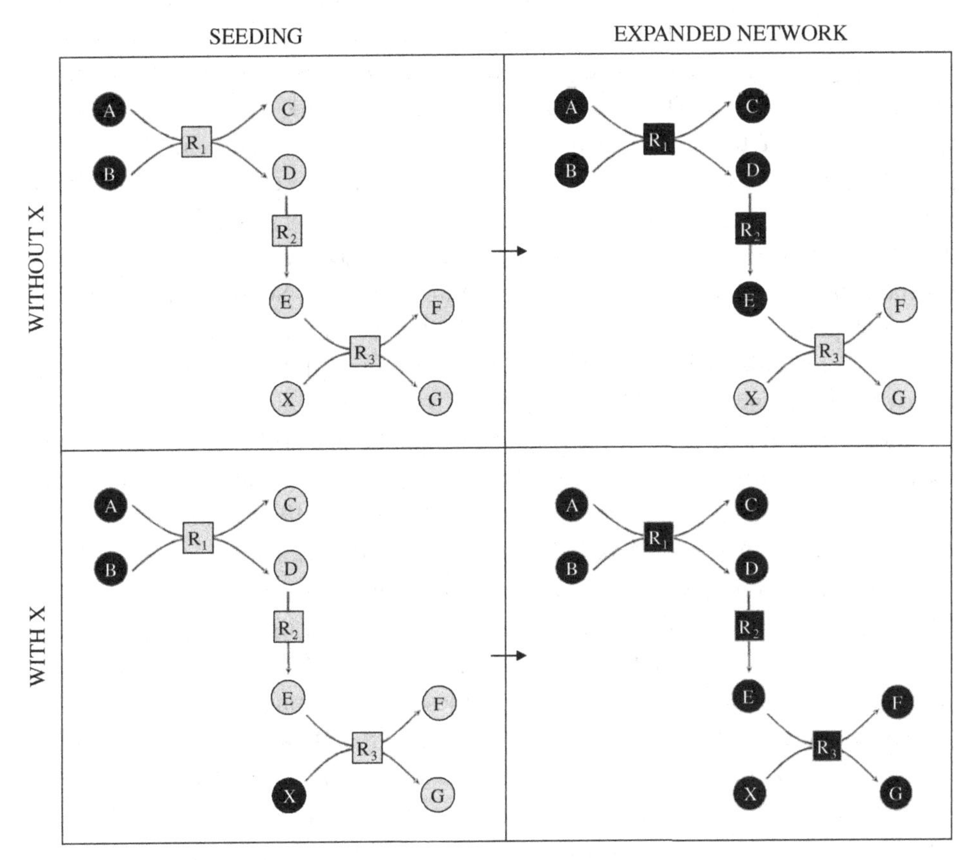

Fig. 6.6. An illustration of how the network expansion algorithm is used to determine the increase in metabolic network complexity (number of metabolites) following the addition of an extra metabolite X. Round nodes represent metabolites, square nodes enzymatic reactions. A network, seeded with a starting set of molecules (A and B, top left), is iteratively expanded following all reactions whose reactants are available. Addition of molecule X to the initial seed (bottom) leads to the generation of a larger feasible network. This process can repeated several times for different seeds.

An analysis of the "meta-metabolome" obtained applying the network expansion algorithm to the set of all known metabolic reactions reveals the existence of discrete regimes of increasing metabolic network complexity, with transitions between regimes contingent upon the presence of key biomolecules. The inclusion of molecular oxygen in biochemical networks constitutes a particularly critical change, as it causes an increase in network complexity unparalleled by any other compound [495].

A statistical approach to prebiotic metabolism

The lack of information becomes even more dramatic when trying to address the fundamental question of how a primitive abiotic Earth environment could gradually give rise to the first organized biochemical systems capable of reproduction and evolution. The idea that a primordial network of chemical reactions could have been the first step towards life has been proposed multiple times [157, 313, 422, 423, 451], often invoking the importance of molecular diversity [313, 546]. As discussed in Sections 6.2 and 6.4, one of the characteristics of metabolism is the presence of very sophisticated catalysts (the enzymes), highly specialized for enhancing the rate of specific reactions. Sophisticated enzymes were not present at the early onset of life, but other forms of catalysis must have acted towards making reactions happen fast enough to establish cycles and homeostatic balance. To address this point it has been suggested that some of the molecules taking part in these proto-metabolic networks could have acted themselves as rudimentary catalysts [157, 313, 550]. Since we don't know what molecular components were available on our planet 3.8 billion years ago, how can we approach this problem? A possible strategy is to define a very simple model of a chemical reaction network, and perform simulations that use parameters drawn from a given probability distribution [549]. The idea is that while we may have very little information on detailed components, it is conceivable that we could come up with reasonable predictions of the shape of these distributions, which may be independent of details. This is what was done in the graded autocatalysis replication domain (GARD) model for prebiotic organization [547, 550]. Consider a chemical reactor in which N chemical species A_1, A_2, ..., A_N can be formed through monomolecular reversible reactions from a common precursor A_0. It is assumed that the monomer concentration A_0 is kept at a fixed value in the reactor, and that an outgoing flux of material is constantly eliminating all the species in proportion to their concentrations A_i. Each molecule can in principle be a catalyst for any of the reactions taking place in the reactor. Mutual catalysis is described by a matrix β, whose element β_{ij} expresses the rate enhancement factor that a molecule of kind j exerts on the reaction by which a molecule of kind i is synthesized.

Catalysis is related to a decrease in the energy barrier for the reaction being accelerated, and therefore forward and backward rates (k_i and k_{-i}) are assumed to be affected equally by the catalyst. This system can be described by the chemical kinetics equations (6.20) which treat molecular populations according to the law of mass action.

$$\frac{dA_i}{dt} = (k_i A_0 - k_{-i} A_i)\left(1 + \sum_{j=1}^{N} \beta_{ij} A_j\right) - \lambda A_i. \tag{6.20}$$

In the GARD model, the elements of the $N \times N$ β matrix of catalytic potencies are sampled from a probability distribution $\Phi(\beta_{ij})$ derived from the Receptor Affinity Distribution (RAD) model [342]. This distribution, originally obtained as a prediction of the affinity of binding between an arbitrary ligand and a large set of receptors, is useful in understanding the statistical nature of molecular recognition in the olfactory and immune systems [510]. Given the statistical and non-specific nature of interactions in a prebiotic molecular mixture, one may expect that binding (and catalysis) in an early Earth environment obeyed a similar model [551].

In the Φ distribution most catalysis values are likely to be very small, signifying almost imperceptible catalytic activity, with occasional higher values, distributed in a graded fashion. The intensity and the organization of the catalytic network connecting the various species A_i was shown to determine the efficiency of self-replication of the GARD system as a whole [547]. Further investigation of similar statistical approaches may help identify parameter distribution constraints necessary for sustaining the spontaneous emergence of a self-reproducing and evolving biochemical system.

Note that the outflow present in the GARD equations (the term $-\lambda A_i$) is formally equivalent to introducing a time-dependent increasing volume, e.g. through enclosure in a growing lipid vesicle. Such flow is also analogous to the growth flux of biomass production described in Section 6.4 for flux balance models, and similarly represents the constant balanced synthesis of several biomass components in fixed proportions. Further elaborations of the GARD model in which the molecular components themselves (the A_i) are vesicle- or coacervate-forming hydrophobic or amphiphilic molecules led to the formulation of the "lipid world" hypothesis for the origin of life [546]. In this scenario, there is no initial specialization in the function of different molecules, as all prebiotic molecular players would concurrently and statistically contribute to structural, information-coding and catalytic roles. Early protocellular assemblies would store and propagate information in their molecular composition (the "compositional genome") [547]. Informational biopolymers such as RNA and DNA could therefore have been an outcome, rather than a prerequisite, of early evolutionary adaptation [547, 549, 551].

6.10 Conclusions

Computational models of metabolism can span multiple spatial and temporal scales. The larger the system (or the more remote in the past), the more difficult it is to know and to take into account all details leading to a certain degree of uncertainty. To some extent, the degree of uncertainty we have to deal with is a natural consequence of the complexity of these networks. Despite the temptation to try and build increasingly detailed models, inclusive of all possible dynamical aspects and kinetic parameters, the establishment of flux balance models and other simplified approaches demonstrates the importance of identifying intermediate levels of description. Such intermediate-level models, by explicitly relying on simplifying assumptions and neglecting specific details, can gain in computational feasibility, and therefore in applicability towards problems still beyond reach experimentally (such as the exhaustive calculation of all double knockouts). In addition, they can allow one to define system-level properties (such as cellular growth), and link them to lower-level variables (such as metabolic fluxes). However, choosing how and where to simplify is not an easy task. Moreover, simplifications come at a price, and the missing input information often translates into limited predictive power. Nonetheless, in general, identifying the right level of description for the challenges lying ahead of us may make the difference between drowning in a sea of details and capturing general principles of biological organization.

7

Hierarchical modularity in biological networks: the case of metabolic networks

ERZSÉBET RAVASZ REGAN

7.1 Introduction

Complexity in biological systems is the rule rather than the exception. Life seems to depend on structures that not only perform a wide variety of functions, but are adaptable and robust at the same time. For a long time the only scientific approach available to study these complex biological systems has been a purely descriptive one. In the second half of the twentieth century molecular biology emerged, along with the development of a variety of experimental methods that allowed an ever-deeper exploration of the constituent parts of a cell, as well as the ways in which these parts assemble. Our chromosomes were shown to be made of tightly packed DNA double helixes that store our genetic code. RNA polymerase, the protein complex responsible for *transcription* of the genetic code to messenger RNA, has been identified, along with many major constituents of the fascinating machinery between genetic code and cellular phenotype. As mRNA molecules leave the nucleus they are met by ribosomes, protein complexes that read the genetic code using groups of three mRNA letters to identify the corresponding amino-acid sequence, and thus *translate* the genetic code into proteins. Proteins are responsible for the majority of biological functions driving a living cell: they orchestrate metabolic reactions, form structural elements like the cytoskeleton, keep track of the extra- and intracellular environment and transmit the signals that constantly reshape gene transcription so that the cell can express precisely the proteins it needs. The "computation" that uses these signals to drive the transcription program is carried out by complex combinatorial interactions between regulatory proteins and specialized DNA regions (promoters) associated to genes.

This brief overview of the way cells use genetic information along with signals from their environment might give the impression that most details of the inner life of the cell are already known. Indeed, many of the most generic processes (transcription, translation, DNA replication, etc.) have been described in detail.

Networks in Cell Biology, ed. M. Buchanan, G. Caldarelli, P. De Los Rios, F. Rao and M. Vendruscolo. Published by Cambridge University Press.

The trouble is that cells have anywhere between 3000 (*E. coli*) to an estimated 40 000 (*H. sapiens*) genes. Most of our knowledge was (and still is) gathered by experts and experiments dedicated to uncovering the secrets of one or very few genes/proteins at a time. Thus, detailed knowledge of some cellular functions with the genes and proteins responsible for them goes hand in hand with the disappointing insufficiency of our knowledge about the function of thousands of genes. This picture, however, has started to change recently.

The development of DNA sequencing was one of the first significant breaks from experiments on a small number of bio-molecules; in 1976 the complete nucleotide sequence of bacteriophage MS2-RNA was published [197]. It was quickly followed by the first DNA-based genome to be sequenced in its entirety (*bacteriophage Φ-X174*) [520], the first free-living organism to be sequenced (*H. influenzae*) [200], most model organisms in bio-medicine [8, 371], and the Human Genome Project [343, 620] (623 organisms are completely sequenced to date[1]). Knowledge of these genomes opened the door for another revolutionary technology, the *DNA microarray chip* [337, 345, 528]: small spots of DNA segments arrayed on a glass (or plastic) plate, each spot selectively binding one of the organisms' messenger RNAs. Microarray chips measure *genome-wide* mRNA expression (and thus indicate protein levels) in cells under a treatment of interest, at a particular developmental stage, or affected by a disease. The system which drives mRNA expression, genetic regulatory interactions, are central to uncovering the program encoded in the genome; it is also one of the least accessible by system-level experiments. Nonetheless, there are ways of probing transcription factor–DNA binding on the scale of the whole cell [356], complemented by a large body of data in curated on-line databases that collect individual experiments reported in the literature [128, 516, 656]. On a different front, experiments that capture protein–protein interactions have been scaled up to cope with entire organisms [196, 629], leading to extensive maps of protein–protein associations for a variety of viruses [198, 413], prokaryotes, like *H. pylori* [488] and eukaryotes, like *S. cerevisiae* [215, 259, 283, 285, 542, 607], *C. elegans* [630] and *D. melanogaster* [221].

The enormous amounts of data coming out of these experiments beg the question: how do we learn from these data, and what kind of questions can they answer? The paradigm of *systems biology* emerged in the wake of system-level experiments: understanding complex biological systems requires understanding and modeling characteristics that are fundamentally determined by the organization of their constituent parts, emergent phenomena created by the interactions of those elements [248, 318, 660].

[1] See `http://www.ncbi.nlm.nih.gov/genomes/static/gpstat.html`

In many cases, if one takes a step back and does not focus on the variation in parts and interactions, many complex systems are made up of assemblages of generic elements and connections that can be framed as a network [15, 52, 63, 72, 76, 153, 154, 436, 439, 466, 538]. By offering a new way to categorize systems of very different origin under a single framework, this approach has uncovered unexpected similarities between the organization of various complex systems, indicating that the networks describing them are governed by generic organizational principles and mechanisms. Networks emerge in biology in many disguises, from food webs in ecology to various biochemical nets in molecular biology, such as protein–protein interaction networks [285, 292, 542, 626], protein domain networks [32, 666, 667] and genetic regulatory networks [417, 558, 593]. Among the most extensive datasets collected from the scientific literature are the metabolic maps of various organisms [307, 456], catalyzing an increasing number of studies on the architecture of the metabolism [189, 294, 627]. Equally complex webs describe human societies, whose nodes are individuals and links represent social interactions [326, 633], the World Wide Web (WWW) [13, 84, 322, 351], where nodes are Web documents connected by URL links, the scientific literature, whose nodes are publications and links are citations [137, 497], or language, made of words and linked by the various syntactic or grammatical relationships between them [152, 192, 564]. The recent availability of system-level data on the network of interactions in large numbers of systems has opened the door for interdisciplinary research in fields where the behavior of the system as a whole is a central question. Recognizing generic organizational principles and order behind diversity and apparent randomness in these different systems has certainly been a surprise along the way.

7.2 Modularity and hubs in biological networks

Two properties of real networks have generated considerable attention. First, many networks are fundamentally modular: one can easily identify groups of nodes that are highly interconnected with each other, but have only a few or no links to nodes outside of the group to which they belong to. Each module is expected to perform an identifiable task, separate from the function of other modules [248, 348, 558, 660]. For example, in protein–protein interaction networks such modules represent protein complexes like the ribosome. This clearly identifiable modular organization is responsible for the high clustering coefficient [637] (the average probability that the two first neighbors of a node are also connected) seen in many real networks. Empirical results indicate that the average clustering coefficient is significantly higher for many real networks than for a random network of similar size [15, 153, 637], and is to a high degree independent of the number of nodes in the network [15]. At the same time, many networks of scientific or technological interest, ranging from biological networks [292, 294, 626, 627] to the World Wide Web [13] have been found to be scale free [53]: there is no well-defined "connectivity scale" that approximates the degree (number of connections) of most nodes in the system. Instead, the distribution of degrees follows an inverse power law with exponents between 2 and 3, indicating that these systems have very large connectivity fluctuations. Most nodes have one or two links, but there are a few hubs with very large degrees. The scale-free property and strong clustering are not mutually exclusive: they coexist in a large number of real networks including metabolic webs, the protein interaction network, the WWW and some social networks [54, 432, 433].

The concept of modularity, the assumption that cellular functionality can be partitioned into a collection of well defined structural elements, is a very popular paradigm of biology [248, 249, 264, 348, 492, 558]. Simultaneously, it is recognized that the thousands of components of a living cell are dynamically interconnected, so that the cell's functional properties are ultimately encoded into a complex intracellular network of molecular interactions [264, 318, 348, 492, 558, 660].

In order to bring modularity, the high degree of clustering and the scale-free topology under a single roof, we have proposed that modules combine with one another in a hierarchical manner, generating what we call a hierarchical network [54, 493, 494]. Hierarchical topology gives a precise and quantitative meaning to network modularity. It indicates that we should not think of modularity as the coexistence of relatively independent groups of nodes. Instead, we have many small clusters that are densely interconnected. These combine to form larger, but less cohesive groups, which combine again to form ever larger and less interconnected clusters. The self-similar nesting of different groups or modules into one another superimposed on the scale free connectivity distribution forces a strict fine structure on real networks. This can be captured in a quantitative manner using a scaling law that describes the dependence of the clustering coefficient on the node degree. We can use this scaling law to look for the presence or absence of hierarchical architecture in real networks.

As a case study, we focus on the topological organization of cellular metabolism, a fully connected biochemical network in which hundreds of metabolic substrates are densely integrated through biochemical reactions. Modular organization within this network (i.e., clear boundaries between sub-networks) is not immediately apparent. The degree distribution $P(k)$ of a metabolic network decays as a power law $P(k) \sim k^{-\gamma}$ with $\gamma = 2.2$ in all studied organisms [294, 627], suggesting that metabolic networks have a scale free topology. This implies the existence of a few highly connected nodes (e.g. pyruvate or coenzyme A), which participate in a very large number of metabolic reactions. With a large number of links, these hubs seem to link all substrates into a single, integrated web in which the existence of fully separated modules is prohibited by definition. Nonetheless, a number of approaches for analyzing the functional capabilities of metabolic networks indicate the existence of separable functional elements [529, 538]. Also, metabolic networks are known to possess high clustering coefficients [627] suggestive of a modular organization. Hierarchical modularity reconciles all the observed properties of metabolic networks within a single framework [494]. Moreover, the hierarchical module structure can be easily uncovered and corresponds to known functional classification of metabolic reactions.

7.3 Scaling of the clustering coefficient: a signature of hierarchy

A hierarchically organized network is made of numerous small, highly integrated modules, which are assembled into larger ones (see an illustration in Fig. 7.1). These larger clusters are less integrated but still clearly separated from each other, and they in turn combine into even larger, less cohesive modules and so on. This type of hierarchy can be characterized in a quantitative manner using the finding of Dorogovtsev, Goltsev and Mendes [151] that in certain deterministic scale-free networks the clustering coefficient of a node with k links follows the scaling law $C(k) \sim k^{-1}$. This scaling law quantifies the coexistence of a hierarchy of nodes with different degrees of clustering. Nodes in the numerous small and cohesive modules have very large clustering coefficients. Nodes that hold together the larger but less cohesive modules have smaller clustering coefficients, indicating that the higher a nodes degree the smaller its clustering coefficient becomes, asymptotically following the $1/k$ law. In contrast, in the Erdős–Rényi random network model [74, 176, 177] or various small world models [440, 442, 637], the clustering coefficient is independent of k [53].

The presence of such a hierarchical architecture reinterprets the role of hubs in complex networks. Hubs, the highly connected nodes at the tail of the power law degree distribution, are known to play a key role in keeping complex networks together, playing a crucial role from the robustness of the network [14, 120, 121] to the spread of viruses in scale free networks [465]. In a hierarchical structure the clustering coefficient of the hubs decreases linearly with their degree. This implies that while the small nodes are part of highly cohesive, densely interlinked clusters, the hubs are not, as their neighbors have a small chance of linking to each other. Therefore, the hubs play the important role of bridging the many

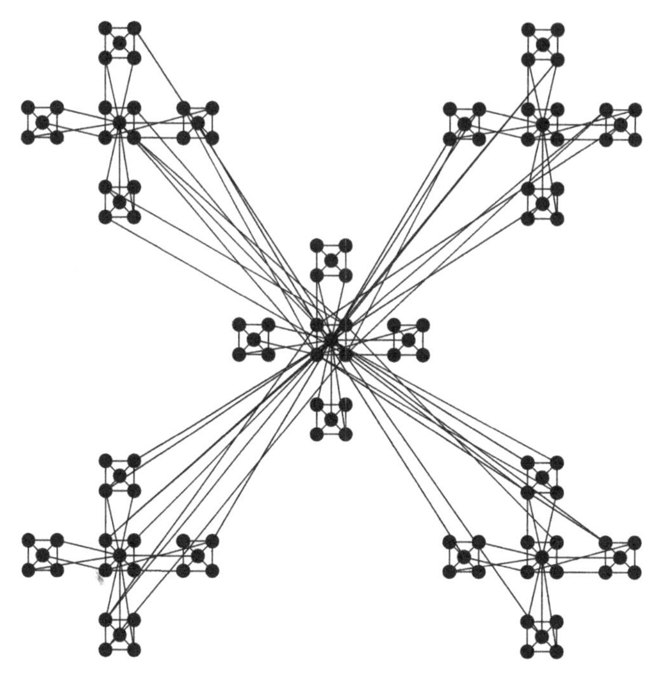

Fig. 7.1. Hierarchical model network ($n = 2$, $p = 3/5$). Its construction starts with a small core of five nodes all connected to each other. In step one ($n = 1$) four copies of the five-node module are made, then a fraction p of the newly copied nodes are picked at random and are connected to a node from the central module (following preferential attachment: the probability that a selected node connects to a node i of the central module is $k_i / \sum_j k_j$, where k_i is the degree of node i and the sum goes over all nodes of the central module.) In the second step ($n = 2$) another four identical copies are created of the 25-node structure obtained thus far, but only a p^2 fraction of the newly added nodes are connected to the central module. Subsequently, in each iteration n the central module of size 5^n is replicated four times, and in each new module a p^n fraction will connect to the current central module, requiring the addition of $(5p)^n$ new links [493].

small communities of clusters into a single, integrated network. Most interesting, however, is the fact that the hierarchical nature of these networks is well captured by the $C(k)$ curve, offering us a relatively straightforward method to identify the presence of hierarchy in real networks (see Section 7.6 for a discussion). The scaling law does not have to have a universal exponent -1; the idea is that larger nodes have low clustering coefficients, their role is to integrate the smaller tight clusters on all scales [494]. Figure 7.2 shows examples of three types of biological networks displaying hierarchical scaling: metabolic, protein–protein interaction and genetic regulatory networks.

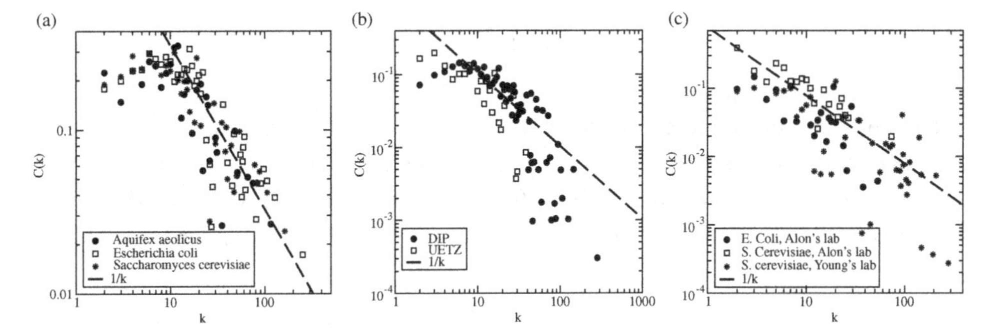

Fig. 7.2. Dependence of the clustering coefficient on node degree in three types of biological networks: (a) metabolic networks of *Aquidex aeolicus* (archaea), *Escherichia coli* (bacterium), and *Saccharomices cerevisiae* (eukaryote) [294, 494]; (b) protein–protein interaction networks from the DIP database [669, 670] and from an extensive yeast two-hybrid experiment [607]; (c) genetic regulatory interaction networks from the RegulonDB database [417, 516, 558] (downloaded from `http://www.weizmann.ac.il/mcb/UriAlon/`) and a system-level experiment in yeast [356].

7.4 Method for finding network modules

A key issue from a biological perspective is to identify the hierarchically embedded modules of biological networks, and understand how the uncovered structure relates to the true functional organization of the system. To this end we used a standard clustering algorithm that uses a similarity measure between nodes to group them onto a hierarchical tree.[2] We define the node-to-node distance through a measure we call topological overlap.

7.4.1 *The topological overlap matrix*

In order to quantify whether two nodes are closely linked into the same local cluster, we introduce the topological overlap matrix, $O_T(i, j)$. Topological overlap of 1 between vertices i and j implies that they are connected to the same vertices, whereas a 0 value indicates that i and j do not share a link, nor links to common substrates among the metabolites they react with. We define the elements of the overlap matrix to be

[2] Other approaches to module detection in metabolic networks are based on the idea that edges along a large number of shortest paths are likely to link different modules of the network [222, 262]. Edges on the largest number of paths were iteratively removed, slowly breaking the network into its functional modules.

$$O_T(i, j) = \frac{\sum_{m=1}^{N} A_{i,m} \cdot A_{j,m} + A_{i,j}}{\min(k_i, k_j) + 1 - A_{i,j}}, \tag{7.1}$$

where $A_{i,j}$ is the adjacency matrix of the network. As the topological overlap matrix is expected to encode the comprehensive functional relatedness of the biological network, we expect that functional modules encoded in the network topology can be automatically uncovered by a standard clustering algorithm.

Adjacency matrix Mathematically, a network can be represented as a table of numbers. This table is called an *adjacency matrix* and it is composed by $n \times n$ entries where n is the number of vertices. In the simplest case of an undirected network, if a link exists between vertex i and vertex j the element a_{ij} of the ith row and jth column is set to 1 while it is 0 otherwise. The same happens to the symmetric element a_{ji}. If instead the graph is directed, only one of these two entries will be set to one (according to the arrow direction). More generally the elements a_{ij} can be real numbers representing (when available) the weights of the edges in the network.

7.4.2 Hierarchical clustering algorithm

We choose the un-weighted average linkage algorithm (or un-weighted pair group method with arithmetic mean) known as UPGMA [166, 569] for our hierarchical clustering method. This algorithm first finds the largest overlap present in the matrix, joins the corresponding substrates u and v to a branching point on the tree, and substitutes them with a "new" cluster $\{u, v\}$. This new unit replaces the original u and v in the overlap matrix. It has an overlap with an arbitrary substrate (cluster) w given by

$$O_T(\{u, v\}, w) = \frac{n_u \cdot O_T(u, w) + n_v \cdot O_T(v, w)}{n_u + n_v}, \tag{7.2}$$

where n_u is the number of components in cluster u. This definition ensures that all original overlap values are represented with the same weight in the overlap value of the joint cluster, hence the method's name, "un-weighted average linkage clustering." Repeating this rule eventually shrinks the overlap matrix to a single unit, corresponding to the root of the hierarchical tree. Thus, we obtain a tree with all the original substrates as its end-leaves, grouped naturally on branches reflecting their hierarchical overlap. When overlap values between clusters are

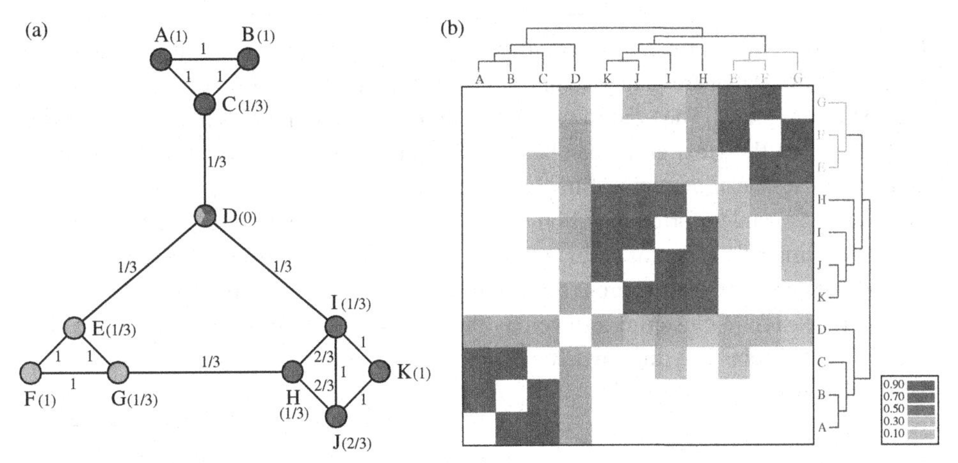

Fig. 7.3. Uncovering the underlying modularity of a complex network. (a) Topological overlap illustrated on a small hypothetical network. On each link, we indicate the topological overlap for the connected nodes, and in parentheses next to each node, we indicate the node's clustering coefficient. (b) The topological overlap matrix corresponding to the small network shown in (a). The rows and columns of the matrix were reordered by the application of an average linkage clustering method to its elements, allowing us to identify and place close to each other those nodes that have high topological overlap. The grayscale denotes the degree of topological overlap between the nodes. The associated tree reflects the three distinct modules built into the model, as well as the fact that the EFG and HIJK modules are closer to each other in the topological sense than to the ABC module.

redundant (i.e. there are at least two groups of clusters with the same overlap value) the program automatically joins the pair found first. The ordering of two branches under a junction is irrelevant, thus arbitrary. The distance between (the height of) two junction levels is defined to be one.[3] We can follow how the clustering algorithm works on a small hypothetical network shown in Fig. 7.3a. The method placed those nodes that have a high topological overlap close to each other (Fig. 7.3b), correctly identifying the three distinct modules built into the network. It also identified the relationship between the three modules, as EFG and HIJK are closer to each other in a topological sense than the ABC module.

[3] The height of a junction could, in principle, contain information about the module represented by the branch under it. For example, the average overlap between metabolites located at the leaves of the branch could determine the height of the junction. However, this additional information does not change the structure of the tree and the way the modules are organized.

Clustering Real networks are often characterized by the presence of groups of vertices (subgraphs) that are tightly connected with each other and loosely connected with the rest of the graph. The goal of clustering algorithms is to find such subgraphs: the vertices within the same group can be considered to belong to a particular cluster (or community in social systems). Simplest null (random) models for graphs and complex networks tend to form rather homogeneous graphs where the structure of clusters is much simpler than in real cases. Different algorithms have been proposed for the determination of these structures (see, for example, the overview in Ref. [134]). The simplest measure of clustering is given by the *clustering coefficient* of every vertex. It measures the probability that two vertices connected with a third are also connected with each other.

7.5 A case study: the *E. coli* metabolic network

To uncover potential relationships between topological modularity and the functional classification of different metabolites we concentrate on the metabolic network of *Escherichia coli*, whose metabolic reactions have been exhaustively mapped and studied [307]. Here we used network data compiled from the WITT database [294, 456, 494]. In order to make our final module map smaller and easier to study, we take advantage of a few peculiarities of metabolism and generate a reduced network before we start the clustering procedure.

7.5.1 Generating the reduced E. coli *metabolic network*

Metabolism relies heavily on the usage of a few substrate pairs, which undergo very generic chemical changes in a large number of reactions of all types. A representative example is the ATP–ADP pair, the cell's energy fuel molecules. As a phosphate group is broken off ATP (adenosine-triphosphate), the energy released from the chemical bond fuels the chemical change of the substrate(s) which ATP reacts with. This mechanism is so generic that ATP and ADP are the greatest hubs of our network: they are linked to a significant fraction of all substrates. A link from ATP, ADP, water, etc., to a metabolite A often carries little biologically relevant information about the function of A. There are many different reactions where other pairs of metabolites help some reactions to take place: exchange of a proton or a methyl group, for example. In order to focus on biologically relevant substrate transformations, we have performed a biochemical reduction of the metabolic network. Our guiding principle was to maintain the main line of substrate

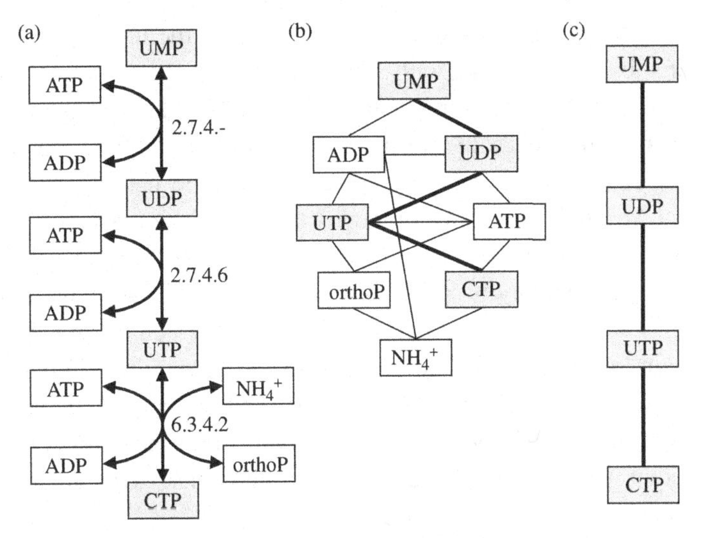

Fig. 7.4. Biochemical reduction of the pathways of the metabolic network. The middle panel shows the full graph theoretic representation of the path way shown in the left panel [294]. The right panel displays the pathway after biochemical reduction.

transformation on each pathway (Fig. 7.4). It is important to note that the reduction process is completely local: it takes place at the level of each reaction, and does not result in the removal of metabolites, but only in the removal of links from the graph representation.

Metabolism Metabolism refers to the ensemble of chemical processes through which living organisms transform resources taken from their environment in the molecules necessary for carrying out cellular functions. Since the products of a chemical reaction are often substrates (i.e. input molecules) of another one, the ensemble of metabolic processes can be conveniently organized as a network. The various chemical compounds present in the reactions are called *metabolites* and reactions are most often catalyzed by specific proteins called *enzymes*. In a metabolic network vertices represent metabolites and edges connect metabolites if they participate in the same reaction.

To further reduce the complexity of the metabolic graph we continue with a two-step topological reduction. Many pathways uncovered by the first reduction are connected to the rest of the metabolic network by a single substrate, or represent a long chain of consecutive substrates that appear as an arc between two substrates, and have no other side connections. Since the topological location of the strings of substrates depends only on one or two multiply connected terminal substrates, we can temporarily remove the long non-branching pathways or replace them with a direct link, without altering the topology of the core metabolism. We define all sets of nodes that can be separated from the network by cutting one link as *hairs*. An arc is an array of nodes connected by only two links to the rest of the metabolism, leading from one well-connected substrate to another. To generate the reduced metabolic network we have removed all hairs from the network and replaced all arcs with a single link, directly connecting the substrates at the two ends of an arc.[4] While the substrates removed during the topological reduction process are biologically important components of the network, their removal does not change the way in which subunits that they were removed from connect to other parts of the metabolism. In this sense they are topologically irrelevant.[5]

7.5.2 Finding the hierarchy of modules

After reducing the metabolic network to a representative core, we proceed to break it up into clusters based on its wiring diagram. The ordering of the overlap matrix according to a substrate's horizontal location on the hierarchical tree leads to Fig. 7.5. This figure provides us with a global topological representation of the metabolism. Groups of metabolites forming tightly interconnected clusters are visually apparent along the diagonal line of the matrix, and upon closer inspection the hierarchy of nested topological modules of increasing sizes and decreasing interconnectedness can also be seen.

To visualize the relationship between topological modules and the known functional properties of the metabolites, we originally color coded the branches of the derived hierarchical tree according to the predominant biochemical class of the substrates it produces, using a standard, small molecule biochemistry-based

[4] Note that we do not repeat the reduction process on the once reduced network. Thus, after the reduced network is ready, it can have arcs and hairs in it. These appear, for example, when two linked nodes both have hair on them, and they both have three links. After the reduction they are left with two links and thus are parts of a newly created arc.

[5] Removing the "hairs" from a network does not alter the way its core nodes connect to each other and form modules. However, the shortening of arcs only makes sense if the network carries some type of conserved flow. In case of the metabolic network two substrates are usually connected via a long arc (pathway) not because of functional distance, but as a result of chemical constraints. Enzymes catalyzing the steps along this pathway typically make small alterations to a molecule. Thus some changes take many steps that are part of the same process of converting A to B. In other types of networks a link may signify a very different relationship: for example, if two proteins in the protein–protein interaction network are connected via a long arc, there is no reason to believe that they are closely related by function. In this network a link means physical binding, the lack of a direct link indicates that the two proteins probably do not work together. As a consequence, the protein interaction network should be analyzed without the shortening of its arcs.

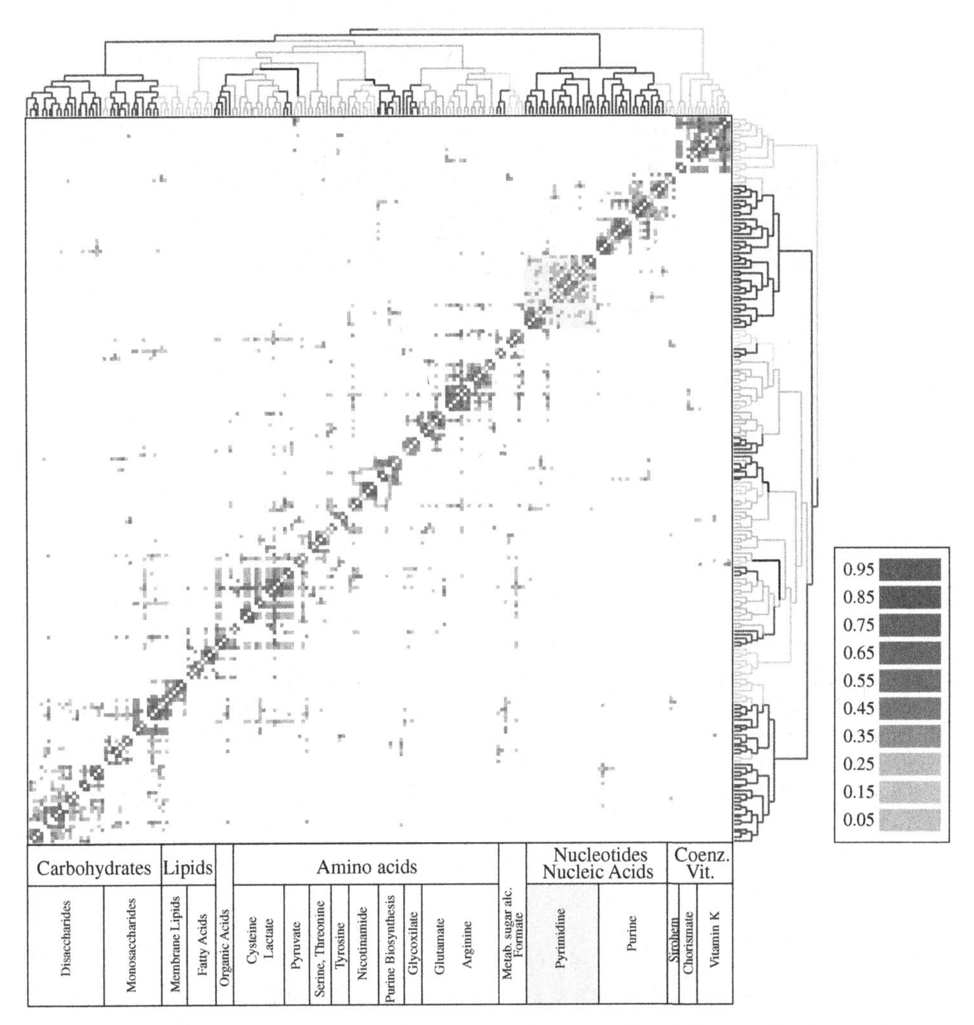

Fig. 7.5. Topological modules in the *E. coli* metabolism: the topologic overlap matrix, together with the corresponding hierarchical tree (top and right) that quantifies the relation between the different modules. Although reproduced here in grayscale only, the branches of the tree were color coded to reflect the predominant biochemical classification of their substrates. The color code of the matrix denoted the degree of topological overlap shown in the matrix. The large-scale functional map of the metabolism, as suggested by the hierarchical tree, is also shown (bottom).

classification of metabolism [456]. The biochemical classes we used to group the metabolites represent carbohydrate metabolism (blue), nucleotide and nucleic acid metabolism (red), protein, peptide and amino acid metabolism (green), lipid metabolism (cyan), aromatic compound metabolism (dark pink), monocarbon

Fig. 7.6. Three-dimensional (3-D) representation of the reduced *E. coli* metabolic network. Although shown here in grayscale, each node was color coded by the functional class to which it belongs. Note that the different functional classes are visibly segregated into topologically distinct regions of metabolism. The oval shaded region denotes the nodes belonging to pyrimidine metabolism, discussed in the main text.

compound metabolism (yellow) and coenzyme metabolism (light orange).[6] The color coding of the hierarchical tree according to biochemical classification of the metabolites proved a very good agreement between the uncovered modular hierarchy and the standard classes of the metabolism. As shown in Fig. 7.5 (grayscale only) and in the three-dimensional representation of the core network in Fig. 7.6 we find that most substrates of a given small molecule class are distributed on the same branch of the tree. Therefore, there are strong correlations between shared biochemical classification of metabolites and the global topological organization of *E. coli* metabolism (Fig. 7.5, bottom).

[6] The functional groups mentioned above are defined by series of reactions, thus many metabolites belong to more than one functional category. The color (category) we choose to represent the metabolite on the tree was the one that matched the color (category) of its neighbors on the tree, indicating that it has stronger connections to this group than any other.

7.5.3 Biochemical pathways in the pyrimidine module

To correlate modules obtained from the graph theory-based analysis to actual biochemical pathways, we concentrate on the pathways of pyrimidine metabolism. The clustering method divides these pathways into four modules, as shown in Fig. 7.7. All highly connected metabolites correspond to their respective biochemical reactions within pyrimidine metabolism, together with those substrates that were removed during the original network reduction procedure, and then re-added. However, it is also apparent that the putative module boundaries do not always overlap with intuitive "biochemistry-based" boundaries. For instance, while the synthesis of UMP from L-glutamine is expected to fall within a single module based on a linear set of biochemical reactions, the synthesis of UDP from UMP leaps putative module boundaries.

7.6 Hierarchy, fractality and the small world of networks

The question of self-similarity or fractal nature has been on the mind of network researchers since the scale-free degree distribution was observed. Are networks like fractals, self-similar on all scales? Hierarchical organization strengthens this image: most networks not only have degrees in a broad range of scales, they are also made of modules of different scales embedded into each other. The problem with complex networks as fractals was that most networks of interest to us are also small world [637]. In a small-world network the number of nodes at a distance l from any given node increases exponentially. One expects this number to grow as a power law in the case of a fractal object: measuring the "mass" within distance l from a point on the object, called the *cluster growing method*, is one way of measuring fractal dimension.

While some deterministic models have been constructed with the idea of fractality in mind [151, 302], the breakthrough in understanding topological self-similarity in complex networks was brought forth by Song, Havlin and Makse [571, 572]. They generalized the standard *box counting method* for measuring fractal dimension of a physical object to complex networks. How does one cover a network with boxes of different sizes? Divide all nodes in groups such that the shortest path between any two nodes in a group is at most l_B long: these are the boxes. Use the smallest number of groups necessary to do this (or a decent approximation) and repeat it for $l_B \in [2, D]$ (D is the diameter of the network). The results of this procedure were quite surprising: many scale-free real-world networks, such as the WWW, actor network and various metabolic networks show a neat fractal scaling between the number of boxes and box size. Thus these networks are self-similar, fractal structures. To prove their point further, the authors used a

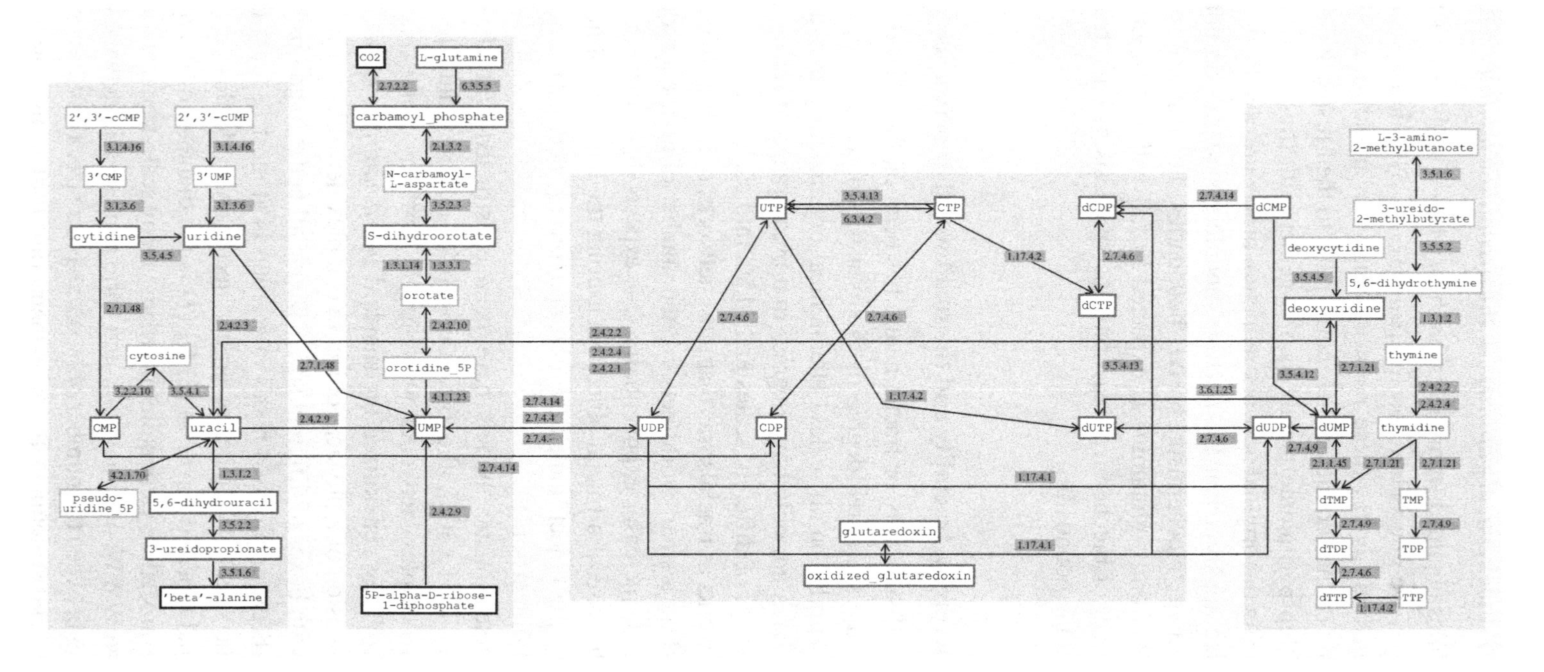

Fig. 7.7. Enlarged view of the substrate module of pyrimidine metabolism, along with a detailed diagram of the metabolic reactions that surround and incorporate it. The arrows show the direction of the reactions according to the WIT metabolic maps [456].

renormalization procedure where they collapsed each box into a node and linked these new nodes to each other if any member of the original boxes had a connection. The networks renormalized this way were also scale free, with the *same* degree distribution exponent, independent of the box size used for renormalization [571].

The question of how these networks are small world *and* self-similar at the same time has also been resolved: the two methods of determining the fractal dimension of an object, box counting and cluster growing are *not* equivalent on complex networks with broad degree distribution. While box counting covers all the hubs only once (they are assigned to one box only), cluster growing finds the hubs for almost any choice of seed node, thus bringing a large part of the network into the cluster with them. This explains the exponential increase in the number of nodes within a distance l, and shows the small-world property of the system. Box counting, on the other hand, can reveal the fractal nature of the network, if present [571].

Song *et al.* [571] uncovered some of the requirements of fractality via proposing a network growth model based on a reverse renormalization. Starting from one point, the network grows by nodes transforming into small clusters in each iteration. The conversion from a node to a small cluster mimics the renormalization process in reverse: the degree of a node grows by multiplication with a scaling factor (thus nodes of previous iterations become the hubs). The newly formed clusters have a diameter of b_B (the box size). They showed that the key feature that influences fractality in the emerging networks is how you connect the clusters: if you always connect them through their hubs (thus hubs are preferentially linked to other hubs leading to assortative mixing), you get a small-world network that is *not* a fractal structure. On the other hand, if the clusters connect via the non-hub (newly created) nodes to each other, the resulting network is a fractal, but it loses its small-world character. Many real-world networks, however, seem to be both fractal and small world; indeed a very small number of hub-to-hub connections in their model can restore the small-world property of the network while preserving its fractal nature.

The relationship between fractality and the small-world nature uncovered by Song *et al.* helps us see through a potential problem with using the $C(k)$ scaling law (which can be measured easily and quickly!) as an indication of hierarchical network architecture. There is a model proposed by Klemm and Eguíluz [324] that obeys the scaling law, it is nonetheless not composed of hierarchically embedded modules. Instead, the topology of the networks generated by the model is similar to a linear chain of locally connected dense clusters [618]. In the Klemm–Eguíluz model a new node joins the network in each time step, connecting to all active nodes in the system. At the same time an active node is deactivated with probability $p \sim 1/k$. Once deactivated, a node (and in general its neighborhood) does not receive more links, and the cluster it sits in is not embedded in larger and looser structures. Rather, a series of power law size-distributed clusters follow each other along a chain. In more general terms: a collection of internally homogeneous modules of different sizes neighboring each other (with just a few links to join them into a network) can also give rise to $C(k) \sim k^{-1}$. However, the Klemm–Eguíluz model leads to a network that is not a small world. The clusters string together along one dimension, instead of being embedded into each other. Thus one must be careful when looking at the $C(k)$ scaling as a signature of hierarchy: it is helpful to know that the network in question is also a small world.

7.7 Conclusions

The use of the above method revealed that the system-level structure of cellular metabolism is best approximated by a hierarchical network organization with seamlessly embedded modularity. In contrast to the intuitive picture of modularity which assumes the existence of a set of modules with a non-uniform size potentially separated from other modules, we find that the metabolic network has an inherent self-similar property: there are many highly integrated small modules, which group into a few larger modules, which in turn can be integrated into even larger modules. This is supported by visual inspection of the derived hierarchical tree (Fig. 7.5), which offers a natural breakdown of metabolism into several large modules, which are further partitioned into smaller, but more integrated submodules. We expect the method to be very useful for automatically uncovering functionally relevant modules in many types of biological networks.

We have found that in a variety of non-biological networks such as synonyms in a language, movie actors [637], the World Wide Web [84, 199, 350, 351] and the Internet [85, 110, 185, 231] represented at autonomous system level (each node is a domain, not just a computer) the clustering coefficient decreases with increasing k, and often scales as $1/k$ [493]. The scaling of the clustering coefficient was subsequently found in a variety of networks: software systems [427], the world trade network [554], the world-wide airport network and the co-authorship network of condmat archive[7] [56, 431]. Not all networks, however, are hierarchical: the power grid and the Internet at router level show no scaling of the clustering coefficient [493]. We suspect that spatial embedding is to blame: in both systems length costs are of concern, and their function is to distribute something to a physical area, thus it does not make much sense to form tight small communities (with many internal wires) in these networks.

[7] `http://arxiv.org/archive/cond-mat`

8
Signalling networks

GIAN PAOLO ROSSINI

8.1 Introduction

Biological systems react to changes in the surrounding environment by adjustment of their properties and functioning. The simplest cases include the capacity of prokaryotes to change the expression levels of specific proteins, as well as their distance from the source of chemical substances. In eukaryotes and multi-cellular organisms, the property of monitoring the environmental conditions and responding to their transformations has attained levels of particular complexity, through the development and evolution of means of supporting the communication among separate districts within the same organism, and among different organisms as well. Two general types of communication are classically described in biological systems. Neuronal communication is the first of them, comprising the networks of fibres connecting the different parts of organisms. Another kind of communication in living systems takes place by chemical signals that are produced and released from some cell sources, diffuse in the environment surrounding the emitting system, being it a liquid or air, and eventually reach the target cells. Two major features distinguish neuronal and chemical signalling: the means supporting the signals and the distance between the source and the target of signals. In neuronal communication the signal is mechanically supported by individual nerve fibres and travels distances related to the size of the organism about, being up to 10 m. The distance between the source of the chemical signal and its target, in contrast, is not limited by the existence of a physical wire connecting the emitting source and its target. These parts of the system, therefore, can be very close, as in the case of autocrine and paracrine communication between contiguous cells of the same tissue, or can be far apart, as in the case of pheromones responsible for male/female attraction in insects, when the distance travelled by a chemical signal can be of the order of 10^4 m, as was beautifully described by Jean-Henri Fabre more than a century ago [181]. The specificity inherent in neuronal communication between the source

Networks in Cell Biology, ed. M. Buchanan, G. Caldarelli, P. De Los Rios, F. Rao and M. Vendruscolo.
Published by Cambridge University Press.

and its target is assured by the physical connection existing between the neuron and the cell it innervates. As far as chemical signals are concerned, the specificity of the communication stems from the structure of the signalling molecule itself and the molecular entity that recognizes and binds it in target cells: the receptor of the incoming signal. Although proper functioning of biological organisms possessing neuronal and chemical signalling demands the integration and contribution of both systems, this chapter will be limited to chemical signalling. Furthermore, in order to keep the complexity of the subject within tractable terms, we will focus our attention onto major signalling pathways in higher vertebrates. We will then outline components and molecular mechanisms of signal transduction, and will discuss cases of cross-talks among distinct pathways, as a basis for the introduction of issues related to signalling networks. We will also discuss some emerging themes and challenges linked to the development of a formal treatment of signalling networks.

8.2 Chemical signalling: many pathways following a few general themes

Signaling networks Signaling networks enable cells to sense changes in their environments and articulate the appropriate responses to them. These networks are made up by collections of interacting signaling pathways, which are cascades of biochemical reactions through which the signals corresponding to the external stimuli are transported to the repository of the genetic code, where they regulate the production of the specific proteins required to orchestrate the overall cellular response.

A chemical can perform some signalling function when a biological system that is exposed to it expresses a molecular component recognizing that chemical in a selective fashion. The recognizing molecule is usually a protein and is termed a receptor. This condition is necessary but not sufficient to have a signalling system. Chemical signalling takes place whenever a specific interaction between the chemical and the receptor triggers some change in the molecular machinery of the biological system, leading to a functional state that differs from the starting conditions of that system. The chain of events mechanistically linking the formation of the chemical–receptor complex and the molecular responses induced in the biological system is the transduction pathway of the signal. The general scheme of a signal transduction pathway in a sensitive biological system then comprises the formation of a signal–receptor complex that turns the activity of one or more molecular effectors on or off, leading to changes in the concentrations, or activity, of molecular components (operators) that perform specific functions in the responding system, whose functional state is then modified. The specificity of signalling systems stems from the fact that any receptor mostly recognizes only one kind of chemical. Although a high specificity in chemical recognition is shared by receptorial components, the signal transduction pathways set in motion by signal–receptor interactions can be essentially comprised within two major groups, according to a key characteristic feature of the receptor. More specifically, the molecular mechanisms of signal transduction can be distinguished into two separate families depending on the cellular location of the receptor, which represents the molecular interface between the exterior of the biological system, from where the signal is originating, and the interior of the responding entity, where the signal brings about its informational properties by setting a different functional state (see Fig. 8.1). When the receptor is located inside target cells, as is the case of classical steroid and thyroid hormone receptor systems, the interaction of the signal with its receptor results in a transducing entity (the ligand–receptor complex) that already is at, or simply moves to the site(s) of its action, where it plays directly an effector role.

If the receptor is embodied in the plasma membrane, at the surface of the target cells, the formation of the signal–receptor complex on the external side of the plasma membrane (the *on* reaction of the transduction pathway) is accompanied by a conformational change of the receptor in the signal–receptor complex. This structural change acts on the molecular machinery at the other side of the plasma membrane, where the signal is then transduced, and effectors of the signalling apparatus are activated, bringing about the molecular response of the system. In

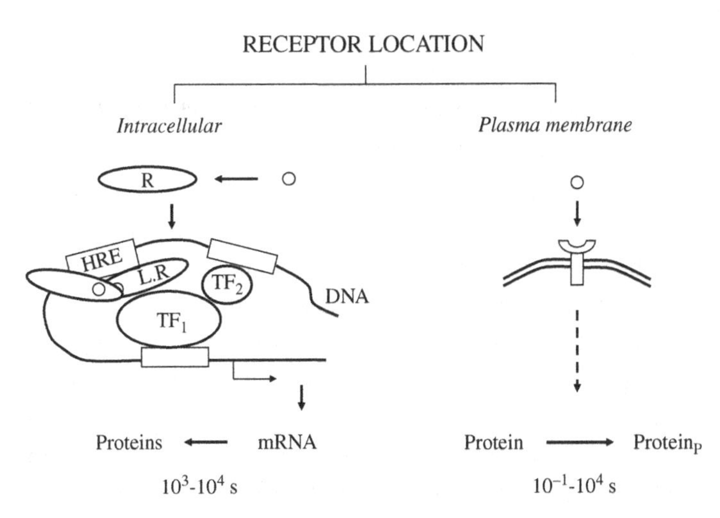

Fig. 8.1. General schemes of signal transduction mechanisms involving receptors located inside target cells and in the plasma membrane of target cells.

signalling pathways involving intracellular receptors, the ligand–receptor complex acts as a modulator of the transcription of selected genes into RNA (see below). Thus, the cellular levels of selected messenger RNAs are changed in the cell and, consequently, the cellular concentrations of the proteins coded by respective genes are affected. When the signal determines a relative increase (positive modulation) of cellular proteins, the response is directly a function of the signalling system. If a negative modulation occurs, instead, the functional response is a function of the ongoing degradation (turnover) of operator proteins, as their biosynthesis is prevented in target cells. In both cases, therefore, responses involve transcriptional events and often ensue in a time frame of hours. When signalling pathways involve membrane receptors, the proximal steps of intracellular transduction most often involve changes in the activity of effector enzymes, which directly regulate the functioning of operator proteins already existing in target cells. In these latter cases, the primary molecular responses stem from post-transcriptional events, and can ensue within fractions of a second from the interaction of the ligand with its receptor.

The simplicity of the two general mechanisms of signal transduction is only apparent, as is implied by the existence of a high degree of specificity in biological systems, with regard to the molecular responses and functional changes triggered by different signals in, and often among their target cells [299]. Before this level of complexity is approached, however, it seems appropriate to outline the major signal transduction pathways that can be grouped according to the cellular location of receptors. The following description simply represents an outline of events, and further details can be found in reviews that will be quoted while describing individual pathways.

8.2.1 The mechanisms of action of ligands interacting with intracellular receptors

This signalling pathway is shared by steroid and thyroid hormones, as well as by retinoids and other low molecular weight natural compounds [70, 394, 601]. Furthermore, it has long been recognized that many proteins existing in organisms can be grouped in the superfamily of steroid and thyroid hormone receptors, based on their structure, although no natural ligand for these proteins has been identified so far, and they are then considered ‘orphan’ receptors [394]. More precisely, the shared structural features of these proteins encompass their organization in three major domains: an amino-terminal portion that is involved in the appropriate regulation of transcriptional responses, a central domain containing two zinc fingers that make direct contact with DNA, and a carboxy-terminal portion that includes the ligand binding domain and is responsible for the trans-activating

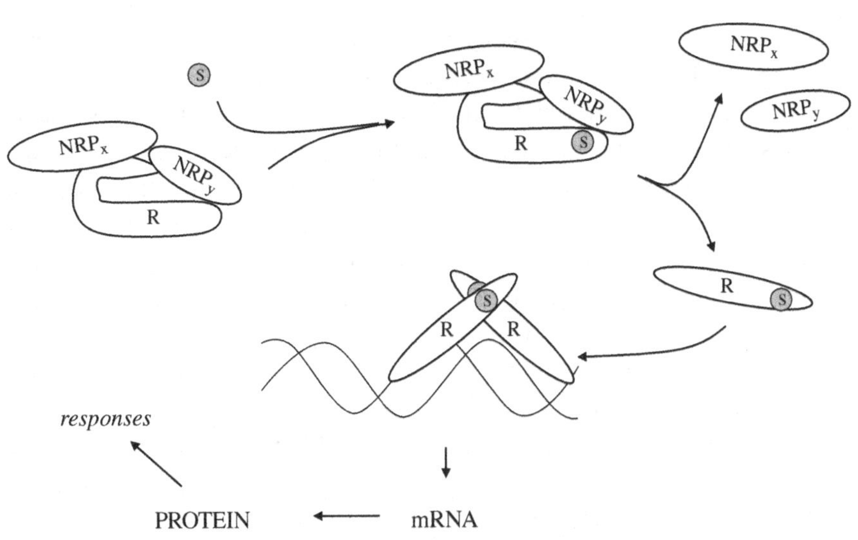

Fig. 8.2. Simplified representation of the mechanism of action of intracellular receptors. The receptor is schematically represented as an heterooligomer containing two different non-receptorial subunits (NRPx, NRPy). Upon ligand (S) binding, the receptor protein undergoes a conformational change involving receptor complex release from the oligomer, and the ligand–receptor complexes then interact productively with the transcription machinery. The protein complexes responsible for the regulation of gene transcription involve receptor dimers and other co-factors, as schematically represented in Fig. 8.1.

functions of these ligand-dependent transcription factors [394, 601]. The classical mechanism of action involving these receptors is depicted in Fig. 8.2. The interaction of a ligand with its receptor brings about a conformational change of the receptor protein, accompanied by the dissociation of proteins from the ligand–receptor complex [10, 203, 601]. These proteins are non-covalently associated with receptors and regulate their properties, but do not interact themselves with the ligand.

Depending on the receptor system, different events may ensue [339]. In the first instance, such as in the case of oestrogen and thyroid hormone receptor systems, receptor proteins are already located inside nuclei, and the binding of the ligand to receptor proteins leads to a conformational change that modifies their interactions with the DNA and other chromatin components. In the case of glucocorticoid signalling, in turn, the receptor protein is believed to be mostly in the cytosolic compartment, and the conformational change induced by the interaction of the ligand with its receptor leads to the translocation of the ligand–receptor complex into the nucleus. Independently of the original cellular location of ligand-free receptors, their interaction with respective ligands results in multiple structural changes that go beyond their dissociation from non-receptorial interacting partners. A first change involves the receptor protein itself, where some portions of the protein become exposed and available for binding to other cellular components. Thus, the zinc fingers of the central domain in the receptors that do not readily bind to DNA when the protein is in a ligand-free form in target cell nuclei become exposed, and can then make contacts with proper nucleotide sequences at the chromatin level. Another set of interactions, in turn, involves the portions of receptor proteins located in the amino-terminal and carboxy-terminal domains of receptors, which assemble with co-factors responsible for proper control of the transcription of ligand-regulated genes, and are therefore endowed with transcription regulatory activity. The co-factors interacting with the receptors can play roles of either activation (co-activators) or inhibition (co-repressors) of transcription [223, 415]. In descriptive terms, the conformational changes induced by ligand binding to its receptor protein result in the assembly (or disassembly) of transcription complexes in the promoter region of regulated genes. Although the details of the complex assembly have not been fully characterized, the process involves the formation of dimers of receptor proteins that interact with the hormone responsive elements (HRE) in the DNA molecule. HREs comprise short DNA stretches, whose sequences and arrangements will eventually dictate the receptorial component(s) that can interact with them. In the case of ‘classical’ steroid hormone receptors, and some other ligands, HREs include a palindromic structure comprising two hexanucleotide sequences separated by three undetermined nucleotides. In other cases, the sequences can be of six or more nucleotides that are arranged in several kinds of repeats, separated by a variable number of undetermined nucleotides [394, 601]. As far as receptor dimers are concerned, these can be made of identical subunits, as is the case of ‘classical’ steroid hormone receptors, or they can be heterodimers, resulting from the interaction of receptors for different ligands [394, 601]. Thus, HREs can be viewed as the sites whereby a ligand–receptor complex is anchored to the proper position at the chromatin level for driving the assembly of protein complexes that will control the transcription of ligand-regulated genes. The assembly of the transcription complexes comprising the ligand–receptor complex, its co-activators/co-repressors and other proteins of the transcription machinery will then determine increased or decreased transcription of specific genes, leading to relative changes in the cellular concentrations of the proteins coded by those genes and, hence, to the functional responses brought about by these proteins [223, 394, 415, 601]. Although the mechanism outlined above represents the core mode of action of the ligands interacting with receptors behaving as transcription factors, these ligands and their receptors can participate in other signalling pathways, as described in Section 8.3.4.

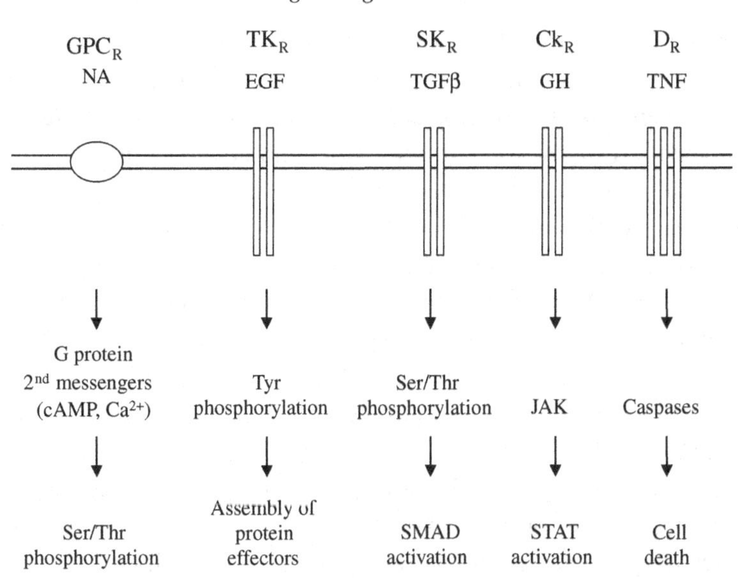

Fig. 8.3. Major signal transduction pathways of plasma membrane receptors. GPCr, G protein coupled receptor; NA, noradrenaline; TK$_R$, tyrosine kinase receptor; EGF, epidermal growth factor; SK$_R$, serine kinase receptor; TGFβ, transforming growth factor β; CK$_R$, cytokine receptor; GH, growth hormone; D$_R$, death receptor; TNF, tumour necrosis factor. See the text for details.

8.2.2 The mechanisms of action of ligands interacting with plasma membrane receptors

The transduction of extracellular signals into target cells by plasma membrane receptors occurs through many pathways, whose molecular steps have a limited degree of similarity. Still, major signalling pathways can be grouped into five general mechanisms, depending on the means by which the information is transferred from the outside to the inside of the target cells by the receptor protein, following its binding to a proper ligand and the conformational change induced by this interaction (see Fig. 8.3).

The first of these transduction mechanisms is used by a large family of membrane receptors, the seven trans-membrane domains receptors, comprising molecules whose polypeptide chain transverses the plasma membrane seven times and is associated with an intracellular protein belonging to another large family of proteins, called G proteins (see Section 8.2.2.1). This protein plays the role of transducer of the extracellular signal into an intracellular signal by interacting, and changing the activity of cytosolic enzymes that most often catalyse the production of intracellular signalling molecules, playing the role of second messengers (such as cyclic AMP, or inositol 1,4,5-trisphosphate). The second messenger then interacts and activates intracellular effectors of the external signal (most often protein kinases). These are enzymes which catalyse the transfer of a phosphate group from a donor molecule (usually ATP) to specific amino acids (serine and threonine in these cases) incorporated into protein molecules, which become phosphorylated. As a consequence of this structural change, the activity of proteins is modified, and this will determine the molecular events representing the response of target cells to the incoming signal. A different mechanism is used by insulin, a protein hormone, and by growth factors, such as epidermal growth factor (see Section 8.2.2.2). In this case, the receptor protein is a protein kinase itself, capable of catalysing the phosphorylation of selected tyrosine residues incorporated into the amino acid sequence of cellular proteins. The binding of the ligand activates the tyrosine kinase activity of receptor proteins, and this leads to phosphorylation of specific tyrosines in the receptor and in other proteins. The phosphorylated tyrosines, in turn, become docking sites for proteins possessing domains that specifically recognize and bind phosphotyrosines. The assembly of multi-protein complexes will then occur, and these multi-component systems will eventually bring about activation of signalling pathways inside the cell. A third type of signal transduction is found when receptors posses serine/threonine kinase activity (see Section 8.2.2.3). In this case, ligand binding to receptors determines the phosphorylation of specific serine residues of effector proteins, the receptor-regulated SMADs (R-SMAD), which move and are retained in nuclei of target cells, where they interact with DNA and regulate the transcription of specific sets of genes. The tyrosine phosphorylation of effector proteins is found also in another signalling pathway, whose receptor proteins are not endowed with tyrosine kinase activity themselves, but interact with tyrosine kinases of the Janus kinase family (JAK) (see Section 8.2.2.4). The interaction of ligands with their receptors then brings about activation of JAK kinases, leading to tyrosine phosphorylation of proteins playing a regulatory role in the transcription of specific genes, termed Signal Transducer and Activators of Transcription (STAT). Thus, tyrosine phosphorylation of STATs determines an increase in the transcription of selected genes whose protein products will eventually bring about the molecular response of the signalling ligand in the target cells. The last pathway comprises the molecular mechanisms responsible for cell death induced by chemical signals, as part of the biological processes controlling proper development and functioning of animal systems (see Section 8.2.2.5). In this case, the ligand interaction with receptors leads to recruitment of proteins and the formation of complexes that include enzymes possessing proteolytic activity, called caspases. The close proximity of multiple caspase molecules leads to activation of their catalytic activity, and the subsequent destruction of cellular proteins resulting in cell demise.

Overall, signalling pathways involving plasma membrane receptors transduce the extracellular signals into sensitive cells by conformational changes in the structure of receptor proteins that recruit several types of effectors on the intracellular side of the membrane. These effectors comprise enzymes catalysing different types of reactions (phosphorylation, hydrolysis and proteolysis, etc.) that change both the intracellular levels of key metabolites as well as macromolecular components, leading to functional rearrangements in target cells. A closer look at key features of the different signalling pathways will provide a more precise picture of the sets of reactions controlling proper functioning of living organisms reacting to defined environmental as well as internal conditions.

8.2.2.1 The signalling pathways of G protein coupled receptors

Signal transduction by G proteins is shared by a very large family of plasma membrane receptors, coded by about 1000 different genes, whose protein products contain seven trans-membrane domains connected by loops on both sides of the plasma membrane, defining the so-called 7TMR family of receptors, or G protein coupled receptors (GPCR) [475]. Many hormones and neurotransmitters, such as glucagon and catecholamines, as well as other bioactive compounds, such as prostaglandins, use signal transduction pathways involving G proteins and the intracellular production/release of second messengers. While the general steps of these pathways are shared by the substances binding to G protein coupled receptors, the specific mechanism of signal transduction used by different receptor systems varies according to the receptor and the G protein interacting with it. The common features of this pathway are found in the initial steps of transduction, when the interaction of the ligand with some extracellular and/or membrane embedded portions of receptor leads to a conformational change on the intracellular parts of the ligand-bound receptor. This conformational change is transmitted to an associated G protein which undergoes a defined set of structural modifications that trigger the intracellular response to the signal (Fig. 8.4). G proteins associated with 7TMRs are composed of three subunits: α, β, γ [241]. Four kinds of α subunit have been described, each of which is typical for any defined receptorial system, and determines the molecular mechanism of cellular responses to the incoming signal. The β and γ subunits are stably associated in heterodimers, and play roles in the signal transduction pathway by both regulating the functioning of the α subunit, and by triggering selected regulatory reactions. The G proteins undergo a cycle of reactions which determines their functioning in signal transduction [241, 365]. In the basal state, G proteins exist as heterotrimers, whose α subunits are associated with GDP. The ligand binding to receptor leads to a conformational change in the α subunit resulting in the release of GDP and association of a GTP molecule. The binding of GTP to the α subunit determines a conformational

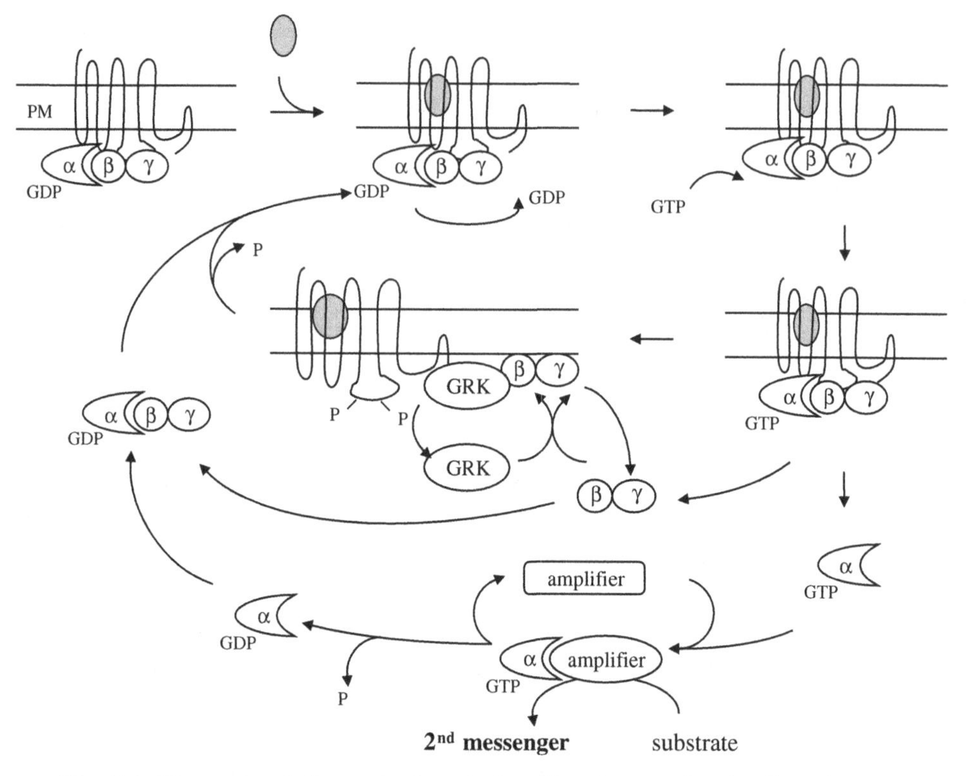

Fig. 8.4. The cycle of G protein functioning. See the text for details and other abbreviations.

change and the resulting structure has a low affinity for the β–γ heterodimers, that dissociate from the α subunit. The GTP-bound subunit, in turn, is in an 'active' conformation and associates with other cytosolic proteins (amplifiers) endowed with specific catalytic activities, that are responsible for the production of intracellular second messengers. The active conformation of α subunits, however, is transient, because the subunit has GTPase activity, and the hydrolysis of GTP to GDP determines a reversion of the structure to that with high affinity for the β–γ heterodimers. Thus the functional cycle of the G protein is terminated by reformation of the GDP-containing α–β–γ heterotrimer. Furthermore, as a consequence of GTP hydrolysis, α subunits lose their capacity to activate the amplifiers, and the signalling of the G protein is interrupted, until another cycle of GDP exchange with GTP restarts signalling through the α subunit. In this signal transduction pathway, $\beta-\gamma$ heterodimers also play some role in the regulation of receptor functioning. The dissociated $\beta-\gamma$ heterodimers, in fact, recruit protein kinases at the intracellular side of the plasma membrane [328, 358]. These kinases (G protein-coupled receptor kinases, GRK) phosphorylate the receptors and inactivate their capacity to transduce the signal, inducing a feed-back mechanism whereby the ligand itself triggers a process that will eventually terminate its own signal (Fig. 8.4). Receptor dephosphorylation by phosphoprotein phosphatases, however, rescues its proper function. As mentioned, four kinds of G proteins exist, with reference to the specific α subunit involved in the heterotrimer, determining different molecular mechanisms of cellular responses. Two of them, the Gs and Gq, are involved in most of the signalling systems, and their signal transduction pathways will be described here.

Signal transduction by cyclic AMP This pathway is outlined in Fig. 8.5 (panel (a)), and is characterized by Gs proteins activating the enzyme adenylate cyclase [241, 273, 365]. The reaction consists in the conversion of ATP to cyclic AMP (cAMP), by the formation of a covalent bond between the phosphate group bound to the C_5 of the ribose moiety of ATP with the C_3 of the same sugar, leading to the cyclic structure characterizing the cAMP molecule and its functional properties. It has been calculated that the time the receptor remains active while bound to one ligand molecule is sufficient to have several cycles of G protein activation and the production of several hundreds molecules of cAMP [365], substantiating the concept of signal amplification in the transduction pathways involving 7TMRs, and the term 'amplifier' given to the enzyme catalysing the reaction producing the second messenger. As a consequence of adenylate cyclase activation, the intracellular concentrations of cAMP will rise, reaching levels that allow cAMP interaction with the protein effector of this pathway, the cAMP-dependent protein kinase (PKA) [590]. This kinase consists of inactive heterotetramers, composed of a core homodimer of regulatory subunits, each of which is associated to one catalytic subunit. Regulatory subunits of PKA contain two sites for binding of cAMP, and the binding of four cAMP molecules to one regulatory dimer leads to the release of the catalytic subunits in an active state. The active PKA then phosphorylates its protein substrates on one or more serine/threonine residues, resulting in changes in the activity of the phosphorylated substrates (either an activation or an inactivation, depending on the protein component and the phosphorylated residue/s). Phosphorylated proteins include enzymes, transcription factors, structural proteins, so that individual cellular responses will involve defined temporal patterns of functional changes including both immediate (post-transcriptional) and delayed (transcriptional) events [410, 522].

Signal transduction by products of phospholipase C A second pathway is triggered by another class of G proteins, the Gq members, that transduce the signal of 7TMRs by interacting and activating some isoforms of phospholipase C (PLCβ)

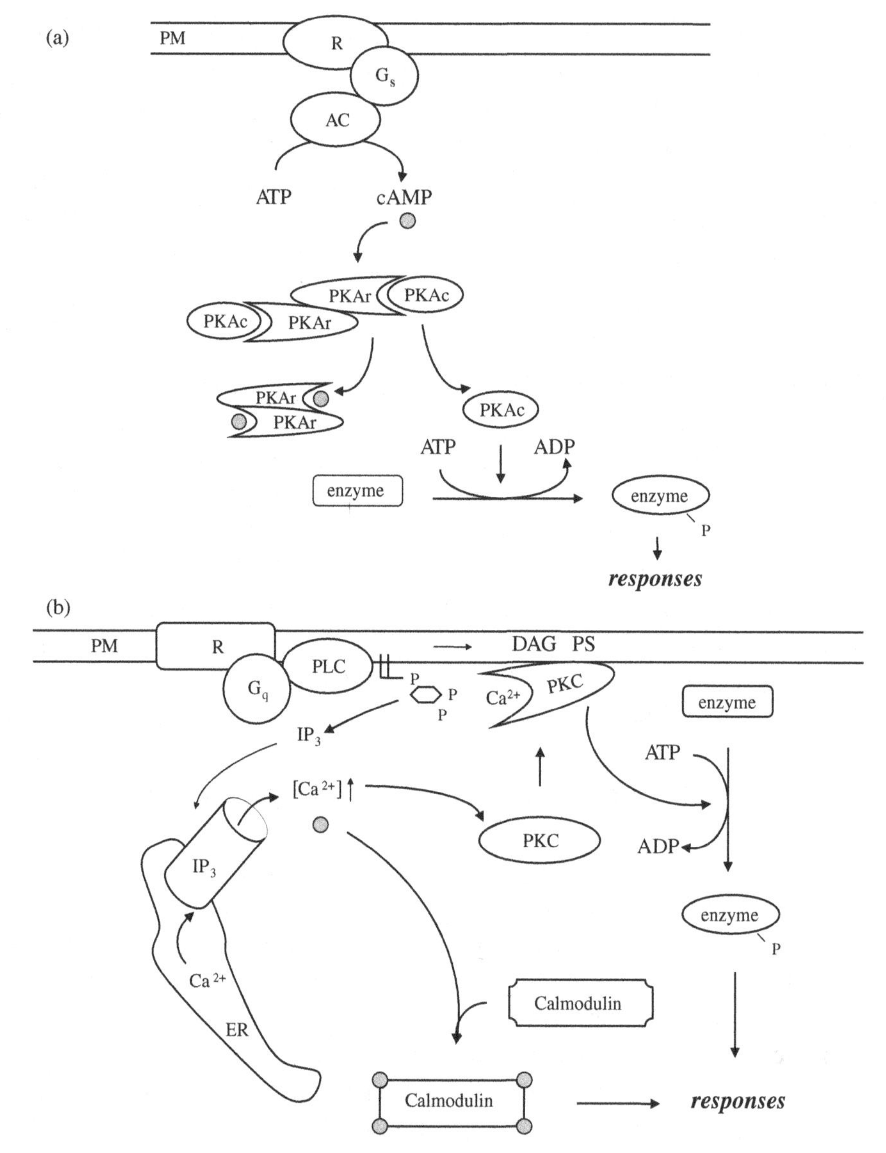

Fig. 8.5. The mechanisms of signal transduction triggered by G protein coupled receptors. Panel (a), the pathway involving cyclic AMP activation of protein kinase A. PKAr, regulatory subunit of protein kinase A; PKAc, catalytic subunit of protein kinase A; PM, plasma membrane. Panel (b), the pathway involving Ca^{2+} mobilization by phosphoinositides and the activation of protein kinase C. DAG, diacylglycerol; ER, endoplasmic reticulum; PLC, phospholipase C; PS, phosphatidylserine; IP_3, inositol 1,4,5-trisphosphate. See the text for details and other abbreviations.

(see Fig. 8.5, panel (b)). These enzymes catalyse the hydrolysis of glycerophospholipids, yielding diacylglycerol (DAG) and the phosphorylated polar head group covalently linked to the C_3 of glycerol [140]. When phosphatidylinositol-4, 5-bisphosphate (PtdIns4,5P_2) is the substrate of the reaction, the polar component released by PLC is inositol-1,4,5-trisphosphate (Ins1,4,5P_3, IP_3) [281, 416]. The following description will embody a general outline of the signal transduction involving DAG and IP_3, but it should be pointed out that it will represent an oversimplification of processes that encompass a notable complexity, and often display marked specificities, owing to both the receptor system and the characteristics of target cells, as will become apparent in some features of the pathway reported below. A first aspect of the complexity inherent in these signalling pathways stems from the fact that both DAG and IP_3 play roles of second messengers. More precisely, DAG interacts and activates some isoforms of protein kinase C (PKC) [80, 443], whereas IP_3 associates with its own receptors, located in the membrane of specialized fractions of the endoplasmic reticulum, triggering the release of calcium ions from these intracellular stores, and leading to a relative increase in the cytosolic calcium concentrations [102, 418]. The events following the release of those intracellular messengers participate in the overall mechanism of signal transduction. The increased concentration of cytosolic calcium ions favours its binding to PKC, that is then recruited at the level of plasma membrane by direct interaction with DAG and another glycerophospholipid, phosphatidylserine. The binding of these three components then results in full activation of the catalytic activity of PKC. As mentioned, this model greatly simplifies the signalling mechanisms, because several isoforms of PKC exist, with different structures and sensitivity to modulators [80, 443]. PKC phosphorylates proteins on serine/threonine. As already described for PKA, the cellular responses triggered by this signalling pathway are brought about by the proteins phosphorylated by the active PKC isoforms in a cell-specific manner. The calcium ions released from intracellular stores by IP_3 have functional roles that go well beyond the activation of PKC. Although a detailed description of the processes set in motion by the increase of cytosolic calcium concentrations is not within the scope of this section, two aspects must be mentioned. On the one hand, it has long been recognized that defined temporal and spatial patterns of calcium mobilization occur in cells upon the production of IP_3 by PLC [418]. Furthermore, the calcium ions act by interacting with a variety of cellular proteins, and calmodulin is probably the most relevant of these cellular effector components. The interaction and functional consequences of calmodulin association with other proteins, in fact is affected by the presence of calcium ions complexed with calmodulin [260]. Thus, the changes in functioning of components controlled by calcium–calmodulin complexes give a contribution to the molecular responses triggered by the ligand interacting with 7TMRs that transduces the incoming signal by activating Gq proteins.

The two signal transduction pathways described above share a high degree of similarity both in their activating reactions by the transducers and regulatory mechanisms. These primarily affect three groups of components: receptors, second messengers, phosphorylated proteins. Regulation of these signalling pathways at the receptor level essentially involves the tuning of the amount of receptor molecules able to trigger signal transduction. After ligand binding and the triggering of responses, these regulatory mechanisms determine the down-regulation and progressve extinction of the *on* reactions. The phosphorylation of the 7TMRs catalysed by the GRKs described above, as well as those brought about by PKA and PKC themselves in their respective pathways, are crucial to the down-regulatory mechanism, and trigger another process. The phosphorylated agonist-bound receptors, in fact, recruit and bind β-arrestins proteins that are free in the cytosol of target cells [328, 358]. Upon associating with receptors, β-arrestins cover part of the binding surface for G proteins on receptors, and prevent any further cycle of signal transduction. Furthermore, the association of β-arrestins with receptors triggers the association of receptors with clathrin, so that receptors are removed from the cell surface by intracellular transport through endocytosis [475]. The removal of receptorial components from the cellular site that is exposed to the incoming signal prevents new cycles of transduction. Indeed, the role of receptor endocytosis is two-fold. On the one hand, the endocytosis can be part of the process the cell uses to destroy receptors. The endocytosed receptors, on the other hand, can be only temporarily maintained at the cell interior without being destroyed, to be eventually recycled back to the cell surface, where dephosphorylation of sites that inactivate receptor activity would yield a functional entity available for new cycles of signal transduction (see Section 8.3.2). The association of β-arrestins with receptors does not represent only a mechanism for the down-regulation of the receptorial machinery, but plays also a key role in some specific responses triggered by the signalling apparatus. These aspects relate to cross-talks of this pathway with another mechanism of signal transduction, and will be described in Section 8.4. The regulation of 7TMR signalling exerted at the level of second messengers is essentially related to their conversion to molecules devoid of the structural features that allow them to interact and activate the effectors of the transduction machinery. As far as cAMP is concerned, the hydrolysis of the phosphate bond with C_3 on the ribose is sufficient to convert the second messenger to 5′-AMP, that has lost the property of activating PKA. The reaction is catalysed by phosphodiesterases (PDE), and the process is inserted in regulatory mechanisms between different signalling systems [125]. An example of these mechanisms will be given in Section 8.3.1. In the case of DAG, in turn, its availability at the level of plasma membrane does not last, as the molecule is readily phosphorylated by DAG kinase, yielding phosphatidic acid that re-enters the metabolic pathway of glycerophospholipids. Furthermore, the conversion of IP_3 into other phosphoinositides is a key element of intracellular calcium homeostasis [281, 416]. Indeed, calcium ions may not be metabolized, and the regulation of calcium-sensitive responses must rely on processes that control the ion concentration in cellular fractions primarily by cellular compartmentalization [102, 418]. The last level of control of responses triggered by ligands interacting with 7TMRs involves the proteins phosphorylated by effector kinases. As mentioned, the response depends on the activity of individual proteins, that is determined by the phosphorylation state of key amino acid residues. Thus regulatory mechanisms essentially involve serine/threonine phosphoprotein phosphatases that catalyse the removal of phosphate groups from the relevant amino acid residues. The regulatory mechanisms essentially follow a feed-back pattern, as described for receptor phosphorylation and conversion of second messengers. Many serine/threonine phosphoprotein phosphatases are expressed in organisms, but our understanding of their individual roles in regulatory mechanisms is far from being fully characterized [648].

8.2.2.2 Signal transduction by receptor tyrosine kinases

The activation of protein tyrosine kinase activity of receptors is the primary event in the transduction of the signal conveyed by many ligands, and is considered the major molecular mechanism controlling cell proliferation [71, 531, 649]. The activation of tyrosine kinase activity of receptors is triggered by ligand binding and is the consequence of changes in the conformation of receptor–ligand complexes [114, 531]. Receptors consist of trans-membrane proteins, whose tyrosine kinase activity is possessed by their intracellular, carboxy-terminal domains. Ligand-free receptors can exist as monomers, as is the case of some growth factor receptors, or as preformed dimers (insulin receptor). Ligand binding triggers the dimerization of monomeric proteins, and the changes in conformation of the intracellular portions of receptor proteins relieve structural constraints thereby allowing the phosphorylation of exposed tyrosines in one receptor molecule by the juxtaposed kinase of the contiguous receptor molecule (see Fig. 8.6). Thus, ligand binding and oligomerization leads to trans and autophosphorylation of receptors in several tyrosine residues [114, 268, 632]. The phosphotyrosines are docking sites for proteins responsible for the transduction of the signal in the cellular response (see Fig. 8.6). The proteins that interact with phosphotyrosines possess SH2 (Src homology 2) and PTB (phosphotyrosine-binding) domains that accommodate the modified amino acid in stable interactions, thereby assembling signalling complexes [468, 674]. The proximity of the docked proteins then leads to changes of their properties and activation of functioning.

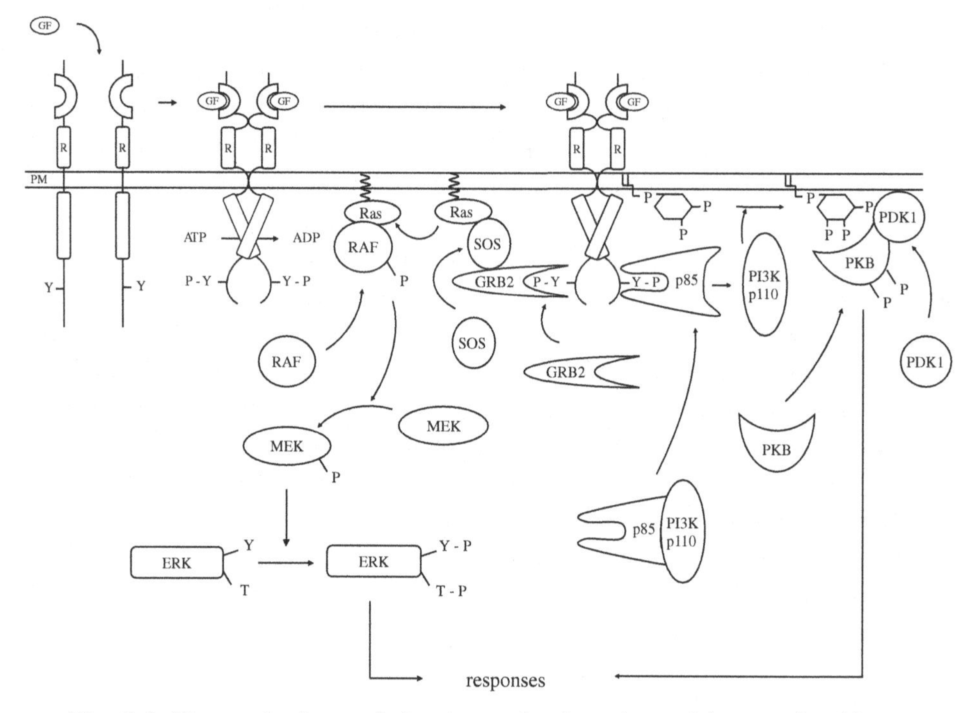

Fig. 8.6. The mechanisms of signal transduction triggered by tyrosine kinase receptors. GF, growth factor; R, receptor. This schematical representation does not include the release of GDP from Ras and its exchange with GTP induced by SOS, as well as the activation of PLCγ by its interaction with phosphotyrosines of activated receptors. See the text for details and other abbreviations.

Although several proteins can bind to tyrosine-phosphorylated growth factor receptors, a different situation exists in the case of the insulin receptor, which recruits only a single class of components, termed insulin receptor substrates (IRS). Thus, the interaction of IRSs with the activated insulin receptor results in phosphorylation of IRSs themselves on tyrosines, and these phosphotyrosines then become docking sites for proteins involved in signalling complexes [588, 679].

Three major pathways participate in signalling by receptor tyrosine kinases, although their relative contributions to responses triggered by individual signals may considerably differ in their target cells, depending on the signal, the system and the cellular context as well [299]. The first of these pathways brings about the activation of the mitogen activated protein kinase/extracellular signal regulated kinases (MAPK/ERKs), which are major determinants in the control of cell proliferation. This activation involves a protein kinase cascade that comprises three different kinase activities and is triggered by signalling complexes where several proteins have been assembled [107, 114, 117, 309, 329]. As far as kinases are concerned, they are often referred to as mitogen-activated protein kinases (MAPK), a term that may be misleading, because it has been used also to indicate protein kinases that do not primarily stimulate cell proliferation but are involved in stress responses (Section 8.2.2.4.). In any case, the three kinases of the cascade are termed MAPKKK (being the protein kinase phosphorylating the kinase that phosphorylates MAPK), MAPKK (being the protein kinase that phosphorylates MAPK), and MAPK itself, that will be indicated as ERK. The signalling complexes assemble on phosphotyrosines of receptors (or of IRS) that bind the adaptor protein GRB2 (growth-factor receptor bound-2) either directly or through its association with a phosphotyrosine-bound Shc (Src-homology-2 containing protein). GRB2 then binds the protein SOS (Son Of Sevenless). SOS is then recruited in the vicinity of the inner leaflet of the plasma membrane, where the protein Ras is located, owing to its covalent bond with a fatty acyl moiety that is embedded in the plasma membrane [243]. Ras belongs to the family of small monomeric G proteins, that is activated by exchange of its bound GDP with GTP, as already described for trimeric G proteins (see Section 8.2.2.1). SOS interacts with Ras, promoting the release of the bound GDP and the association of Ras with GTP, thereby activating Ras [243]. As a consequence, activated Ras can bind the first kinase of the cascade leading to ERK activation. The protein kinase activated by Ras, the MAPKKK, is termed RAF, and is activated by phosphorylation of multiple serine and threonine residues, by a complex mechanism that is still partially unresolved and that could include autophosphorylation [647]. In any case, activation of the kinase activity of RAF results in an enzyme that phosphorylates the MAPKK of the cascade, the MAPK/ERK kinases (MEK). MEK is a dual protein kinase that is then capable of phosphorylating proteins on both tyrosine and threonine, and becomes activated after phosphorylation of two serine residues catalysed by RAF. MEK then catalyses the phosphorylation of ERK on a threonine and a tyrosine residue, separated by one unrelated amino acid, and the dual phosphorylation of ERK activates its catalytic properties [117]. ERKs represent major effectors of receptor tyrosine kinase signalling pathways, and their activation leads to phosphorylation of target proteins in serine/threonine, with consequent changes in the functioning of enzymes (some ERK substrates are protein kinases themselves), structural proteins and transcription factors in target cells [250, 649].

A second pathway of receptor tyrosine kinase signalling involves the assembly of phosphatidylinositol 3-kinase (PI3K) on phosphotyrosines. PI3K is a dimeric enzyme, comprising a regulatory (85 kDa) and a catalytic (110 kDa) subunit [335, 395]. The regulatory subunit associates with phosphotyrosines and the interaction relieves the inhibition exerted by the regulatory subunit on the catalytic subunit of PI3K, while recruiting the active enzyme to the plasma membrane, where it catalyses the phosphorylation of phosphatidylinositols in the C3 position of inositol, particularly in phosphatidylinositol 4,5-bisphosphate [281, 416]. The activation of receptor tyrosine kinase signalling, therefore, leads to a substantial increase in the membrane levels of phosphatidylinositol 3,4,5-trisphosphate (PtdIns3,4,5P_3). Because each phosphate group has two negative charges, increased PI3K activity leads to increased density of negative charges in some areas of the internal side of the plasma membrane. These clusters of negative charges then bind proteins harbouring the so-called plekstrin homology domain (PH domain), that includes sequences of amino acids containing positively charged side chains (lysine, arginine) [468]. Two proteins are then recruited at the level of the plasma membrane: the protein kinase B (PKB/Akt) and the phosphoinositide-dependent protein kinase 1 (PDK1). Recruitment of PKB at the plasma membrane then results in its phosphorylation on a serine and a threonine residue, and this leads to activation of the kinase activity. It is recognized that threonine phosphorylation of PKB is catalysed by PDK1, but it is still debated whether serine phosphorylation is due to PDK1 or a separate kinase [395, 527, 613]. In any case, PKB activation is the result of receptor tyrosine kinase signalling, leading to serine/threonine phosphorylation of a number of proteins, including enzymes and transcription factors, in target cells [335, 395]. Indeed, the activation of PKB is considered the major determinant of insulin signalling, particularly with regard to its metabolic effects [119, 395]. Moreover, PKB activation is important also in the inhibition of cell death by apoptosis [106, 395].

A third mechanism involved in receptor tyrosine kinase signalling is the activation of some PKC enzymes [309]. In this particular case, some isoforms of phospholipase C (PLCγ) associate with phosphotyrosines on ligand-bound growth factor receptors and become activated, leading to the production of DAG and IP_3, and the consequent activation of PKC and calmodulin-dependent enzymes (see Section 8.2.2.1).

Attenuation of responses in this signal transduction pathways includes the different molecular steps of the pathway already approached with G protein coupled receptors. Because phosphotyrosines are key elements in receptor tyrosine kinase signalling, the removal of phosphate groups from tyrosines, catalysed by phosphotyrosine phosphatases, is of primary importance [531, 689]. Down-regulation of receptors by endocytosis, however, also contributes to attenuation of responses triggered by receptor tyrosine kinases [114, 531].

8.2.2.3 The signalling pathway of SMADs

Signal transduction through this pathway is used by signalling molecules belonging to the transforming growth factor β family [532, 559]. The process is triggered when two molecules of the ligand, associated as homodimer, bind to their receptors. Two types of receptor exist: type II, possessing a constitutive serine/threonine protein kinase activity, and type I receptor molecules, which acquire their kinase activity after type II receptors have transphosphorylated them in several serine/threonine

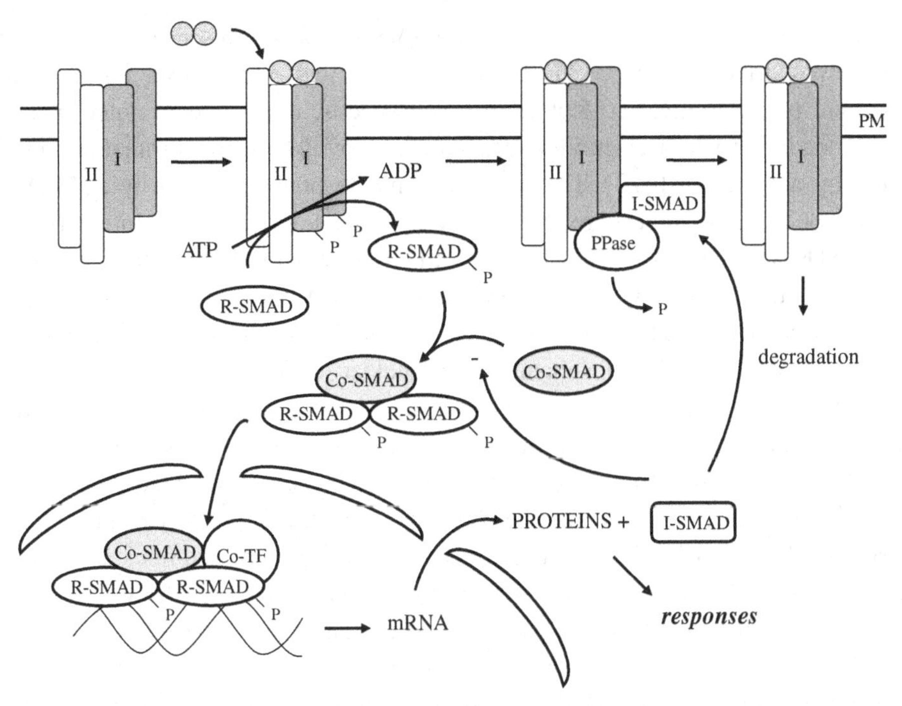

Fig. 8.7. The mechanism of signal transduction involving SMADs. Co-TF, co-factors of the transcription machinery; PM, plasma membrane. See the text for details and other abbreviations.

residues located in their carboxy-terminal, intracellular domain (see Fig. 8.7). Ligand binding to heterotetrameric receptors then leads to phosphorylation of type I receptors by their type II partners and to the subsequent interaction of the intracellular domain of type I receptors with SMAD proteins, which are next phosphorylated by these receptor kinases in two serine residues [532, 559]. Three classes of SMAD proteins exist [532, 672]. Regulatory SMADs (R-SMADs), such as SMAD1, SMAD2, and others, are phosphorylated, and the phosphorylation controls their functional properties. Other SMADs include co-SMADs, such as SMAD4, that are not phosphorylated by the receptor kinases, but interact with R-SMAD and contribute to their functioning. Inhibitory SMADs (I-SMADs), in turn, represent the third class, and include proteins that participate in the negative regulation of this signalling pathway. In basal states SMADs continuously shuttle between the cytosol and nuclei, and the phosphorylation of R-SMADs leads to nuclear transfer of these proteins [532, 559, 672]. The phosphorylated R-SMADs form heterotrimeric oligomers, composed of two R-SMAD molecules and one co-SMAD subunit, and these trimers then move into nuclei, where they bind to DNA and determine changes in the transcription of specific genes [532, 559, 672]. The SMAD trimers have low affinity for their binding sites on DNA (SMAD binding element, SBE), and do not appear to assemble the basal transcription machinery themselves. In turn, the SMAD trimers interact with other DNA binding proteins and transcriptional co-factors in a cell-specific and context-related manner [406], leading to transcription regulation that involves also chromatin remodelling by acetylation [672]. The transcriptional response induced by SMADs is dependent on their nuclear retention, due to phosphorylation and interaction of R-SMADs with co-SMADs. In line with this mechanism, the negative regulation of the signalling pathway involves modes that affect these two processes. Thus, signal extinction occurs as a consequence of dephosphorylation reactions catalysed by phosphoprotein phosphatases at a nuclear level [532, 672]. As far as the interaction of R-SMAD with co-SMADs is concerned, the process leading to disruption of productive trimers follows the transcriptional response, that includes the induction of inhibitory SMADs (I-SMADs). Thus, the SMAD6 produced as part of the cellular response to TGFβ in target cells, competes with phosphorylated R-SMADs for binding to co-SMADs, leading to decreased nuclear levels of R-SMADs-co-SMAD oligomers. As ongoing phosphorylation of R-SMADs would maintain signalling through this pathway, extinction of the *on* reaction of the process, by down-regulation of R-SMAD phosphorylation at the receptor level, should accompany their nuclear dephosphorylation. I-SMADs appear to participate to this process, by inducing down-regulation of R-SMAD phosphorylation through type I receptor dephosphorylation as a consequence of recruitment of a phosphoprotein phosphatase (type 1 ser/thr phosphoprotein phosphatase) at the receptor level, and by receptor degradation due to recruitment of ubiquitin ligases [672]. Ubiquitination and degradation of R-SMADs also contribute to extinction of the signal transduction through this pathway.

8.2.2.4 The JAK-STAT pathway

Cytokines, such as interferons and interleukins, as well as classical hormones, such as prolactin and growth hormone, trigger their responses in target cells by interacting with plasma membrane receptors devoid of enzymatic activity, but capable of activating tyrosine kinases of the Janus kinase family (JAK) that are associated with the intracellular domain of receptor proteins [344, 366, 530]. The binding of the ligands to receptor proteins leads to receptor oligomerization, whereby JAK molecules are brought in close proximity, leading to trans-phosphorylation of kinases themselves in specific tyrosine residues and the release of their catalytic activity (see Fig. 8.8). The activated JAKs then phosphorylate receptor proteins in tyrosine, providing binding sites for proteins containing SH2 domains. The proteins that recognize and bind these phosphotyrosines represent transcription factors that move into nuclei, thereby transducing the hormone signal intracellularly, and are then termed Signal Transducers and Activators of Transciption (STAT) [373].

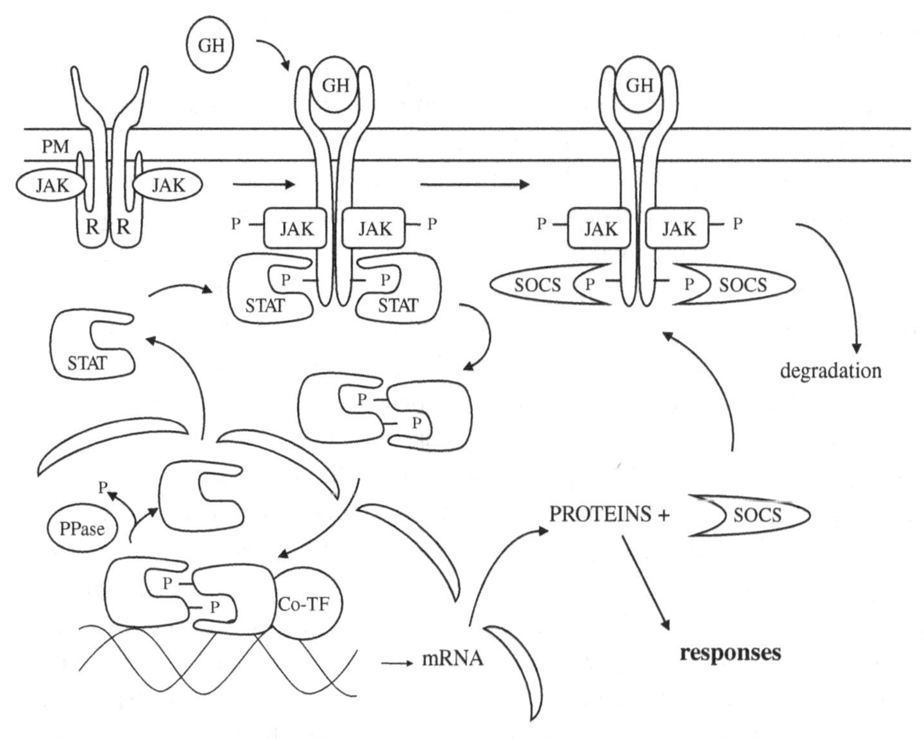

Fig. 8.8. The JAK-STAT signalling pathway. Co-TF, co-factors of the transcription machinery; GH, growth hormone; PM, plasma membrane; PPase, phosphoprotein phosphatase; R, receptor. See the text for details and other abbreviations.

The structure of STATs includes a central DNA binding domain, separated from a SH2 region adjacent to a tyrosine residue (tyrosine activation residue around 700) by a linker region, followed by a carboxy-terminal transcriptional activation domain (TAD). The SH2 domain of STATs binds the phosphotyrosine on receptors and the receptor-bound JAKs then phosphorylate the tyrosine activation residue of STATs. The tyrosine phosphorylated STATs then dissociate from the receptors and two molecules interact through their SH2 and phosphotyrosine residues in a head-to-tail arrangement. The homodimeric, activated, STATs then translocate inside the nucleus of target cells, where they interact with DNA sequences in the promoters of regulated genes. These STAT binding sites are organized as palindromes, and two STAT dimers are associated with some natural promoters, so that tetramers appear to maximally activate transcription [366, 425, 530]. The transcriptional response triggered by STATs can be affected by the interaction of STATs with co-factors and the covalent modification of STATs themselves [366, 530]. Furthermore, the serine phosphorylation of STATs has been found to enhance gene transcription, a response that can be observed also when STAT proteins have not undergone tyrosine phosphorylation, pointing to a role of these components in transduction pathways used by other in signals in responsive cells. Thus, it is expected that significant cross-talks between the JAK-STAT and other signalling pathways can occur. As in SMAD signal transduction, the control of the JAK-STAT signalling pathway is exerted through regulation of nuclear retention of STATs, covalent modifications and degradation of components of the signalling apparatus, as well as by protein modulators of receptors and STATs. The nuclear location of STATs, for instance, depends on tyrosine phosphorylation of the homodimers, whose subunits are recycled back to the cytoplasm upon phospho-tyrosine dephosphorylation [366, 530]. Indeed, the STATs appear to be continuously shuttling between the nucleus and the cytoplasm upon stimulation of this pathway, as a consequence of tyrosine phosphorylation catalysed by JAKs at the receptor level and phospho-tyrosine dephosphorylation in nuclei. In more general terms, the signalling pathway as a whole is affected by phospho-tyrosine dephosphorylation, as the removal of phosphate groups from phospho-tyrosines of receptor proteins and JAKs by phospho-tyrosine phosphatases SHP1, SHP2 and PTP1B in the cytoplasm, is involved in signal down-regulation [366, 530]. The interaction of protein components of the pathway with other proteins is another factor contributing to the overall control of the signalling apparatus. The transcriptional response triggered by the incoming signal in this pathway includes the expression of protein modulators, such as the supressors of cytokine signalling (SOCS). Proteins of this group possess an SH2 domain, and inhibit signalling by binding to phospho-tyrosines of receptor proteins, JAKs and STATs [366, 373, 425, 530]. Furthermore, SOCS have been also involved in targeting of STATs for proteasomal degradation [366, 373, 530].

8.2.2.5 Death receptor pathways

Controlled cell death contributes to the normal functioning of organisms, where it plays roles in proper development, as well as tissue remodelling and homeostasis [79, 258, 683, 690]. Several cytokines participate in the control of cell death pathways, including tumour necrosis factor (TNF), TNF-related apoptosis inducing ligands (TRAIL), and FAS ligand [472, 526]. In this pathway, the ligands are preassembled as homotrimers, so that interaction with their receptors brings about the oligomerization of receptor proteins (see Fig. 8.9). The intracellular portion of these proteins consists of the so-called death domain, and receptor oligomerization then recruits protein components that bind to these domains, with the formation of the death-inducing signalling complex (DISC). This complex comprises receptor-associated adaptor proteins (Fas-associated death domain protein, FADD), that interact with other proteins through a death effector domain (DED) located at their amino termini. The major protein interacting with the DED of FADD is the

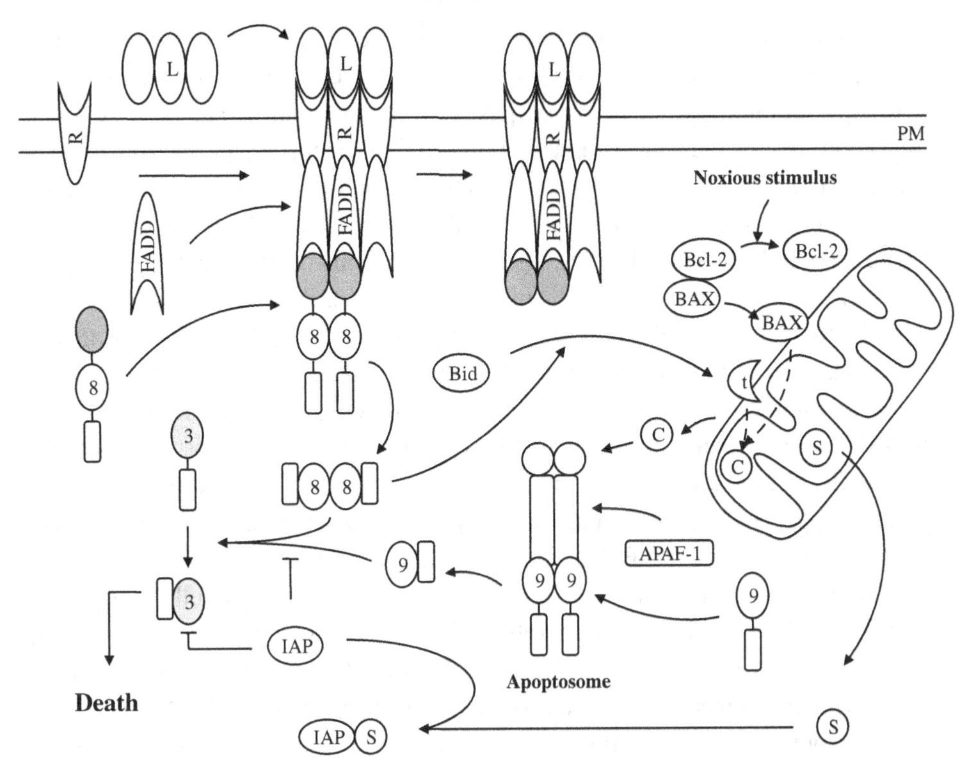

Fig. 8.9. The death pathways. The extrinsic pathway is activated by an incoming signal (L), whereas the intrinsic pathway is triggered by a noxious stimulus altering normal cell functioning. For simplicity, the apoptosome depicted in this scheme contains two cytochrome C and APAF-1 subunits, although it is an heptameric complex. 3, caspase 3; 8, caspase 8; 9, caspase 9; C, cytochrome C; R, receptor; S, SMAC/DIABLO; t, tBid. See the text for details and other abbreviations.

pro-enzymatic, inactive form of the caspase 8 (or caspase 10) protease. Caspases 8 and 10 belong to a class of cysteine proteases that selectively catalyse proteolysis of their substrates after an aspartate position [50, 338, 502]. This class of enzymes includes many different members (more than ten in mammals), that are usually grouped as either 'initiator' or 'effector' caspases. The distinction between these two types of caspases stems from the role they play in the death response (apoptosis), so that enzymes involved in the triggering of responses are initiator caspases, whereas those responsible for cleavage of cellular proteins, as part of cell destruction, are effector caspases. The initiator caspase molecules recruited at the DISC do not possess catalytic activity, which is gained following their assembly as homodimers at receptor level. The close proximity of inactive pro-enzymes is believed to cause a conformational change of zymogens that allows autocatalytic attacks, leading to the separation of the protease portion of caspase 8 from the

DED, and the cleavage of the protease, giving rise to the two subunits of the active enzyme, and the release of catalytically active heterotetramers containing two p18 and two p10 subunits of the active caspase 8 in the cytoplasm. The active caspase 8 then catalyses the cleavage of inactive pro-enzymatic forms of effector caspases, such as caspase 3 and caspase 7, leading to the formation of the catalytically active dimers, which will eventually bring about the proteolysis of cellular proteins and execute the death response leading to cell demise. This death response, called the extrinsic response because it is triggered by an extracellular signal, is often accompanied by the activation of an intrinsic death response, which can be also activated by noxious agents in affected cells. The mitochondria play a major role in the intrinsic apoptotic pathway, whose activation involves proteins of the Bcl-2 family [106, 678]. The proteins belonging to this family are classified as anti-apoptotic (such as Bcl-2 and Blc-X_L) and pro-apoptotic (Bax, Bak, Bid) components. The death response in the intrinsic pathway is triggered by changes in the interactions among components of the Bcl-2 family, owing to covalent modifications (phosphorylation) of existing molecules or an imbalance between their cellular levels. In general, it is believed that anti-apoptotic components associate with pro-apoptotic counterparts, preventing them from perturbing mitochondrial membrane integrity. Thus, a triggering event would lead to increases in the intracellular levels of pro-apoptotic components interacting with mitochondria, resulting in the formation of pores, and the consequent release of the mitochondrial proteins cytochrome c and 'second mitochondria-derived activator of caspases' (SMAC/DIABLO) [472, 502]. The released cytochrome c would then lead to the assembly of a multimeric protein complex (apoptosome) in the cytoplasm, containing multiple molecules of three proteins: cytochrome c, the apoptotic protease activating factor 1 (APAF-1) and the pro-enzymatic forms of another initiator caspase, caspase 9 [50]. The assembly of these components in the apoptosome then leads to the activation of caspase 9, and the active caspase 9 eventually activates effector caspase 3 and 7 that execute cell death [50, 338, 502]. The role of SMAC/DIABLO in this activation process, in turn, is related to its affinity for cytosolic proteins that function as inhibitors of both initiator and effector caspases, and are then termed 'inhibitors of apoptosis proteins' (IAP) [272, 519]. Thus, by direct association of SMAC/DIABLO with IAP, endogenous inhibitors of caspases are inactivated and cell demise is brought about by activated effector caspases. The molecular link between the extrinsic and the intrinsic pathways of apoptosis is represented by one member of the Bcl-2 family, the Bid protein. Active caspase 8, in fact, attacks intact Bid molecules, releasing a truncated fragment (tBid) that induces the release of cytochrome c from the mitochondria (most likely by interfering with anti-apototic members of the Bcl-2 family), and then activates the intrinsic apoptotic pathway [106, 472, 678].

Although cell death represents the key event of these pathways, other molecular processes are set in motion by ligand interaction with death receptors. In particular, when TNF interacts with its receptors, signal transduction comprises complex arrays of components, and the ultimate response does not invariably result in cell death. In turn, TNF responses often include the inhibition of cell death and induce the selective regulation of cellular processes, involved in proper cell functioning, including cell survival and proliferation [459, 628]. The details of the molecular mechanisms of these responses are far from being clarified, but the wealth of information on TNF action reveals extensive cross-talks with other signal transduction and regulatory pathways, as will be considered in Section 8.3.3.

8.3 Cross-talks among signal transduction pathways

The major signalling mechanisms described above consist mainly of linear pathways. As we learn more on signal transduction pathways used by the many chemical signals in organisms, however, this kind of arrangement of components in the signalling apparatus is becoming an oversimplification. Indeed, signal transduction by linear transduction pathways most likely represents an exception rather then a rule in biological systems. Many examples of molecular links between different pathways, as well as branching points of signalling systems leading to multiple effector molecules are known. A few of them will be described in this section, to provide a general frame for relevant features of cross-talks in signal transduction. In some cases, the cross-talk can be recognized as a feed-back mechanism responsible for the proper control of functionally antagonist signalling systems, whereby the pathway triggered by a first signal is progressively attenuated and eventually overcome by the events set in motion by a different agent, as is the case of the glucagon/insulin pairs in higher vertebrates. In other instances, no feed-back or controlling mechanisms are apparent, and the cross-talks mostly represent a means for the diversification and specialization of responses based on combinations of common signalling pathways [299]. Some of the cross-talks among signalling pathways are described below and are summarized in Fig. 8.10.

8.3.1 Insulin effects on components of the signalling pathway involving cAMP

Insulin and glucagon play opposed roles in controlling the metabolism in higher vertebrates. The general effect of insulin is anabolic, with stimulation of biosynthetic reactions, including glycogen, lipid and protein synthesis, whereas

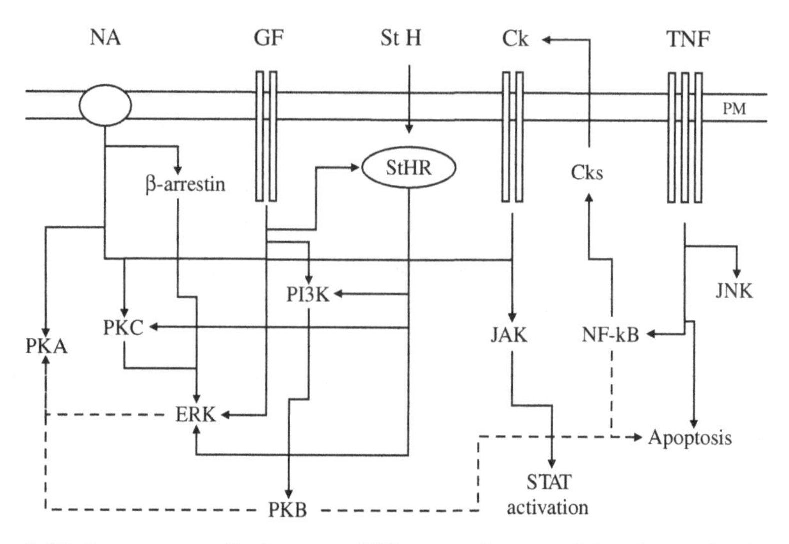

Fig. 8.10. Some cross-talks between different pathways of signal transduction. In this scheme the solid lines represent activating mechanisms, whereas the dashed lines represent inactivating mechanisms. Cks, cytokines; GF, growth factor; NA, noradrenaline; StHR, steroid hormone-receptor complex. See the text for details and other abbreviations.

glucagon stimulates glycogenolysis and lipid catabolism. At a molecular level, the effects of glucagon are triggered by increased cellular concentrations of cAMP and PKA activation, whereas the opposite holds true when insulin is the dominant signal in the organism [143]. The molecular bases of insulin antagonism on cellular functions stimulated by increased cellular levels of cAMP and PKA activation impinge onto many different components and involve several processes displaying notable target organ specificity. The cross-talks among the two signalling pathways approached here will be confined only to two aspects of signal transduction and the resulting cellular effects. As far as signalling is concerned, the antagonistic effect exerted by insulin on processes stimulated by PKA activity involves a decrease in the cellular levels of cAMP, owing to increased PDE activity. This type of effect would stem from the activation of serine/threonine kinases induced by insulin in target cells (possibly ERK, or another protein kinase activated by ERK) that would catalyse the phosphorylation and thereby the enhancement of the activity of some PDE isoforms [125]. A separate mechanism contributes to the effect of insulin on glycogen metabolism, leading to stimulation of glycogen synthesis coupled to inhibition of glycogenolysis in target cells. In this case, insulin enhances the synthetic process, while opposing the molecular changes underlying the stimulation of glycogenolysis that have been induced by increased cellular levels of cAMP. A simplified representation of the mechanism controlling glycogen metabolism

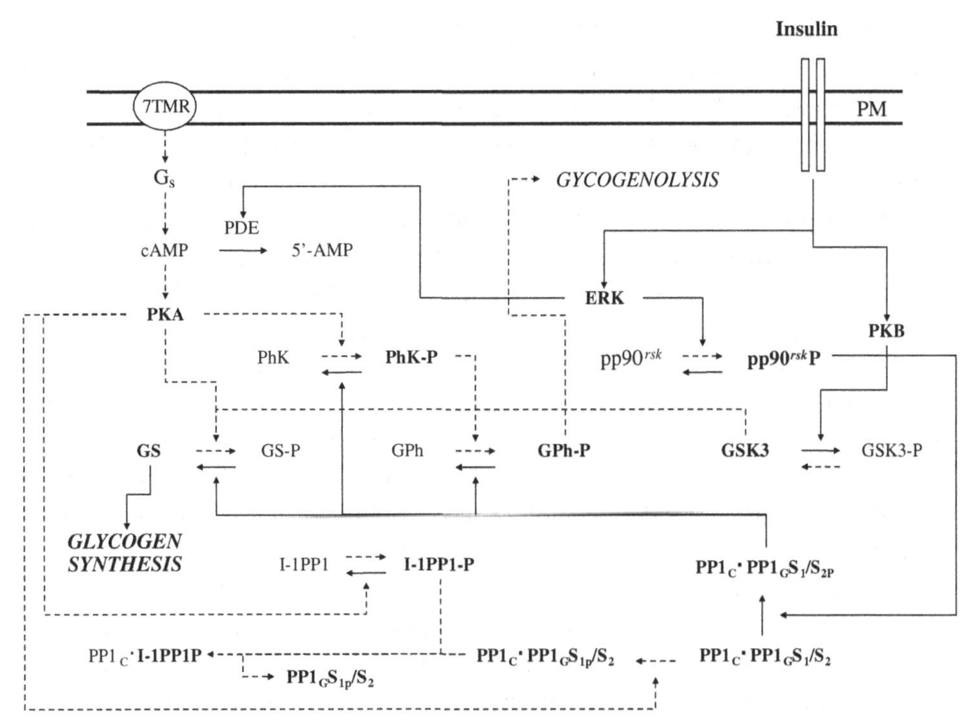

Fig. 8.11. Functional antagonism between insulin action and signalling through activation of adenylate cyclase. The mechanism activated by insulin and the consequent activated reactions/functions, are indicated in bold characters and by the solid lines. The mechanism activated as a consequence of cAMP sinthesis, that are inhibited by insulin signalling and result in inhibition of reactions/functions, are indicated by regular characters and by the dashed lines. 7TMR, seven trans-membrane receptor; Gs, protein G stimulating the adenylate cyclase; GPh, glycogen phosphorylase; GS, glycogen synthase; I-1PP1, inhibitor 1 of type 1 serine/threonine phosphoprotein phosphatase; PhK, glycogen phosphorylase kinase; PP1c, catalytic subunit of type 1 serine/threonine phosphoprotein phosphatase; PP1g, regulatory subunit of type 1 serine/threonine phosphoprotein phosphatase; S, phosphorylation sites in the regulatory subunit of type 1 serine/threonine phosphoprotein phosphatase. See the text for details and other abbreviations.

is reported in Fig. 8.11, showing that PKA activation leads to phosphorylation of glycogen phosphorylase and glycogen synthase. The first kind of phosphorylation is responsible for enzyme activation, whereas the second one leads to loss of catalytic activity, and the overall effect is the stimulation of glycogenolysis coupled to inhibition of glycogen synthesis [143]. The effect of insulin is achieved by stimulation of phosphate removal from the enzymes involved in glycogen metabolism, owing to enhanced activity of type 1 serine/threonine phosphoprotein phosphatase (PP1), associated with a decreased glycogen synthase phosphorylation

[118, 119, 505, 518, 648]. The PP1 enzyme is responsible for dephosphorylation of glycogen synthase (increasing its activity), as well as of phosphorylase kinase and glycogen phosphorylase, that become inactivated. Thus, increasing the activity of PP1 results in inhibition of glycogenolysis, accompanied by stimulation of glycogen synthesis. The effect of insulin on PP1 is mediated by a modification of the phosphorylation state of one of its regulatory subunits (the G subunit) [118, 648]. PP1 exists in oligomeric forms, comprising the catalytic subunit and the regulatory subunit G, which is associated with glycogen and targets the PP1 enzyme to its sites of action. The association of the catalytic subunit of PP1 and its activity are controlled by the phosphorylation state of the G subunit. The G subunit has two separate sites, containing serine, whose phosphorylation determines opposite effects (see Fig. 8.11). Site 1 is phosphorylated by pp90^{rsk}, that is a protein kinase activated by ERK, whereas site 2 is phosphorylated by PKA. The phosphorylation of site 2 induces the dissociation of the catalytic subunit of PP1 from the G subunit.

While this dissociation determines a loss of PPase activity, the inactivation of released PP1 is achieved by its interaction with a selective inhibitor of the phosphatase (I-1), that is phosphorylated by PKA. The association of the catalytic subunit of PP1 with I-1 is promoted by phosphorylation of the inhibitor. Thus, signals increasing the cellular concentration of cAMP trigger the activation of PKA that leads to activation of the enzymes responsible for glycogenolysis, as well as the inactivation of glycogen synthase and PP1. The stimulation of PP1 by insulin is caused by ERK activation, with the consequent phosphorylation of site 1 of the G subunit of PP1, and the stabilization of the catalytically active PP1 oligomers at the level of glycogen molecules, where the enzymes involved in glycogen metabolism are located. Thus, insulin stimulation of target cells induces the activation of PP1, causing dephosphorylation of glycogen phosphorylase, which is inactivated, and of glycogen synthase, which is activated, leading to stimulation of glycogen synthesis and inhibition of glycogenolysis. The decreased phosphorylation of glycogen synthase induced by insulin is further supported by a decreased rate of its phosphorylation, owing to the activation of PKB induced by insulin (see Section 8.2.2.2). PKB phosphorylates glycogen synthase kinase 3 (GSK3), which becomes inactivated by this phosphorylation [119]. GSK3 has multiple cellular substrates, including glycogen synthase, whose phosphorylation by GSK3 is accompanied by a loss of catalytic activity [505]. Thus, the phosphorylation of GSK3 by active PKB in insulin-stimulated cells favours the dephosphorylated state of glycogen synthase and, hence, the stimulation of glycogen synthesis [119, 518]. Overall, the cross-talks between signalling pathways involving PKA activation and those activating ERKs and PKB delineate a set of coordinated processes that essentially determine feed-back control loops.

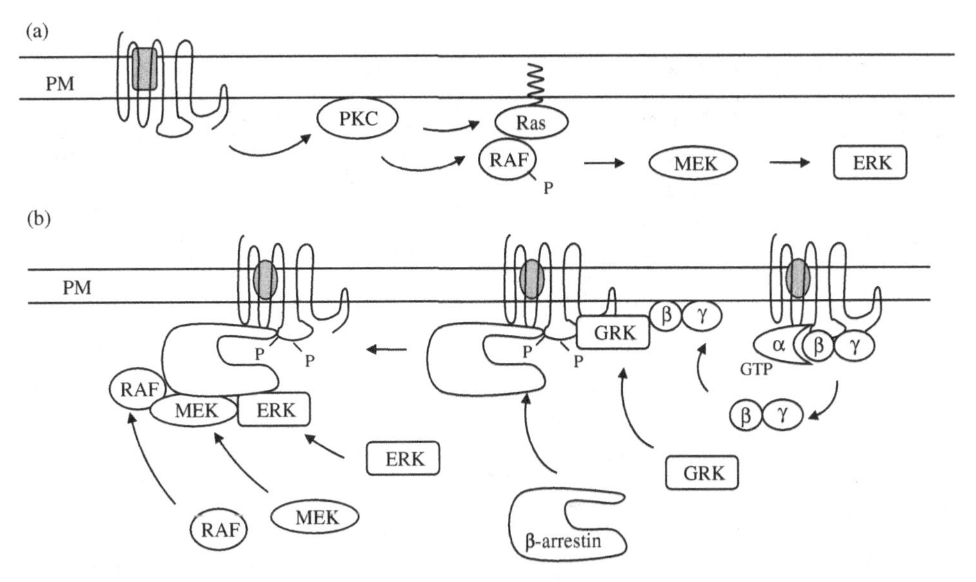

Fig. 8.12. Activation of ERK by signalling through protein G-coupled receptors. Panel (a) (from left to right), activation of ERK by signalling involving PKC activation. Panel (b) (from right to left), activation of ERK through signalling involving recruitment of β-arrestin by phosphorylated receptors. See the text for details.

8.3.2 Activation of ERK by signal transductions acting through G protein coupled receptors

The kinase cascade leading to ERK activation is not triggered only by ligand binding to receptor tyrosine kinases. Multiple mechanisms of ERK activation participating in signalling pathways involving G proteins are known [358, 372], and two of them are outlined in Fig. 8.12. The first mechanism (Fig. 8.12, panel (a)) involves the activation of some PKC isoforms that are believed to lead, either directly or indirectly through Ras activation, to RAF phosphorylation and the consequent activation of ERK [80, 235, 309, 358]. Another mechanism (Fig. 8.12, panel (b)) of ERK activation would be triggered by phosphorylation of G protein-coupled receptors, that then recruit the cytosolic proteins β-arrestins. These proteins can play the role of docking sites for components of the kinase cascade activating ERKs, and the assembly of the RAF-MEK-ERK system is the basis of ERK activation [358].

8.3.3 The TNF activation of NF-kB and its role in the inhibition of apoptosis

It has long been recognized that TNF does not only promote cell death, but is also a potent pro-inflammatory component promoting cell proliferation in certain

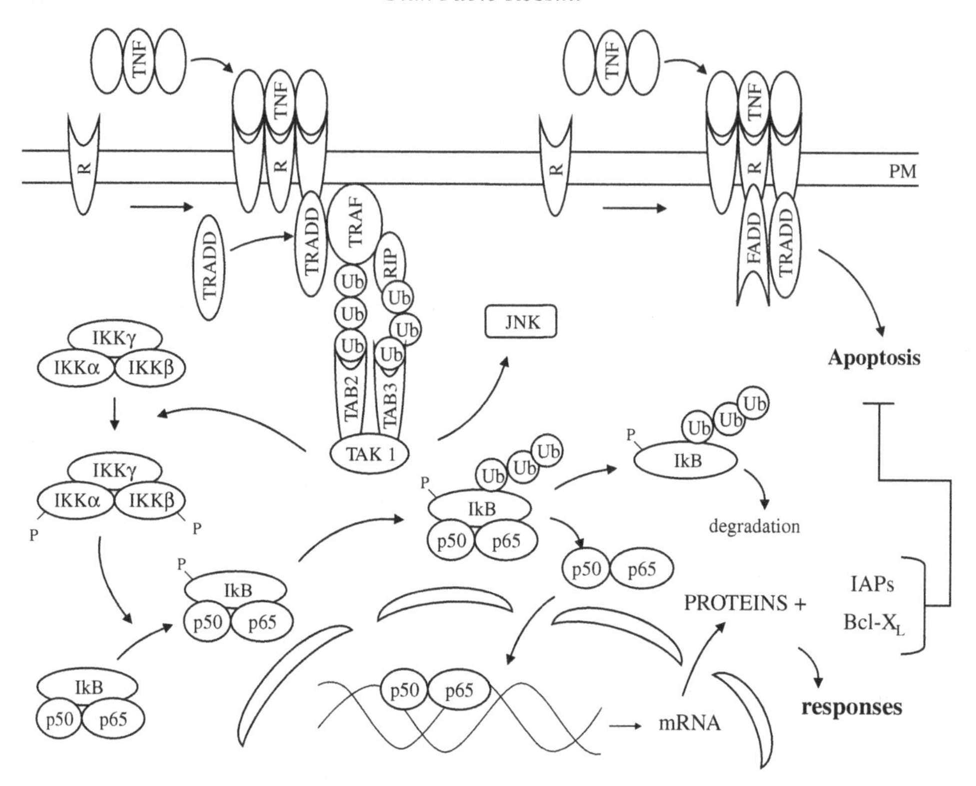

Fig. 8.13. The molecular mechanism of NF-kB activation by tumour necrosis factor. R, receptor; Ub, ubiquitin. See the text for details and other abbreviations.

systems [628, 668]. This effect of TNF is exerted through the activation of a group of transcription factors termed nuclear factors kappa B (NF-kB) [459, 628, 668]. The inactive forms of these proteins reside in the cell cytosol as trimers, composed of an NF-kB homodimer, formed by p50 and p65 subunits, that is associated to an inhibitor of kappa B (IkB) (see Fig. 8.13, lower left). This inhibitor blocks NF-kB function by interacting with its DNA binding domain, thereby preventing its nuclear translocation and binding to DNA sequences of NF-kB responsive genes. TNF activates NF-kB by a process that is triggered by its binding to the TNF receptor [358, 632].

The formation of TNF–receptor complexes induces the assembly of multi-protein structures, through the death domains residing in the intracellular portion of trimeric receptors, with the recruitment of TNF receptor-associated death domain protein (TRADD), the TNF receptor-associated factors (TRAF) and a receptor interacting protein (RIP). The activation of NF-kB is triggered by the phosphorylation of IkB, that results from the activation of a protein kinase cascade triggered by the TNF receptor signalling platform. The current view is that TRAF catalyses the covalent binding of ubiquitin to both RIP and itself [111, 668]. Polyubiquitins then recruit the first component of the kinase cascade, by the association with the TAK1 protein kinase, through the involvement of TAK1 binding proteins (TAB) 2 and 3. TAK1 then phosphorylates the regulatory subunit of the IkB kinase complex (IKK). This complex is composed of two distinct IkB kinases (IKK-α and IKK-β) associated to their regulatory subunit (IKK-γ/NEMO). The phosphorylation of IkB by the IKK complex leads to ubiquitination and IkB degradation, with the release of the NF-kB dimers that enter the nucleus and stimulate the transcription of NF-kB responsive genes [628]. The protein platform assembled at the level of the TNF receptor can also trigger death responses through the formation of another TNF receptor signalling complex (Fig. 8.13, upper right). In this latter case, the platform recruits FADD and caspase 8 (see Section 8.2.2.5, and Fig. 8.9). Apoptosis, however, can be inhibited when NF-kB is activated, because it enhances the transcription of several anti-apoptotic components [628]. In particular, NF-kB induces the transcription of several caspase inhibitors, including the caspase 8 modulator FLIP, and several IAP components (cIAP and XIAP), as well as anti-apoptotic Bcl-2 family members, such as Bcl-X_L. The increased transcription of genes coding for anti-apoptotic components has a relevant role in TNF responses, due to a third transduction pathway that is activated by TNF [459, 668], involving a serine/threonine protein kinase termed stress-activated protein kinase/c-Jun N-terminal kinase (SAPK/JNK). The JNK kinases belong to the MAPK family, and are believed to have a role mostly in stress responses of cells exposed to noxious stimuli, leading to cell death [136]. Indeed, a sustained activation of JNK by TNF results in apoptosis, overcoming the anti-apoptotic action of NF-kB. The relevance of JNK activation by TNF, with regard to cross-talks among signal transduction pathways, is linked to the dual role of this kinase as an effector of either cell death or differentiation [136, 459].

Transduction pathways triggered by TNF then establish multiple cross-talks. By activating NF-kB, in particular, an inhibitory action is exerted on the apoptotic machinery, Furthermore, NF-kB responsive genes include several cytokines, that activate the JAK-STAT pathways, establishing another cross-talk among signalling mechanisms, Finally, by the activation of JNK, TNF could either determine cell death or contribute to the control of specialized cellular functions.

8.3.4 Non genomic actions of steroid hormones and cross-talks with receptor tyrosine kinases

The canonical mechanism of action of steroid hormones and other agents regulating cellular functions through binding to intracellular ligand-activated transcription factors are exerted at nuclear level, through changes in the expression of selected genes (Section 8.2.1). It has long been recognized, however, that steroid hormones can trigger non-genomic effects in target cells [376, 445]. The key features of these 'non-canonical' responses involve their occurrence even if gene transcription and protein synthesis are blocked by specific inhibitors, and their detection in seconds/minutes after hormonal challenge of sensitive systems. Furthermore, in non-genomic responses the ligand behaviour as agonist or antagonist does not always reproduce that established for transcriptional responses, indicating that a separate class of receptorial components might be involved in the transduction pathways responsible for non-genomic actions of steroid hormones [376, 387, 445, 638]. Indeed, a new class of receptors has been uncovered [370] that is distinct from the ligand-activated nuclear receptor superfamily [394]. Still, a cytoplasmic and/or plasma membrane location of the canonical 'nuclear' receptors has been postulated for many non-genomic actions of steroid hormones [445, 638]. The major non-genomic actions reported for steroid and thyroid hormones involve the activation of PI3K, ERK/MAPK and PKC [376, 445]. Other cross-talks between steroid hormones and signal transduction pathways of membrane receptors, however, have been recorded, including changes in intracellular calcium and cAMP levels [376, 445]. The cross-talk between steroid hormones and ERK cascades, in particular, has been found in multiple systems where it would take place through several transduction mechanisms. Furthermore, the activation of transcriptional responses by nuclear steroid hormone receptors in their ligand-free state has been reported to be induced by EGF and other growth factors, through phosphorylation of steroid hormone receptors at specific sites [642].

8.4 Signalling networks, system organization and modelling

From a general biological perspective, chemical signalling is a means to regulate and coordinate the molecular elements performing different functions at a system level, whether this is a single cell, an organ in a living organism, an individual, or else higher levels of interactions among individuals of the same or different species. The materials described in the previous sections have outlined the features of the constitutive elements of major signalling networks in living organisms, as we know them at the moment, and the few examples of cross-talks among signal transduction pathways have provided insights onto reactions and molecular mechanisms connecting different pathways, mostly focused at a cellular level. The large body of available data provides mostly phenomenological explanations of parts composing the systems and their behaviour, whereas a mechanistic description of the whole system that might support deterministic predictions of its behaviour has not been achieved. Thus, available knowledge can often explain what has been recorded, rather than allow an observer to make quantitative inferences on how the system will evolve within a spatio-temporal frame, and modelling of signalling networks remains a major challenge.

The emphasis on the importance of integrating experimental approaches with modelling and advancements in our conceptual frameworks [248] is a relevant aspect of our understanding of signalling networks. Modelling is recognized as a multi-step challenge, where model construction is only the first part [17]. When signalling networks are considered, model construction relies on prior knowledge of the molecular transformations and mechanisms of the pathway itself [17, 287]. Indeed, model reconstruction comprises the identification of the molecular components participating in the signalling pathway and the reactions involving these components, as well as the arrangement of those reactions to give the molecular process that links an input with its output in the system [460].

Whenever modelling of signalling networks is concerned, the first step of component and reaction identification is less straightforward than might be expected, because the use of equations to describe the dynamics of the system [17, 68] demands precise estimates of values that can represent the conditions existing in the living cell. General considerations on the components involved in signalling networks show that the list is amazingly large, as many of them can have multiple molecular structures (for instance, phosphorylated and dephosphorylated states of the same protein), displaying different functional states [460]. Furthermore, the steps of signalling pathways are compartmentalized in the cell, so that the measurement of a concentration must take into consideration both the heterogeneous distribution of the relevant component into the cell and its dynamic states [332]. Several factors then emerge that have relevance with regard to the mechanism whereby a system reacts to an incoming signal. These factors include signal amplitude and duration of the input [299]. As signalling is a general means to control system functioning, and this is often accompanied by a net difference in the type of processes brought about by the cell (for instance the opposition glycogen lysis/synthesis approached in Section 8.3.1), an efficient machinery is capable to distinguish between signal and noise, and this is accomplished by mechanisms of signal transduction that are triggered only when the incoming signal is of sufficient amplitude and duration [299].

The description of the mechanisms of signal transduction by definition of the sequence of reactions making up the pathway has been the most common recognizable part of classical investigations on signal transduction, and has led to recognition that linear pathways are not free-standing entities but are parts of larger networks [299], as exemplified by the few cases described in Section 8.3. The emerging picture is a large network of signalling pathways including molecular components that receive signals from multiple inputs (junctions) and capable to transduce the incoming signal to different effectors (nodes), leading to a variety of outputs [299]. While this level of complexity requires a basic knowledge of network topology, the structure of components and their cause–effect relationships, defining the mechanisms that govern the specificity of outputs with regard to inputs becomes a particularly relevant issue. In fact, the identification of junctions and nodes is accompanied by the recognition that the signalling networks are mostly structured by *modules/motifs* that organize the molecular machinery of the signalling apparatus.

The concept of the module stems from the recognition that biological functions may not be reduced to an individual molecule but arise from the interactions among many components, and modules then represent critical levels of biological organization [248], that would essentially catch what is essential to modelling of biological systems [444]. The concept of the module has a wide scope that encompasses topological and functional issues. Both terms have been introduced in a topological context, to identify recurrent patterns of interconnections [417, 558], and groups of highly interconnected nodes [55]. The importance to correlate those topological entities with functional roles, however, was immediately apparent [55], and the recognition that modules include groups of components and reactions that perform an identifiable function is now part of the concept [317, 460]. While modules and motifs have been used as small subunits for network analysis [124], it should be recognized that the boundaries of modules can be ill-defined. Thus, both the different phosphorylated/unphosphorylated states of a protein kinase, as well as the protein kinase cascade including that kinase as a component can be regarded as functional modules [317].

The loose boundaries of modules do not hamper the possibility of modelling signalling pathways themselves, but would rather determine the level of detail of the model [17]. In any case, the recognition that signalling networks are organized in modules, and the definition of the connectivity among modules is considered the basis for the specificity of cellular responses that use the same or very similar sets of modules. Many combinations of modules can be identified in different spatio-temporal conditions/patterns, and these would be responsible for cell-specific responses under defined conditions [299, 316, 460]. Thus, transient and sustained activation of ERK are the basis for the proliferative and differentiative responses triggered by EGF and NGF, respectively, in PC12 cells [316, 398].

The recognition of the importance of modules in modelling of signalling networks provides a major impetus for their identification, and the high-throughput technologies developed in recent years have been giving major contributions in the collection of information and data for reconstruction of signalling networks and their modules [55, 301]. Because of the need to base modelling on empirical data and constants [17, 287], a recent study by the Alliance for Cellular Signalling (AfCS) seems of particular importance with regard to the methodology used to gain knowledge about signalling networks and the cross-talks existing when a system is challenged with multiple signals. In that study, the interaction of different signalling pathways was analysed in a large-scale investigation, by the use of RAW 264.7 macrophages that were exposed to 22 receptor-specific ligands both alone and in all pairwise combinations [428]. The system was initially validated by analysing the outputs of single ligands, and the detection of expected results with regard to final and intermediate outputs of individual signals. Interestingly, evaluations of the 231 pairwise combinations of ligands showed that these have context-dependent roles in modulating cellular outputs and that the signalling network comprises a few modular transduction units linked by a limited set of interaction agents [428]. Thus, a study that has been built on the measurement of a defined set of molecular parameters has led to characterization of signalling networks at a wider level.

8.5 Conclusions and outlook

In recent years, studies on signalling networks have been characterized by both bottom up and top down approaches aimed at a better description of mechanisms (cause–effect relationships) and aimed at models suitable for quantitative predictions about the behaviour of systems challenged by chemical signals. If those studies and the approaches used to obtain modelling of signalling networks are put into perspective, it seems appropriate to depict the overall process as an integrative and iterative activity capitalizing contributions from both experimentalists and theoreticians [460]. Indeed, the impressive advancement of our comprehension of these complex biological processes appears to stem from the fruitful integration of results obtained with both reductionist and systemic approaches.

Appendix A

Complex networks: from local to global properties

D. GARLASCHELLI AND G. CALDARELLI

A.1 Introduction

In this appendix we provide basic concepts aimed at introducing the formalism of networks. We first introduce graphs and make simple examples, and then discuss the topological properties of networks. Other general presentations of network structure can be found in review articles [12, 72, 153, 436, 583] and books [88, 89, 98, 636].

A *network* (or *graph* in a more mathematical language) is defined as a set of N **vertices** (or *nodes*) connected by **links** (or *edges*). Links, and consequently the whole graph, can be either *directed* (*oriented*), if a direction is specified as in Fig. A.1a, or *undirected* (*not oriented*), if no direction is specified as in Fig. A.1b. More precisely, undirected links are rather *bidirectional* ones, since they can be traversed in both directions. For this reason an undirected graph can always be thought of as a directed one where each undirected link is replaced by two directed links pointing in opposite directions (see Fig. A.1c). A link in a directed network is said to be *reciprocated* if another link between the same pair of vertices, but with opposite direction, is there. Therefore, an undirected network can be regarded as a special case of a directed network where all links are reciprocated. The links of a network may also carry a number, referred to as the *weight* of the edge, representing the strength of the corresponding interaction. In such a case one speaks of a *weighted network*. In the present appendix we do not consider weighted networks explicitly.

A.1.1 Adjacency matrix

All the topological information can be compactly expressed by labelling each vertex with an integer number $i = 1 \ldots N$ and defining the $N \times N$ **adjacency matrix** of the graph, whose entries indicate whether a link is present between two vertices.

Networks in Cell Biology, ed. M. Buchanan, G. Caldarelli, P. De Los Rios, F. Rao and M. Vendruscolo. Published by Cambridge University Press.

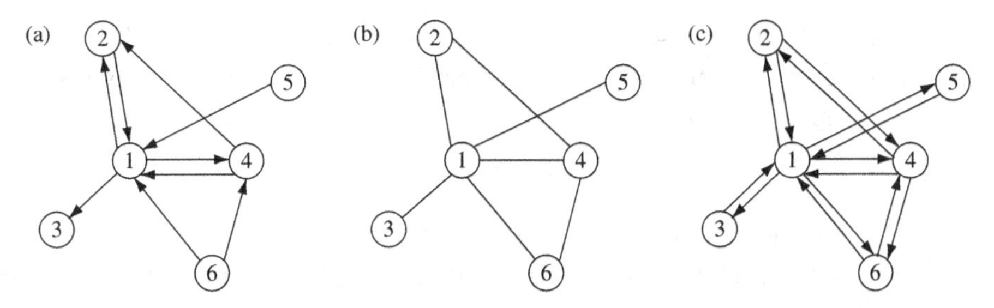

Fig. A.1. Simple examples of networks, each with $N = 6$ vertices. (a) A directed graph. Here the edges between vertices 1 and 2 and between 1 and 4 are reciprocated. (b) An undirected graph, which is also the undirected version of graph (a). (c) The directed version of graph (b). Here all egdes are reciprocated.

For directed networks we denote the adjacency matrix by A and define its elements $\{a_{ij}\}$ as follows:

$$a_{ij} \equiv \begin{cases} 1 & \text{if a link from } i \text{ to } j \text{ is there} \\ 0 & \text{else.} \end{cases} \tag{A.1}$$

For undirected networks we prefer to denote the adjacency matrix by B, and define its elements $\{b_{ij}\}$ as

$$b_{ij} \equiv \begin{cases} 1 & \text{if a link between } i \text{ and } j \text{ is there} \\ 0 & \text{else.} \end{cases} \tag{A.2}$$

Note that the choice of the symbol b_{ij} for undirected networks is somewhat non-standard. Outside the present appendix, when there is only one unambiguously defined kind of network (either directed or undirected), the symbol a_{ij} is generally preferred for both cases. Also note that a non-zero diagonal element of the adjacency matrix ($a_{ii} = 1$ or $b_{ii} = 1$) corresponds to a link starting and ending at the same vertex, which is denoted *self-loop* or simply *loop*. In this appendix we assume no self-loops in the network.

For undirected networks $b_{ij} = b_{ji}$, while in general $a_{ij} \neq a_{ji}$ for directed networks ($a_{ij} = a_{ji} = 1$ if and only if the edges between i and j are reciprocated). For instance, the adjacency matrices a_{ij} and b_{ij} corresponding to Fig. A.1a and Fig. A.1b respectively are given by

$$a_{ij} = \begin{pmatrix} 0 & 1 & 1 & 1 & 0 & 0 \\ 1 & 0 & 0 & 0 & 0 & 0 \\ 0 & 0 & 0 & 0 & 0 & 0 \\ 1 & 1 & 0 & 0 & 0 & 0 \\ 1 & 0 & 0 & 0 & 0 & 0 \\ 1 & 0 & 0 & 1 & 0 & 0 \end{pmatrix} \qquad b_{ij} = \begin{pmatrix} 0 & 1 & 1 & 1 & 1 & 1 \\ 1 & 0 & 0 & 1 & 0 & 0 \\ 1 & 0 & 0 & 0 & 0 & 0 \\ 1 & 1 & 0 & 0 & 0 & 1 \\ 1 & 0 & 0 & 0 & 0 & 0 \\ 1 & 0 & 0 & 1 & 0 & 0 \end{pmatrix}. \tag{A.3}$$

As mentioned above, an undirected network can be regarded as a (fully reciprocated) directed one; in this case, the adjacency matrix a_{ij} of the resulting directed network is simply given by

$$a_{ij} \equiv b_{ij}, \tag{A.4}$$

where b_{ij} is that of the original undirected network. In this particular case a_{ij} is a symmetric matrix. Note that this mapping can be reversed in order to recover the original undirected network: from Fig. A.1b we can always obtain Fig. A.1c, and vice versa. By contrast, the mapping of a directed network onto an undirected one – where an undirected link is placed between vertices connected by *at least one* directed link – is also possible, even if in general it cannot be reversed due to a partial loss of information. For instance, the graph shown in Fig. A.1b is the undirected version of that shown in Fig. A.1a. From Fig. A.1a we can obtain Fig. A.1b, but then we cannot go back to Fig. A.1a unless we are told how to. The relation between the adjacency matrix of the undirected network (B) and that of the original directed one (A) is

$$b_{ij} \equiv a_{ij} + a_{ji} - a_{ij}a_{ji}, \tag{A.5}$$

where the last term on the right hand side prevents the occurrence of non-unit entries corresponding to doubly linked vertices. Note that this relation can be tested on the matrices of Eq. (A.3).

A.1.2 Regular graphs

As a starting example, we briefly mention one of the simplest kinds of networks, namely the class of **regular graphs**. Regular graphs are networks where each vertex is connected to the same number z of neighbours in a highly ordered fashion. In Fig. A.2 we show three examples of regular (undirected) graphs: a periodic chain with first- and second-neighbour interactions ($z = 4$), a two-dimensional lattice with only first-neighbour interactions ($z = 4$) and a *fully connected* graph (where each vertex is connected to all the others: $z = N - 1$). Chains and square lattices are examples of the more general class of D-dimensional discrete lattices, used whenever a set of elements is assumed to interact with its first, second, … and lth neighbours embedded in some (D-dimensional) geometric space (*nearest-neighbour approximation*). In this case, each vertex is connected to $z = 2Dl$ other vertices. Fully connected graphs are instead used when infinite-range interactions are assumed, resulting in the so-called *mean-field* approximation (since in general the 'field', or interaction, experienced by any vertex is the average of the one produced by all others) where $z = N - 1$. The highly ordered structure of these graphs translates into particular regularities of their adjacency matrices: for instance, the

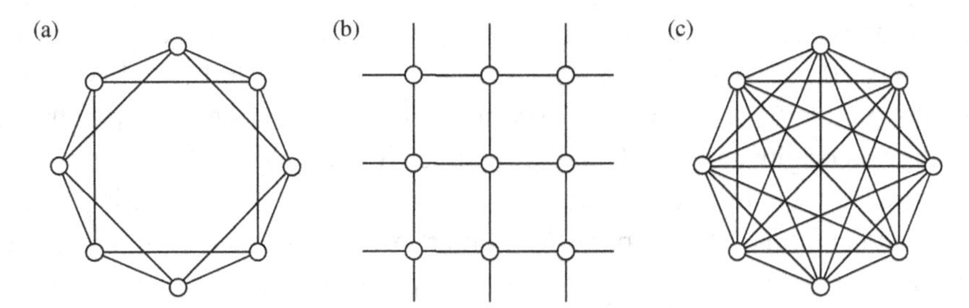

Fig. A.2. Examples of 'familiar' undirected graphs in the physics literature. (a) Periodic one-dimensional chain (ring) with first- and second-neighbour interactions. (b) Two-dimensional lattice with only first-neighbour interactions. (c) Fully connected graph (mean-field approximation). All these graphs are regular since no 'disorder' is introduced.

adjacency matrices for the graphs in Fig. A.2a (assuming any cyclic labelling of vertices) and Fig. A.2c are given respectively by:

$$b_{ij} = \begin{pmatrix} 0 & 1 & 1 & 0 & 0 & 0 & 1 & 1 \\ 1 & 0 & 1 & 1 & 0 & 0 & 0 & 1 \\ 1 & 1 & 0 & 1 & 1 & 0 & 0 & 0 \\ 0 & 1 & 1 & 0 & 1 & 1 & 0 & 0 \\ 0 & 0 & 1 & 1 & 0 & 1 & 1 & 0 \\ 0 & 0 & 0 & 1 & 1 & 0 & 1 & 1 \\ 1 & 0 & 0 & 0 & 1 & 1 & 0 & 1 \\ 1 & 1 & 0 & 0 & 0 & 1 & 1 & 0 \end{pmatrix}, \quad b_{ij} = \begin{pmatrix} 0 & 1 & 1 & 1 & 1 & 1 & 1 & 1 \\ 1 & 0 & 1 & 1 & 1 & 1 & 1 & 1 \\ 1 & 1 & 0 & 1 & 1 & 1 & 1 & 1 \\ 1 & 1 & 1 & 0 & 1 & 1 & 1 & 1 \\ 1 & 1 & 1 & 1 & 0 & 1 & 1 & 1 \\ 1 & 1 & 1 & 1 & 1 & 0 & 1 & 1 \\ 1 & 1 & 1 & 1 & 1 & 1 & 0 & 1 \\ 1 & 1 & 1 & 1 & 1 & 1 & 1 & 0 \end{pmatrix}.$$

Regular networks can be built deterministically, or in other words without introducing 'disorder'. This is by far the simplest choice for a network, but it severely restricts the extremely large number of possibilities for its topology. Traditional assumptions such as the nearest-neighbour or the mean-field one cannot be seriously considered as good choices for many real networks whose edges are not simply determined by the positions of vertices in the embedding space. Importantly, the predictions for the outcomes of dynamical processes defined on regular graphs cannot be extended to the behaviour of the same processes on real networks. This motivates the introduction of more complicated models, some of which will be presented in Appendix B.

A.1.3 Levels of topological complexity

The above considerations indicate that, at least in general, a detailed description of networks is necessary in order to characterize their topology. This ultimately results in the need to define a set of topological properties that can be measured on any network. Several reviews already exist in the literature [12, 52, 88, 89, 98, 153, 436, 583, 636] presenting network properties from various viewpoints. Here we follow an original approach and present network topology progressively, from local to global properties. More explicitly, we first consider the properties specified by the first neighbours of a vertex ('first-order' properties), then those specified by its first and second neighbours ('second-order' properties), and so on until we come to those relative to the whole network ('global' properties). We will also emphasize the differences arising between directed and undirected networks.

A.2 First-order properties

By 'first-order' properties we mean the set of topological quantities that can be specified by starting from a vertex and considering its first neighbours. This information is captured by simply considering single elements of the adjacency matrix, or linear functions of them.

A.2.1 Degree

In an undirected network, the simplest first-order property is the number k_i of neighbours of a vertex i, or equivalently the number of links attached to i. The quantity k_i is called the **degree** of vertex i. In terms of the adjacency matrix $B = \{b_{ij}\}$, the degree can be defined as

$$k_i \equiv \sum_{j \neq i} b_{ij}, \tag{A.6}$$

where the requirement $i \neq j$ indicates that self-loops, if present, do not contribute to the sum. In a directed network, it is possible to distinguish between the **in-degree** k_i^{in} and the **out-degree** k_i^{out} of each vertex i, defined as the number of in-coming and out-going links of i respectively. In this case, if $A = \{a_{ij}\}$ is the adjacency matrix the degrees read

$$\begin{aligned} k_i^{in} &\equiv \sum_{j \neq i} a_{ji} \\ k_i^{out} &\equiv \sum_{j \neq i} a_{ij}. \end{aligned} \tag{A.7}$$

In an undirected network the vector $\{k_i\}_{i=1}^N$ of vertex degrees is called the **degree sequence**. A very important quantity for the empirical description of the first-order topological properties of a network is the (normalized) histogram of the values $\{k_i\}_{i=1}^N$, or the **degree distribution** $P(k)$, representing the fraction of vertices with degree k, or equivalently the probability that a randomly chosen vertex has degree k. In a directed network it is possible to introduce the *in-degree sequence* $\{k_i^{in}\}_{i=1}^N$ and the *out-degree sequence* $\{k_i^{out}\}_{i=1}^N$. Correspondingly, it is possible to define the *in-degree distribution* $P^{in}(k^{in})$ and the *out-degree distribution* $P^{out}(k^{out})$.

The empirical behaviour of the degree distribution of real networks is probably the initial reason why they have attracted extremely high interest from the scientific community over the past few years. Indeed, it turns out that, for a large number of networks, the degree distribution displays the *power-law* form

$$P(k) \propto k^{-\gamma} \tag{A.8}$$

with $2 \leq \gamma \leq 3$ [12, 98]. Directed networks often display the same qualitative behaviour for the in-degree and/or the out-degree:

$$P^{in}(k^{in}) \propto (k^{in})^{-\gamma_{in}} \qquad P^{out}(k^{out}) \propto (k^{out})^{-\gamma_{out}}, \tag{A.9}$$

where γ_{in} and γ_{out} have in general different values ranging between 2 and 3 [12, 98]. For the practical purpose of plotting empirical degree distributions and estimating their exponents, the *cumulative distributions* are commonly used:

$$P_>(k) \equiv \sum_{k' \geq k} P(k') \qquad P_>^{in}(k^{in}) \equiv \sum_{k' \geq k^{in}} P^{in}(k') \qquad P_>^{out}(k^{out}) \equiv \sum_{k' \geq k^{out}} P^{out}(k'). \tag{A.10}$$

In this way the statistical noise is reduced by summing over k'. If the actual degree distribution has the power-law behaviour of Eqs. (A.8) or (A.9), then the cumulative distributions are again power-laws but with a different exponent:

$$P_>(k) \propto k^{-\gamma+1} \qquad P_>^{in} \propto (k^{in})^{-\gamma_{in}+1} \qquad P_>^{out} \propto (k^{out})^{-\gamma_{out}+1} \tag{A.11}$$

The power-law behaviour is witnessed by a straight-line trend of the degree distribution, when plotted in log–log scale.

Power-law distributions are very important from a general point of view since they lack a typical scale [389, 438]. More exactly, they are the only distributions satisfying the scaling condition

$$P(ak) = f(a)P(k) \tag{A.12}$$

and their funtional form is therefore unchanged, apart for a multiplicative factor, under a rescaling of the variable k. Because of this absence of a characteristic scale, power-law distributions are also called **scale-free** distributions. An important consequence of the scale-free behaviour is the non-vanishing occurrence of rare events, or in other words the presence of 'fat tails'. On the contrary, the tails of other distributions usually decay at least exponentially. In the context of networks, the scale-free behaviour means that there are many low-degree vertices but also few high-degree ones, which can be connected to a significant fraction of the other vertices. The presence of very large degrees is only algebraically suppressed, giving rise to a whole hierarchy of connectivities, from small to large. This has a remarkable dynamic effect: if some kind of 'information' travels on the network, once it has reached a high-degree vertex it then propagates to almost the entire graph, resulting in an extremely fast communication process. By contrast, note that regular networks introduced in Section A.1 have a delta-like degree distribution of the form $P(k) = \delta_{k,z}$ where z is the degree of every vertex (see Fig. A.2). For D-dimensional lattices with lth-neighbours interactions, $z = 2Dl$ and (for large networks and small D and l) no vertex is connected to a significant fraction of the others. For fully connected graphs, $z = N - 1$ and every vertex is connected to all the others. In any case, no hierarchy is present and the network is perfectly homogeneous.

A.2.2 *Link density*

It is possible to consider the **average degree** $\bar{k}$ as a single quantity characterizing the overall first-order properties of a network, and then compare different networks with respect to it. In an undirected graph the average degree can be simply expressed as

$$\bar{k} \equiv \frac{\sum_i k_i}{N} = \frac{\sum_i \sum_{j \neq i} b_{ij}}{N} = \frac{2L^u}{N}, \tag{A.13}$$

where

$$L^u \equiv \sum_{i=1}^{N} \sum_{j<i} b_{ij} = \frac{1}{2} \sum_{i=1}^{N} \sum_{j \neq i} b_{ij} \tag{A.14}$$

is the total number of undirected links in the network. For directed networks, it is easy to see that the average in-degree $\bar{k}^{in}$ equals the average out-degree $\bar{k}^{out}$, and both quantities can be expressed as

$$\begin{aligned} \bar{k} \equiv \bar{k}^{in} &= \frac{\sum_i k_i^{in}}{N} \\ \equiv \bar{k}^{out} &= \frac{\sum_i k_i^{out}}{N} \\ &= \frac{\sum_i \sum_{j \neq i} a_{ij}}{N} = \frac{L}{N}, \end{aligned} \tag{A.15}$$

where

$$L \equiv \sum_{i=1}^{N} \sum_{j \neq i} a_{ij} \tag{A.16}$$

is the total number of (directed) links expressed in terms of the adjacency matrix a_{ij}. We chose a different notation for L^u and L to avoid confusion when an undirected graph is regarded as directed, with two directed links replacing each undirected one. In that case the mapping described by Eq. (A.4) allows to recover Eq. (A.16) consistently from Eq. (A.14), and our notation yields $L = 2L^u$ as expected. Note that in terms of the degree distribution the average degree $\bar{k}$ reads

$$\bar{k} = \sum_{k'} k' P(k') \qquad \bar{k} = \sum_{k'} k' P^{in}(k') = \sum_{k'} k' P^{out}(k') \tag{A.17}$$

for undirected and directed networks respectively. The number of links is an interesting property by itself, being a measure of the 'density' of connections in the network. In order to compare networks with different numbers of vertices, the number of links is usually divided by its maximum value in order to obtain what is called the *link density* or **connectance**. In an undirected network, the maximum number of links (with self-loops excluded) is given by the total number of vertex pairs, which is $N(N-1)/2$ if the number of vertices is N. Therefore the connectance is defined as

$$c^u \equiv \frac{2L^u}{N(N-1)}. \tag{A.18}$$

By contrast, in a directed network the maximum number of links is given by twice the number of vertex pairs (since each pair can be occupied by two links with opposite directions) and the connectance is therefore defined as

$$c \equiv \frac{L}{N(N-1)}. \tag{A.19}$$

For regular graphs $k_i = \bar{k} = z$ and therefore $L^u = Nz/2$ and $c^u = z/(N-1)$. Therefore in the limit $N \to \infty$ of large network size D-dimensional lattices display $c^u = 2Dl/(N-1) \propto N^{-1} \to 0$, while fully connected graphs always display $c^u = 1$. The case $c^u \to 0$ is often referred to as the 'sparse graph' limit, while $c^u \to const$ is the 'dense graph' limit. Real networks are obviously of finite size, therefore we can speak of the 'infinite size' limit only as a formal extrapolation. In this sense, one finds that most real networks are sparse, while few are dense.

A.3 Second-order properties

By 'second-order' topological properties we denote those properties that depend not only on the connections between a vertex and its nearest neighbours, but also on the structure relating a vertex with the 'neighbours of its neighbours'. Therefore the computation of these properties involves products of *two* adjacency matrix elements $b_{ij} b_{jk}$.

A.3.1 Degree correlations

An important example of second-order structure is given by the *degree correlations*: is the degree of a vertex correlated with that of its first neighbours? The more complete way to describe second-order topological properties is to consider the two-vertices **conditional degree distribution** $P(k'|k)$ specifying the probability that a vertex with degree k is connected to a vertex with degree k'. In the trivial case with no correlation between the degrees of connected vertices, the second-order properties can be obtained in terms of the first-order ones, or in other words the conditional probability must be equal to the (unconditional) probability that *any* vertex is connected to a vertex of degree k':

$$P(k'|k) = \frac{k'}{\bar{k}} P(k'). \tag{A.20}$$

However, as we will show in the following, real networks display a more complicated behaviour and are characterized by non-trivial degree correlations which make the form of $P(k'|k)$ deviate from Eq. (A.20).

Estimating the empirical form of the conditional probability directly from real data is difficult since $P(k'|k)$ is a two-parameter curve and is affected by statistical fluctuations (however, two-parameter plots of this type have been studied for instance by Maslov *et al.* [405]). A more compact description that also partly averages out the statistical noise is given by defining the **average nearest neighbour degree** (*ANND* in the following) of a vertex i as the average degree of the neighbours of i. For an undirected graph, the *ANND* can be denoted by k_i^{nn} and simply defined in terms of the adjacency matrix as

$$k_i^{nn} \equiv \frac{\sum_{j \neq i} b_{ij} k_j}{k_i} = \frac{\sum_{j \neq i} \sum_{k \neq j} b_{ij} b_{jk}}{\sum_{j \neq i} b_{ij}}. \tag{A.21}$$

Then it is possible to average k_i^{nn} over all vertices with the same degree k and plot it against k to obtain the one-parameter curve $\bar{k}^{nn}(k)$. The slope of this curve gives direct information on the nature of the degree correlations: if $\bar{k}^{nn}$ is an increasing function of k then degrees are positively correlated (high-degree vertices are on average linked to high-degree ones) and the network is said to be **assortative** (or to display *disassortative mixing*), while if $\bar{k}^{nn}$ decreases with k then degrees are negatively correlated and the network is said to be **disassortative** (or to display *disassortative mixing*). These possible trends are shown in Fig. A.3. In the uncorrelated or 'neutral' case the *ANND* is instead independent of k. Note that for regular networks (see Section A.1) $k_i^{nn} = z \forall i$ and degrees are perfectly corre lated, however the $\bar{k}^{nn}(k)$ plot reduces to the single point (z, z). Real networks are found to be either assortative or disassortative, and they never seem to be uncorrelated. This means that the first-order topological properties such as the degree

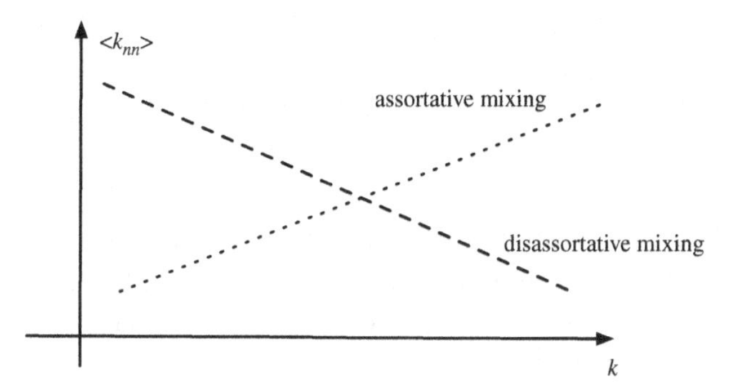

Fig. A.3. Assortative or disassortative mixing can be identified when considering the behaviour of the curve $\bar{k}^{nn}(k)$.

distribution, even if non-trivially interesting by themselves, still do not capture the whole complexity of real networks. We note that the quantity $\bar{k}^{nn}$ can be expressed in terms of the conditional probability $P(k'|k)$ as

$$\bar{k}^{nn}(k) = \sum_{k'} k' P(k'|k). \tag{A.22}$$

From the above expression we recover the expected constant trend for the uncorrelated networks described by Eq. (A.20), which inserted into Eq. (A.22) yields $\bar{k}^{nn} = \overline{k^2}/\bar{k}$ independently of k. In some cases the $\bar{k}^{nn}(k)$ curve is particularly interesting since it displays the empirical form

$$\bar{k}^{nn}(k) \propto k^{\beta}. \tag{A.23}$$

Relations similar to (A.21) hold for directed networks as well. More specifically, it is possible to define the *average nearest neighbour in-degree* $k_i^{nn,in}$ and the *average nearest neighbour out-degree* $k_i^{nn,out}$:

$$k_i^{nn,in} \equiv \frac{\sum_{j\neq i}\sum_{k\neq j} a_{ji}a_{kj}}{\sum_{j\neq i} a_{ji}} \qquad k_i^{nn,out} \equiv \frac{\sum_{j\neq i}\sum_{k\neq j} a_{ij}a_{jk}}{\sum_{j\neq i} a_{ij}} \tag{A.24}$$

and correspondingly the $\bar{k}^{nn,in}(k^{in})$ and the $\bar{k}^{nn,out}(k^{out})$ curve. However, it is also possible to regard the directed network as undirected using the mapping described by Eq. (A.5) and then consider the undirected *ANND* defined in Eq. (A.21) and the corresponding curve $\bar{k}^{nn}(k)$.

A.3.2 Assortativity coefficient

As for the first-order properties, it is possible to define single quantities characterizing the overall second-order properties of the network as a whole. For instance, Newman [434, 435] introduced the **assortativity coefficient** as the

Pearson correlation coefficient between the degrees at either ends of an edge, which for an undirected network reads

$$r_a \equiv \frac{\overline{kh} - \bar{k}\bar{h}}{\overline{k^2} - \bar{k}^2}, \tag{A.25}$$

where k and h are the degrees of the two vertices at the ends of an edge, and the bar indicates an average over all the edges in the network. By regarding each link as an independent contribution to r_a, it is possible to evaluate the statistical error on r_a, or its standard deviation σ_{r_a}, as

$$\sigma_{r_a}^2 = \sum_{l=1}^{L^u} \left[r_a^{(l)} - r_a \right]^2, \tag{A.26}$$

where $r_a^{(l)}$ is the value of r_a displayed by the network if the lth link is removed [435]. Similar expressions can be derived for directed networks. Newman showed that, consistently with the analysis of the *ANND* curve, real networks are always either assortative ($r_a > 0$) or disassortative ($r_a < 0$). Interestingly, social networks turn out to be assortative, while other systems such as biological networks, the WWW and the Internet turn out to be disassortative [434, 435].

A.3.3 Reciprocity

We conclude our discussion of the second-order properties with the notion of **reciprocity**, which is a characteristic of directed networks. As anticipated in Section A.1, a link from a vertex i to a vertex j is said to be reciprocated if the link from j to i is also present. The number $L^{\leftrightarrow}$ of reciprocated links can be defined in terms of the adjacency matrix a_{ij} as

$$L^{\leftrightarrow} \equiv \sum_{i=1}^{N} \sum_{j \neq i} a_{ij} a_{ji}. \tag{A.27}$$

It is interesting to compare the above expression with Eq. (A.16). As expected, while each non-zero element a_{ij} gives a contribution to the number of directed links, only reciprocated pairs of vertices such that $a_{ij} a_{ji} = 1$ contribute to $L^{\leftrightarrow}$. Since $0 \leq L^{\leftrightarrow} \leq L$ it is possible to define the reciprocity r of the network as

$$r \equiv \frac{L^{\leftrightarrow}}{L} \tag{A.28}$$

so that $0 \leq r \leq 1$. The measured value of r allows to assess if the presence of reciprocated links in a network occurs completely by chance or not. To see this, note that r represents the average probability of finding a link between two vertices already connected by the reciprocal one. If reciprocated links occurred by chance, this probability would be simply equal to the average probability of finding a link between *any* two vertices, which is the connectance c. Therefore if r equals

$$r^{rand} = c = \frac{L}{N(N-1)} \tag{A.29}$$

then the reciprocity structure is trivial, while if $r > c$ (or $r < c$) then reciprocated links occur more (or less) often than predicted by chance.

There are two limitations with the above definition of reciprocity. First, r has only a *relative* (with respect to r^{rand}) meaning and does not carry complete information by itself. Secondly, and consequently, the definition does not allow a clear ordering of different networks with respect to their actual degree of reciprocity. Indeed, r^{rand} is larger in a network with larger link density, and it is therefore impossible to compare the values of r for networks with different density, since they have distinct reference values. In order to avoid the aforementioned problems, as an improved definition of reciprocity (denoted as ρ to avoid confusion with r) Garlaschelli and Loffredo [212] proposed the correlation coefficient between the entries of the adjacency matrix of the directed graph:

$$\rho \equiv \frac{\sum_{i=1}^{N}\sum_{j\neq i}(a_{ij}-\bar{a})(a_{ji}-\bar{a})}{\sum_{i=1}^{N}\sum_{j\neq i}(a_{ij}-\bar{a})^2} = \frac{r-c}{1-c}, \tag{A.30}$$

where the equalities $\bar{a} \equiv \sum_{i\neq j} a_{ij}/N(N-1) = L/N(N-1) = c$, $\sum_{i=1}^{N}\sum_{j\neq i} a_{ij}a_{ji} = L^{\leftrightarrow}$ and $\sum_{i=1}^{N}\sum_{j\neq i} a_{ij}^2 = \sum_{i=1}^{N}\sum_{j\neq i} a_{ij} = L$ have been used. The requirement $i \neq j$ means that self-loops, if present, are ignored. The correlation-based reciprocity ρ avoids the conceptual problems mentioned above, since it is an *absolute* quantity which directly allows to distinguish between *reciprocal* ($\rho > 0$) and *antireciprocal* ($\rho < 0$) networks, with mutual links occuring more and less often than random, respectively. The neutral or *areciprocal* case corresponds to $\rho = 0$. Note that if all links occur in reciprocal pairs one has $\rho = 1$, as expected. The use of the above quantity shows that real networks are always characterized by a non-trivial degree of reciprocity. Moreover, networks of the same kind result ordered in classes according to their reciprocity, so that the reciprocity structure is a key aspect of the topology of real networks [212].

A.4 Third-order properties

The third-order topological properties of a network are those which go 'the next step beyond' the second-order ones, since they involve the structure of the connections between a vertex and its first, second *and third* neighbours. The computation of third-order properties involves products of *three* adjacency matrix elements

$b_{ij}b_{jk}b_{kl}$. In the general language of conditional degree distributions, the relevant quantity for an undirected network is now the three-vertices probability $P(k', k''|k)$ that a vertex with degree k is simultaneously connected to a vertex with degree k' and to a vertex with degree k''.

A.4.1 Clustering

The most studied third-order property of a vertex i is the **clustering coefficient** C_i, defined (for an undirected graph) as the number of links connecting the neighbours of i to each other, divided by the total number of pairs of neigbours of i (therefore $0 \leq C_i \leq 1$). In other words, C_i is the connectance (see Section A.2) of the subgraph defined by the neighbours of i, and can therefore be thought of as a 'local link density'. It can also be regarded as the probability of finding a link between two randomly chosen neighbours of i. A simple example is shown in Fig. A.4. It is easy to see that, if b_{ij} is the adjacency matrix of the graph, then the number of interconnections between the neighbours of i is given by $\sum_{j\neq i}\sum_{k\neq i,j} b_{ij}b_{jk}b_{ki}/2$. The clustering coefficient C_i is then obtained by dividing this number by the number of possible pairs of neighbours of i, which equals $k_i(k_i-1)/2$ if k_i is the degree of i. It follows that

$$C_i \equiv \frac{\sum_{j\neq i}\sum_{k\neq i,j} b_{ij}b_{jk}b_{ki}}{k_i(k_i-1)} = \frac{\sum_{j\neq i}\sum_{k\neq i,j} b_{ij}b_{jk}b_{ki}}{(\sum_{j\neq i} b_{ij}-1)\sum_{j\neq i} b_{ij}}. \tag{A.31}$$

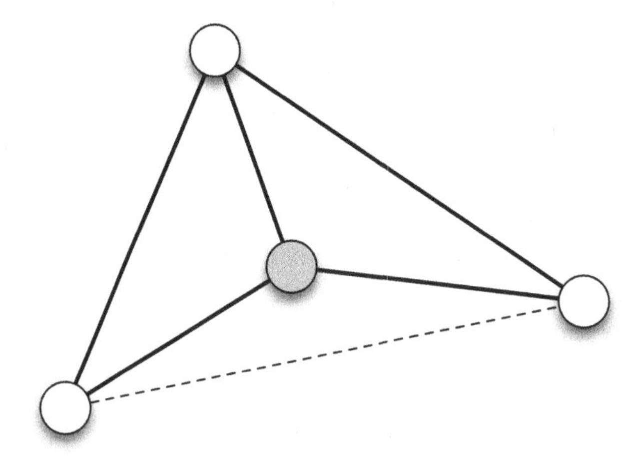

Fig. A.4. The bold edges present in graph form two triangles insisting on the grey central vertex. If the missing (dashed) edge were present, the clustering coefficient of the grey vertex would equal the maximum value 1. In the case where the dashed edge is absent, the clustering coefficient is 2/3.

It is easy to show that for regular graphs (see Section A.1) each vertex has the same clustering coefficient: $C_i = 1$ in a fully connected graph, while $C_i = (3z - 6D)/(4z - 4D)$ in a regular lattice with $z = 2Dl$ (see Fig. A.2). The latter expression is zero if $3z = 6D$ (corresponding to $l = 1$) and tends to $3/4$ if $3z \gg 6D$ (corresponding to $l \gg 1$), which is a large value for the clustering coefficient whose maximum value is 1.

A statistical way to consider the clustering properties of real networks, used for instance by Ravasz and Barabási [493], is similar to that introduced for the degree correlations. By computing the average value of C_i over all vertices with a given degree k and plotting it versus k, it is possible to obtain a $\bar{C}(k)$ curve whose trend gives information on the scaling of clustering with degree. Remarkably, the analysis of real networks reveals that in many cases the average clustering of k-degree vertices decreases as k decreases, and that this trend is sometimes consistent with a power-law behaviour of the form

$$\bar{C}(k) \propto k^{-\tau}. \tag{A.32}$$

The decrease of C_i with the degree k_i is a topological property often referred to as **clustering hierarchy** [493], since it signals that the network is hierarchically organized in such a way that low-degree vertices are surrounded by highly interconnected neighbours forming dense subnetworks, while high-degree vertices are surrounded by loosely connected neighbours forming sparse subnetworks. Dense subnetworks can be thought of as 'modules' in which the whole graph is subdivided. Low-degree vertices are more likely to *belong* to such modules, while high-degree ones are more likely to *connect* different modules together. By contrast, note that for regular networks the $\bar{C}(k)$ curve, as the $\bar{k}^{nn}(k)$ one, reduces to a single point which now is (z, C_i).

It is also possible to compute the **average clustering coefficient** $\bar{C}$ over all vertices:

$$\bar{C} \equiv \sum_{i=1}^{N} C_i. \tag{A.33}$$

This quantity represents the average probability to find a link between two randomly chosen neighbours of a vertex (clearly $0 \leq \bar{C} \leq 1$). The empirical analysis of real networks reveals that they are always characterized by a 'large' value of $\bar{C}$. For some real networks the rescaled quantity $\bar{C}/c^u$ displays an approximate linear dependence on the number of vertices N:

$$\bar{C}/c^u \propto N. \tag{A.34}$$

As a comparison, note that regular graphs obviously display $\bar{C} = C_i$, and therefore $\bar{C}/c^u = 1$ for fully connected graphs, $\bar{C}/c^u = 0$ for regular lattices with $l = 1$, and $\bar{C}/c^u = \frac{3(N-1)(z-2D)}{4z(z-D)} \propto N$ for regular lattices with $l > 1$. Therefore the latter display the qualitative linear scaling of $\bar{C}/c^u$ with N observed for real networks. As regular lattices with $l > 1$, real networks are on average highly clustered.

A.4.2 *Directed motifs*

Clearly, the clustering coefficient of a vertex is the normalized number of triangles (closed loops of three vertices) originating at that vertex. For directed networks, there are various possible patterns for the triangles originating at a vertex, depending on the directions of the edges. More generally, in directed networks there are many more possible subgraphs of three vertices than in the undirected case. These subgraphs, which completely characterize the third-order topology of directed graphs, are called **motifs**. Motifs turn out to be fundamental components of many biological networks. They play the role of modular circuits, each one with a clear associated function. The topology of the network often results from the way these functions are integrated together in the overall operations supported by the system. Motifs deserve a separate discussion, which is provided in Section C.2.

A less informative description of the third-order structure of directed graphs can be obtained by regarding the edges as undirected and computing the clustering coefficient on the undirected version of the network. In this case Eq. (A.31) holds for directed networks as well, with b_{ij} given by Eq. (A.5).

A.5 Global properties

Although it is in principle possible to proceed with the analysis of fourth-order properties and so on, the study of higher-order properties of real networks generally goes directly to the *global* ones, which are those that require the exploration of the entire network to be computed. Since in a network with N vertices the longest path required to go from a vertex to any other contains at most $N-1$ links, or N if one is interested in loops of length N, it follows that global properties involve products of at most N adjacency matrix elements:

$$\underbrace{b_{i_1 i_2} b_{i_2 i_3} \ldots b_{i_{N-1} i_N} b_{i_N i_1}}_{N \text{ factors}} . \tag{A.35}$$

Global properties often have the most important effect on processes taking place on graphs, since they are responsible for the way information 'spreads' over the network and for the possible emergence of any sort of collective behaviour of vertices. Here we consider a few (out of the many) examples of global network properties: the *connected components*, the *shortest distance*, the *betweenness centrality*, and the *community structure*.

A.5.1 Connected components

Two vertices in an undirected network are said to belong to the same **cluster**, or *connected component*, if a path exists connecting them through a finite number of links. The *size* of a cluster is the number of vertices present in it. Note that for each of the regular networks shown in Fig. A.2 all vertices belong to the same cluster. For directed networks, it is in general possible that a path going from a vertex i to a vertex j exists, while no path from j to i is there. In other words, it is possible to define the *in-component* of vertex i as the set of vertices from which a path exists to i, and the *out-component* of i as the set of vertices to which a path exists from i. Finally, two vertices i and j are said to belong to the same *strongly connected component* if it possible to go both from i to j and from j to i. There is in principle no limit on the number and sizes of connected components in a graph. However, an empirical property of most real networks is the presence of one largest component containing a huge fraction of the vertices, plus a certain number of much smaller components with the few remaining vertices. This means that the spread of information on real networks is efficient, since starting from a vertex in the largest component it is possible to reach a large number of other vertices in the same component. The presence of the largest component is interesting also for theoretical reasons, since it is related to the occurrence of a *phase transition* in models where links are drawn with a specified probability.

A.5.2 Shortest distance

Another important property, which clarifies better the communication properties in a network, is the **shortest distance** between vertices. For each pair of vertices i and j in an undirected graph, it is possible to define their shortest distance d_{ij} as the minimum number of links that must be crossed to go from i to j. An example is shown in Fig. A.5. Note that for directed graphs in general $d_{ij} \neq d_{ji}$; however, in this case the average distance is generally computed on the undirected version of the network. It is then possible to define the **average distance** $\bar{d}$ over all pairs of vertices. Note that if two vertices belong to different clusters, then the distance between them, and consequently $\bar{d}$, is formally infinite. To prevent this outcome, $\bar{d}$ is usually computed only over those pairs of vertices belonging to the same cluster. Similarly, the *diameter* of a network is the largest, rather than the average, value of the shortest distance between two vertices in the entire network. The empirical behaviour of $\bar{d}$ and of the diameter is very important. It turns out that, even in a network with an extremely large number of vertices, the average distance and the diameter are always very small. Even if with a number of exceptions, it is found

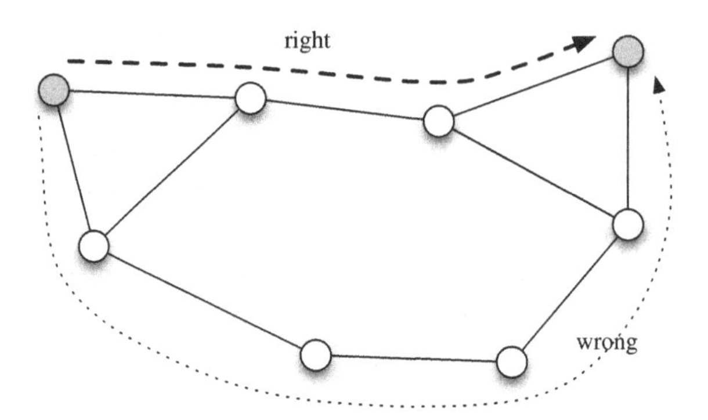

Fig. A.5. The shortest distance between the two grey nodes is three, and equals the minimum number of edges separating them. The right path along which this distance can be computed is the shortest one (dashed), while longer paths (such as the dotted one) are wrong paths.

that both quantities scale at most logarithmically with the number of vertices N. In particular, in most cases $\bar{d}$ scales with N according to the approximate law

$$\bar{d} \approx \frac{\ln N}{\ln \bar{k}}. \tag{A.36}$$

This property is usually taken as a quantitative statement of the so-called **small-world** effect. Its importance lies in the remarkable deviation from the behaviour of regular graphs in any Euclidean dimension D, which instead display $\bar{d} \propto N^{1/D}$ and are therefore characterized by a much larger average distance. The small-world effect is sometimes defined (in a 'stronger' sense) as the *simultaneous* presence of a small average distance and a large average clustering coefficient. Both properties are empirically observed in most real-world networks. Moreover, in some cases an **ultra-small-world** effect is observed, corresponding to an extremely weak size dependence of the distance described by $\bar{d} \propto \ln(\ln N)$.

A.5.3 Betweenness centrality

There are various strategies for defining a measure of importance for the vertices in a graph. Most of them are based on different versions of the concept of *centrality*: a node is more important if it is close to many other nodes. A widely used choice is the notion of **betweenness centrality**. The betweenness of an edge or vertex in a network measures the number of shortest paths between all the possible couples of vertices which pass through it [204]. As shown in Fig. A.6, the node that is crossed the most times is also the most central. Whenever two or more paths exists, the contribution given to a particular edge or vertex will be 1 divided by the number of possible paths. Symbolically, we can write the betweenness b_i of a vertex i as

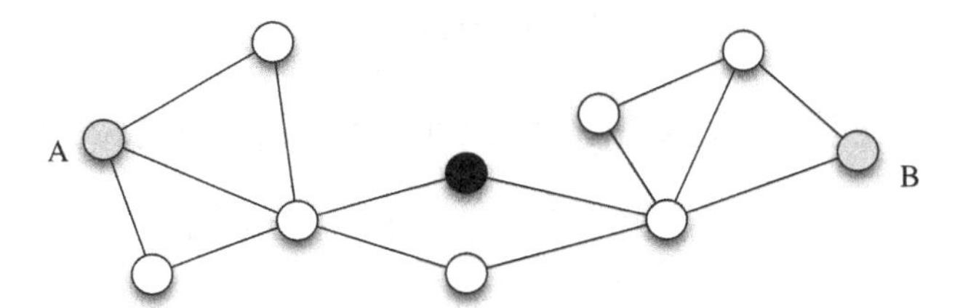

Fig. A.6. The betweenness of the central black vertex is computed by considering all shortest paths (distances) between all the possible pairs of vertices. For the two grey vertices in the figure there are two different shortest paths, each of them contribute with a weight of 1/2.

$$b_i = \sum_{j \neq k \neq i} \frac{\mathcal{N}_{jk}(i)}{\mathcal{N}_{jk}}, \tag{A.37}$$

where the sum runs from 1 to N (total number of vertices), and one must take care that j is different from k and both different from i. The quantity $\mathcal{N}_{jk}$ gives the total number of shortest paths between j and k, while $\mathcal{N}_{jk}(i)$ is the number of such paths passing through i.

A.5.4 Community structure

Other important subgraphs that can be identified in a network are **communities** of densely connected vertices. A community can consist of any number of vertices (from a few ones to a large fraction of the network), and a network can therefore be partitioned into heterogeneously sized communities. There is no unique definition of a community, and even when a single definition is adopted, there are various methods to identify the communities of a particular network. The most popular approaches invoke the concepts of link density and betweenness centrality to define and detect communities in large networks. In any case, communities are strongly non-local, as they often require the whole knowledge of network topology. The problem of community definition and identification will be discussed in more detail in Section C.3.

Appendix B

Modelling the local structure of networks

D. GARLASCHELLI AND G. CALDARELLI

B.1 Introduction

In this appendix we review various theoretical models that have been proposed in order to reproduce some of the empirically observed properties of real networks. We consider only the models that focus on the local topological properties, in particular (in the language of Appendix A) on the first- and second-order properties. As a result, the higher-order properties of the networks generated by the models considered here are the result of local rules alone. Nonetheless, suitable local rules are often enough in order to reproduce most of the observed complexity of real networks. Moreover, it is believed that most real networks are indeed shaped by local rules alone, as higher-order mechanisms requiring the knowledge of the entire network are in most cases unfeasible.

The models presented here share a common aspect: the deviation of real networks from regular graphs is modelled through the introduction of some 'disorder' according to suitable stochastic rules. All the models described below (and largely most models in the literature) are therefore stochastic models. As a consequence they are also *ensemble* models, since they define a whole set of possible realizations of a network, rather than a single graph. Ensemble averages give the expected value of any topological property. They will be denoted by angular brackets $\langle \cdots \rangle$ to avoid confusion with averages over the vertices of a single graph, which are instead denoted by a bar as in Appendix A.

B.2 The random graph model

Introduced by Erdős and Rényi in 1959 [176], the *random graph* (RG in the following) model is the simplest stochastic model of an undirected network. Despite its simplicity and inadequacy in reproducing most empirical topological properties, this model remains an instructive reference for many other models, which are often

Networks in Cell Biology, ed. M. Buchanan, G. Caldarelli, P. De Los Rios, F. Rao and M. Vendruscolo.
Published by Cambridge University Press.

defined as generalizations of it. The RG model simply assumes a fixed number N of vertices. Undirected links between different pairs of vertices are drawn independently of each other, and with the same probability q. Note that the choice of the symbol q is somewhat non-standard, as the connection probability is usually denoted by p in the literature on the RG model. However, in the present appendix we need to distinguish clearly undirected graphs from directed ones. For this reason, we prefer to use q for the probability to draw an undirected edge, and p for the probability to draw a directed edge (for instance if one is considering a directed variant of the RG model). As q varies from 0 to 1, the random graph ranges from an empty to a completely connected network. As a simple example, in Fig. B.1 we show four realizations of the RG model with $N = 10$ corresponding to different values of q.

The expected topological properties of the RG model can be computed quite easily as functions of q [12, 74, 176]. We first consider the expected first-order properties (see Section A.2): since there are $N(N-1)/2$ possible pairs of vertices, each 'occupied' by a link with independent probability q, the expected number of edges is

$$\langle L^u \rangle = q\frac{N(N-1)}{2}, \tag{B.1}$$

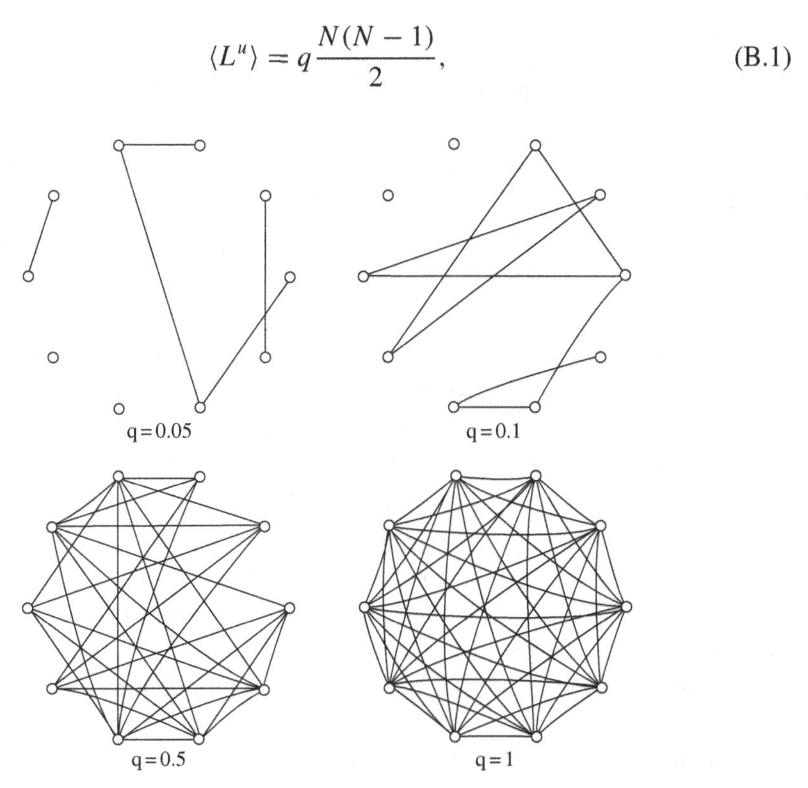

Fig. B.1. Four realizations of a random graph with $N = 10$ for different values of the connection probability q.

and by a direct comparison with Eq. (A.18) we find for the expected connectance $\langle c^u \rangle = q$. Similarly, the expected mean degree is

$$\langle \bar{k} \rangle = 2\frac{\langle L^u \rangle}{N} = q(N-1). \qquad \text{(B.2)}$$

If we want to model a real undirected network with a given number of links, we can generate an ensemble of random graphs such that the ensemble average $\langle L^u \rangle$ equals the *desired* (observed) value $\tilde{L}^u$. To do this, it suffices to choose $q = \tilde{c}^u$, where $\tilde{c}^u = 2\tilde{L}^u/N(N-1)$ is the *desired connectance*. Even if the individual graphs in the ensemble will have different values of L^u, the ensemble average $\langle L^u \rangle$ will equal the desired value $\tilde{L}^u$. Therefore the RG model succeeds in generating an ensemble of networks with any desired connectance, number of links or mean degree.

The other first-order properties are instead in sharp contrast with what is empirically observed in real-world networks. Each vertex has the same expected degree, which is given by the fraction q of the other $N-1$ vertices to which it is on average connected:

$$\langle k_i \rangle = q(N-1) \quad \forall i. \qquad \text{(B.3)}$$

The expected degree distribution $P(k)$ (see Section A.2) of the model can be computed analytically due to the simple assumptions of equiprobability and independence of different links. It can be shown [12, 74] that $P(k)$ has a binomial form, therefore approaching a Poisson distribution for large N:

$$P(k) = \binom{N-1}{k} q^k (1-q)^{N-1-k} \approx e^{-qN}\frac{(qN)^k}{k!} = e^{-\langle \bar{k} \rangle}\frac{\langle \bar{k} \rangle^k}{k!}, \qquad \text{(B.4)}$$

where we have assumed $N-1 \approx N$ for large N. The form of the degree distribution of various realization of the RG model is shown in Fig. B.2.

The Poisson form of the degree distribution is the major drawback of the RG model. The Poisson distribution, as the Gaussian one, has *exponentially decaying* tails which are in striking contrast with the much 'heavier' power-law tails of empirically observed degree distributions described by Eq. (A.8) and discussed in Section A.2. In particular, the RG model underestimates the number of vertices with large degree. As a consequence, the failure of the RG model points out that the 'disorder' observed in real networks cannot be reproduced by means of a completely random model, and requires the introduction of more complicated stochastic rules.

Also, the second- and third-order properties (see Sections A.3 and A.4) of the RG model are in constrast with empirical data. The expected *average nearest neighbour degree* can be computed as

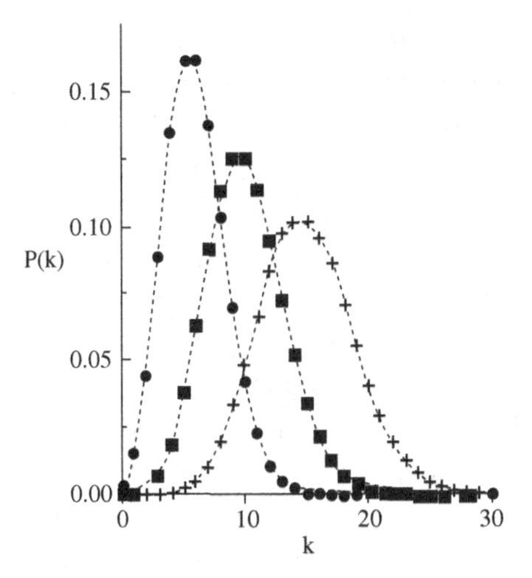

Fig. B.2. Degree distribution $P(k)$ in three realizations of the random graph model corresponding to different values of q.

$$\langle k_i^{nn} \rangle = \frac{\sum_{j \neq i} q \langle k_j \rangle}{\langle k_i \rangle} = q(N-1) \quad \forall i \tag{B.5}$$

and therefore the $\bar{k}^{nn}(k)$ curve is flat, in striking contrast with the empirical trends exemplified in Fig. A.3.

Similarly, since the *clustering coefficient* C_i of a vertex equals the probability of finding a link between a pair of neighbours of i (see Section A.4), and since in the RG model this probability equals q for any pair of vertices, the expected clustering coefficient $\langle C_i \rangle$ is the same for each vertex i. It is therefore equal to its expected mean value $\langle \bar{C} \rangle$:

$$\langle C_i \rangle = \langle \overline{C} \rangle = q \quad \forall i. \tag{B.6}$$

Therefore the $\bar{C}(k)$ curve is flat too and the RG model does not display the clustering hierarchy observed in most real networks, corresponding to a decreasing trend often of the form $\bar{C}(k) \propto k^{-\tau}$, as discussed in Section A.4. Moreover, it turns out that, once q is fixed in order to reproduce the empirical connectance of a real network, the resulting expected value $\langle \bar{C} \rangle$ is in general much smaller than the observed one. To see this, note that the RG model predicts $\bar{C}/c^u = 1$ independently of N, while this quantity presents the already discussed approximate linear dependence on N in real networks. In other words, the edges in real networks are arranged in such a way that there is a higher level of mean clustering than if they were drawn uniformly between all vertex pairs, and the RG model cannot reproduce simultaneously the connectance and the average clustering of real networks. This important point is the basic motivation for the introduction of the *small-world* model that shall be presented in Section B.3.

While the low-order properties of the RG model are trivial, the global ones are very interesting, since they highlight some instructive aspects of the model. In some cases, they are also in partial accordance with the empirical results. The cluster structure of the model crucially depends on the parameter q: when q is small, there are many small clusters in the network, while as q increases a *phase transition* occurs and a very large cluster forms, which spans the entire network when $q = 1$. More quantitatively, in the infinite size limit $N \to \infty$ the fraction of vertices in the largest cluster turns out to be zero if $q < q_c$ and *finite* if $q > q_c$, where $q_c \sim 1/N$ marks the *critical point* of the phase transition. In the $q > q_c$ phase, the largest component is also called the *giant component*, and its onset is referred to as the *percolation* transition. When $q < q_c$ the size of the various clusters is exponentially distributed, and for $q > q_c$ the non-giants components have exponentially distributed sizes too. Remarkably, right at the percolation transition $q = q_c$ the cluster size distribution has instead a power-law form, meaning that clusters of all sizes are present in the network.

The expected average distance $\langle \bar{d} \rangle$ of the RG model is interesting too. We recall that this quantity represents the average number of links separating two vertices. Therefore it can be computed approximately by noting that, since the expected average number of first neighbours of a vertex is $\langle \bar{k} \rangle$, then that of second neighbours is $\langle \bar{k} \rangle^2$ and that of lth neighbours is $\langle \bar{k} \rangle^l$. The average distance $\langle \bar{d} \rangle$ must be such that all the N vertices are on average reached in $\langle \bar{d} \rangle$ steps, or in other words $N \approx \langle \bar{k} \rangle^{\langle \bar{d} \rangle}$, yielding

$$\langle \bar{d} \rangle \approx \frac{\log N}{\log \langle \bar{k} \rangle} \tag{B.7}$$

in accordance with the empirical result reported in Eq. (A.36). Therefore the RG model displays the small-world effect discussed in Section A.5, even if in the 'weaker' sense since the clustering coefficient is small.

We finally mention that the RG model can be defined for directed networks as follows. Each pair of vertices i and j is considered twice, a first time drawing a link from i to j with probability p and a second one drawing a link from j to i with the same probability p. The results are straighforward generalizations of the undirected case. The expected average in- or out-degree is $\langle \bar{k} \rangle = (N-1)p$ and the expected connectance is $\langle c \rangle = p$. The in- and out-degree distributions have a Poisson form, and all vertices have the same expected average nearest neighbour degree. The clustering coefficient and the average distance can be measured on the undirected version of the network through the mapping defined by Eq. (A.5). We note that the undirected version of a directed RG model with connection probability p is formally equivalent to the undirected RG model with connection probability

$$q = 2p - p^2, \tag{B.8}$$

since the probability of having *at least one* directed link between two vertices i and j equals that of having a link from i to j, plus that of having a link from j to i, minus the probability p^2 of having both links simultaneously. Note that the latter also determines the expected number $\langle L^{\leftrightarrow} \rangle$ of reciprocated links, which is a fraction p^2 of the $N(N-1)$ possible ones: $\langle L^{\leftrightarrow} \rangle = p^2 N(N-1)$. Then the expected reciprocity is

$$\langle r \rangle = \frac{\langle L^{\leftrightarrow} \rangle}{\langle L \rangle} = \frac{p^2 N(N-1)}{pN(N-1)} = p, \tag{B.9}$$

which is equal to the expected connectance. Therefore the directed RG model displays a trivial reciprocity structure, in constrast with real networks as discussed in Section A.3.

B.3 The small-world model

In Sections A.4 and A.5 we reported a series of empirical results showing that real networks display a small average distance (like random graphs) and a large average clustering coefficient (like regular lattices with $l > 1$), and we denoted the coexistence of both properties as the 'strong' small-world effect. In order to reproduce this somehow intermediate behaviour of real networks between regular lattices and random graphs, Watts and Strogatz introduced in 1998 the so-called *small-world* (SW in the following) model [637]. The idea of the model is to start with a regular lattice and then to introduce disorder by randomly 'perturbing' it. More precisely, each link of a D-dimensional lattice is *rewired* with probability p in such a way that one of its ends is moved to a randomly chosen vertex (provided that self-loops and multiple links between two vertices are avoided). This procedure is shown in Fig. B.3 for a ring ($D = 1$) with first- and second-neighbours interactions ($l = 2$).

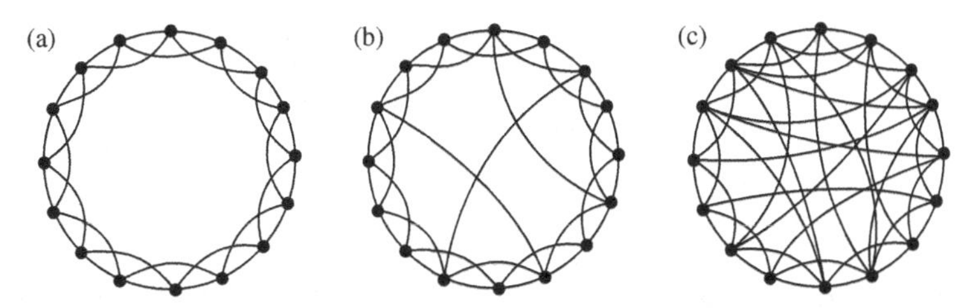

Fig. B.3. Example of the SW model built on a regular ring ($D = 1$ and $l = 2$) for three different values of p: (a) $p = 0$ (maximally ordered graph), (b) $0 < p < 1$ (small-world network), (c) $p = 1$ (maximally random graph).

For $p = 0$ no link is rewired and the network is the original regular lattice. In this case each vertex has the same degree $z = 2Dl$ and both the average distance and the average clustering coefficient are 'large'. At the opposite extreme ($p = 1$) all edges are randomly rewired and the network is similar to a random graph with the same number of links and average degree $\bar{k} = z$. As expected, in this case both the average distance and the average clustering coefficient are

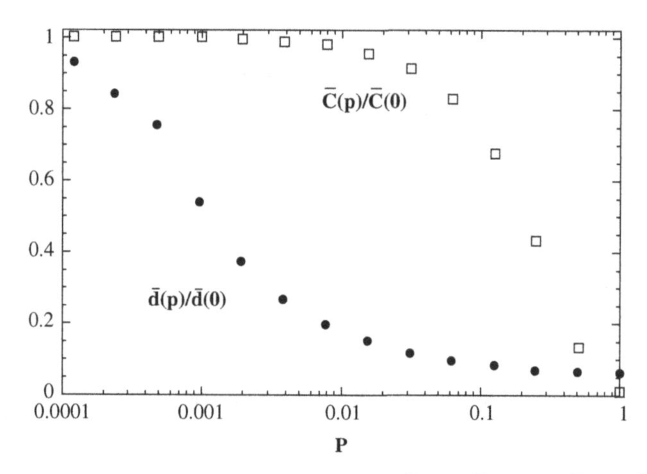

Fig. B.4. Dependence of the rescaled quantities $\bar{C}(p)/\bar{C}(0)$ and $\bar{d}(p)/\bar{d}(0)$ on the rewiring probability p in various realizations of the SW model (after ref. [637]).

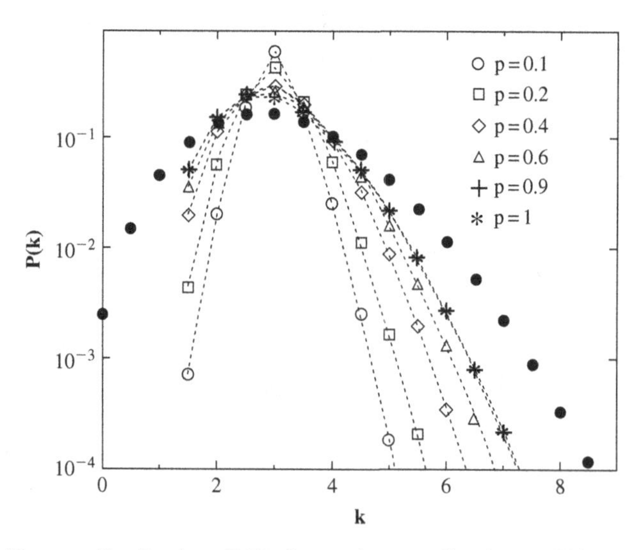

Fig. B.5. Degree distribution $P(k)$ for various realizations of the small-world model corresponding to $\langle k \rangle = 3$ and different values of p. The full circles represent the degree distribution for a random graph with the same average degree (after ref. [57]).

'small'. The intriguing point is that, for intermediate values of p, there exists a region where the average clustering is large and the average distance is small. The dependence on p of the rescaled quantities $\bar{C}(p)/\bar{C}(0)$ and $\bar{d}(p)/\bar{d}(0)$ is illustrated in Fig. B.4. The 'small-world' region is present for a wide range of the parameter p. Therefore the model provides an interpolation between regular lattices and random graphs which successfully reproduces the small-world effect observed in real networks.

Despite this satisfactory behaviour of $\bar{C}$ and $\bar{d}$, the SW model fails in reproducing other important topological properties, especially the degree distribution $P(k)$. The form of $P(k)$ is shown, for various values of p, in Fig. B.5. Recall that for $p = 0$ the regular lattice has a delta-like degree distribution. As p is increased towards 1, the degree distribution 'broadens' around the mean value $\bar{k} = z$, but no power-law tails are observed. As a comparison, the Poisson degree distribution of a RG model with the same average degree is shown. Despite its incompleteness, the SW model remains instructive as a reference for the intermediate behaviour of real networks between 'order' and 'randomness'. It is exactly this property that makes real networks difficult to be captured by simple models.

B.4 Evolving models

Although throughout the present work we are mainly interested in networks with a fixed number of vertices, which are probably most frequent in biology, we briefly mention here another important class of models exploiting the idea of network evolution through the addition of vertices.

B.4.1 Preferential attachment

Most evolving models are based on the idea of *preferential attachment*. We describe here the earliest and most influential of such models, which was proposed by Barabási, Albert and Jeong in 1999 [53]. In the Barabási–Albert (BA in the following) model, as it is known, a network with initially m_0 vertices grows through successive timesteps. At each timestep, a new vertex is added to the network, together with $m \leq m_0$ new links originating from it and pointing to m preexisting vertices chosen with a probability proportional to their degree. In other words, the choice of the m partners is not random, and high-degree vertices are more likely to 'attract' future connections. The model is aimed at reproducing real growing networks such as the WWW and collaboration networks, whose topology can be reasonably assumed to evolve gradually in such a way that more 'visible' high-degree vertices are more successful in developing new connections. It is possible to show, using various analytical techinques [12, 153], that the expected long-term form of the degree distribution resulting from this process is

$$P(k) \propto k^{-3} \tag{B.10}$$

independently of the values of the parameters m_0 and m. Therefore the BA model succeeds in reproducing qualitatively the power-law degree distribution of real networks (see Section A.2). However, the only possible value of the exponent is -3. Moreover, the model has other limitations, including the absence of degree correlations [434, 464] and of clustering hierarchy [493]. A variety of extensions of the BA model have been proposed (many of them are reviewed in refs. [12, 153]) in order to refine it. An interesting example is the class of growing models with *nonlinear preferential attachment*, where the probability that a preexisting vertex receives a link from the newly introduced one is no more proportional to its degree k, but in general to the power k^β with $\beta > 0$. One of the motivations for introducing nonlinear preferential attachment is that various empirical analyses suggest [293, 431] that some real networks such as the Internet and collaboration networks indeed grow by preferential attachment, sometimes linearly but more in general with a power-law dependence on vertex degree. It is, however, interesting to note that, while real scale-free networks driven by nonlinear preferential attachment are observed, the BA model has been shown [12, 153] to generate scale-free networks only in the linear case $\beta = 1$.

B.4.2 Vertex copying

Another class of evolving models is based on the *vertex copying mechanism* [323, 570, 617]. In this case too, a new vertex with a certain number of links is added to the network at each timestep, but the target vertices for the newly introduced links are now partly 'copied' from the links of an already existing vertex. This results in a sort of 'speciation' that generates a vertex that partly inherits the links of an 'ancestor', and partly undergoes a 'mutation' selecting new neighbours. Vertex copying mechanism are suitable to reproduce the WWW (as an alternative mechanism different from the preferential attachment one), where it is reasonable to assume that new pages are created by 'copying' a preexisting page with all its hyperlinks and then partly modifying it [323]. Another promising application is in protein networks [570, 617], since genes that code for specific proteins can duplicate during their evolutionary development and subsequently differentiate due to selection. This results in the consequent duplication and mutation of the set of interactions specified by the original gene. The vertex copying mechanism has been shown to reproduce various network properties including the scale-free degree distribution [323, 570, 617].

B.5 The configuration model

With this section we turn back to static models with a fixed number N of vertices. We start with the so-called *configuration model* [405, 419, 441], which can be regarded as a generalization of the RG model presented in Section B.2 in the following sense. The ensemble of networks generated by the RG model is completely random except the expected connectance, which is specified by fixing the connection probability q. In a similar manner, the ensemble of networks generated by the configuration model is completely random except the degree sequence, which is specified from the beginning.

To be more precise, let us first consider the model for the case of an undirected network. Each vertex i is assigned a desired degree k_i and then the ensemble of all graphs compatible with the resulting degree sequence is constructed by randomly drawing links between vertices. It can be shown [419] that every possible topology of a graph with the given degree sequence occurs in the ensemble with equal probability. The degree sequence $\{k_i\}_{i=1}^N$ can be picked from any desired degree distribution $P(k)$, for instance a scale-free one with desired exponent. Note that, once the desired degree sequence is specified, the number of links $L^u = \sum_i k_i/2$ is consequently fixed.

For directed networks, the configuration model is easily generalized by assigning each vertex i a given in-degree k_i^{in} and out-degree k_i^{out}, drawn from the desired distributions $P^{in}(k^{in})$ and $P^{out}(k^{out})$. It is also possible to specify a joint degree distribution $P(k^{in}, k^{out})$ for the probability that a vertex has a desired in- and out-degree simultaneously. Fixing the degree sequences also fixes the number of directed links $L = \sum_i k_i^{in} = \sum_i k_i^{out}$. These links are randomly drawn in such a way that the desired in- and out-degree sequences are both realized.

The aim of the configuration model is to check whether the higher-order structure is consistent with that of real networks, once the first-order properites alone are fixed. If this is the case, one must conclude that higher-order properties are a mere outcome of the specified form of the degree distribution, being consistent with a random assignment of links preserving the degree sequence. For this reason, the configuration model does not aim at testing explicit hypotheses on how networks organize into a given structure. It is rather a *null model* generating a suitably randomized ensemble of graphs once some low-level information is assumed as an input. The configuration model is extremely insightful, since it shows that the degree sequence alone determines higher-order patterns, especially disassortativity and clustering, which are often very close to the empirically observed ones. We briefly consider the two main implementations of the model.

B.5.1 The local rewiring algorithm

A random assignment of links, under the constraint that each vertex has a specified degree, gives rise systematically to undesired self-loops and multiple links between two vertices. Therefore in principle the graph ensemble should not be compared with real networks that miss such kinds of links, since this comparison could highlight patterns that are merely the result of the differences between the two specified topological classes. For this reason, rather than generating a graph ensemble separately, and then comparing the latter with a real network, Maslov, Sneppen and Zaliznyak [405] proposed starting directly with the real network. Then, they proposed a rule to generate a randomized ensemble of networks with the same degree sequence of the original one. This mechanism, which is the so-called *local rewiring* algorithm, avoids the occurrence of multiple links and self-loops in the randomized networks.

The local rewiring algorithm consists in the iteration of the elementary step shown in Fig. B.6a and Fig. B.6b for undirected and directed networks respectively: two edges are randomly chosen from the initial graph G_1 and the vertices at their ends are exchanged in such a way that the new graph G_2 has the same degree sequence of the initial one. If the 'new' edges already exist in the network, the step is aborted and two different edges are chosen. In this way an ensemble of randomized networks is generated, having the same degree sequence of the original one and no multiple links or self-loops. Therefore this ensemble is different from the version of the configuration model where such links are present. In particular, since two vertices cannot be connected more than once, here the presence of links between high-degree vertices is suppressed, determining a certain degree of 'spurious' disassortativity that is not caused by an 'active' anticorrelation between vertex degrees. This remarkable point highlighted by Maslov, Sneppen and Zaliznyak led them to show that much of the disassortativity observed in the Internet

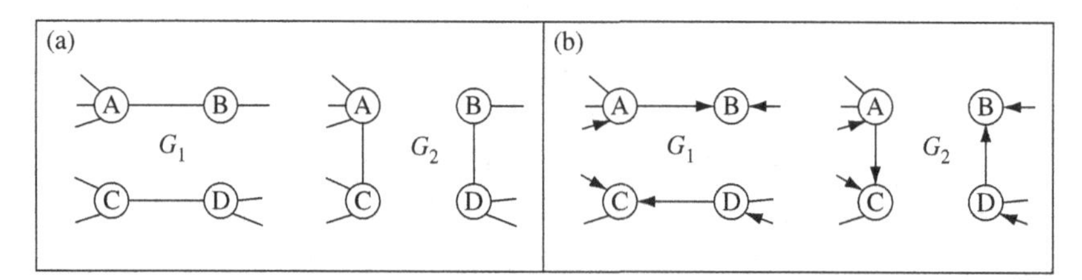

Fig. B.6. Elementary step of the *local rewiring* algorithm for (a) undirected and (b) directed networks. Two pairs of edges (here $A-B$ and $D-C$) are randomly chosen from graph G_1 and the vertices at their ends are exchanged to obtain the edges $A-C$ and $D-B$ in graph G_2. Note that the degree of each vertex is unchanged (in the directed case, the in- and out-degrees are separately conserved).

(see Section A.3) can be accounted for in this way, while other patterns such as the clustering properties are instead genuine [405].

B.5.2 Analytical approaches

A number of expected quantities of the configuration model can be computed by making use of a *generating function formalism* [441], including the clustering coefficient, the average distance, the critical point of the phase transition where a giant component forms, and the size of the non-giant components. It is also possible to go one step further and fix not only the degree sequence, but also the degree correlations [434, 435]. In this case too, various expected quantities can, in principle, be computed analytically and compared with real data. In some cases, the expected properties partially fit some of the empirical ones.

There is also another approach, obtained by allowing the degrees to fluctuate about an average value, and thus fixing the *expected* degree of each vertex. Using the jargon of statistical mechanics, this change of perspective corresponds to modify the graph ensemble from canonical to grand-canonical. This approach was first explored by Chung and Lu [113], who considered an undirected network where each vertex i is assigned a *desired degree* $\tilde{k}_i$ (which also fixes the desired number of links $\tilde{L}^u = \sum_i \tilde{k}_i/2$). They assumed that a link between two vertices i and j is drawn with probability

$$q_{ij} = \frac{\tilde{k}_i \tilde{k}_j}{2\tilde{L}^u}. \tag{B.11}$$

The reason for the above choice is that the ensemble averages of the degrees converge to their desired values:

$$\langle k_i \rangle = \sum_j q_{ij} = \tilde{k}_i \frac{\sum_j \tilde{k}_j}{2\tilde{L}^u} = \tilde{k}_i. \tag{B.12}$$

Therefore if the desired degrees are picked from a distribution $P(\tilde{k}_i)$ the expected degrees will be distributed according to the same distribution (however, as we show below, we are not free to choose *any* desired form of the distribution). The factorizable form of q_{ij} also implies that no degree correlations are introduced:

$$\langle k_i^{nn} \rangle = \frac{\sum_j q_{ij} \tilde{k}_j}{\tilde{k}_i} = \frac{\sum_j \tilde{k}_j^2}{\sum_j \tilde{k}_j}, \tag{B.13}$$

which is independent of i and has the expected form $\overline{k^2}/\bar{k}$ valid for uncorrelated networks (see Section A.3).

The model can be formulated for directed graphs as follows: each vertex i is assigned simultaneously a *desired in-degree* $\tilde{k}_i^{in}$ and a *desired out-degree* $\tilde{k}_i^{out}$, and a directed link from i to j is drawn with probability

$$p_{ij} = \frac{\tilde{k}_i^{out}\tilde{k}_j^{in}}{\tilde{L}}, \tag{B.14}$$

where $\tilde{L} = \sum_i \tilde{k}_i^{in} = \sum_i \tilde{k}_i^{out}$ is the desired number of directed links. This choice ensures that $\langle k_i^{in}\rangle = \tilde{k}_i^{in}$ and $\langle k_i^{out}\rangle = \tilde{k}_i^{out}$ for all vertices, generalizing Eq. (B.12).

The Chung–Lu model avoids by construction the occurrence of multiple links and self-loops, since each pair of (distinct) vertices is considered only once. However, to ensure $0 \leq q_{ij} \leq 1$ and $0 \leq p_{ij} \leq 1$ for all i, j in Eqs. (B.11) and (B.14), one is forced to consider only those degree sequences satisfying the constraint

$$\tilde{k}_i \leq \sqrt{2\tilde{L}^u} = \sqrt{\sum_{j=1}^{N} \tilde{k}_j} \tag{B.15}$$

and similarly $\tilde{k}_i^{in} \leq \sqrt{\tilde{L}}$ and $\tilde{k}_i^{out} \leq \sqrt{\tilde{L}}$ for directed graphs. We can regard this requirement from a different point of view: a connection probability $q_{ij} > 1$ corresponds physically to the presence of multiple links between i and j, and this possibility is avoided only by imposing the above constraint. Therefore the problem of the occurrence of multiple links is circumvented by restricting the possible degree sequences to those satisfying Eq. (B.15).

Unfortunately, the above constraint is very strong, and is often violated by real power-law degree distributions where few very large degrees are present. This drawback of the Chung–Lu model led Park and Newman [463] to modify it in such a way that no restriction on the desired degree sequence is imposed, and at the same time no multiple edges are generated. Park and Newman started from the general problem of finding the form of the connection probability q_{ij} that generates an ensemble of graphs with no multiple links and such that two graphs with the same degree sequence are equiprobable, in the original spirit of the configuration model. Remarkably, they found that this condition is not satisfied by Eq. (B.11). Indeed, the correct expression turns out to be [463]

$$q(x_i, x_j) = \frac{z x_i x_j}{1 + z x_i x_j}, \tag{B.16}$$

where x_i controls the expected degree of vertex i, and $z > 0$ is a parameter controlling the expected number of links such as the factor $(2\tilde{L}^u)^{-1}$ in Eq. (B.11). The quantities $\{x_i\}_{i=1}^N$ play a role similar to that of the desired degrees $\{\tilde{k}_i\}_{i=1}^N$ in the Chung–Lu model, even if in this case they turn out to be in general different from the expected degrees $\{\langle k_i\rangle\}_{i=1}^N$ and are therefore denoted by a different symbol. For the same reason, we use the notation $\sigma(x)$ to indicate the statistical distribution of the values of x. The above expression is of fundamental importance. It ensures that $0 \leq q_{ij} \leq 1$ with no restriction on the distribution $\sigma(x)$. However the requirement of no multiple links, which is implicit in Eq. (B.16), implies a deviation of the expected degrees from the assigned values of x:

$$\langle k_i \rangle = \sum_j q_{ij} = \sum_j \frac{zx_i x_j}{1 + zx_i x_j}. \tag{B.17}$$

We can inspect the above formula by noting that the expected degree of a vertex with a given value of x depends on x alone:

$$\langle k(x) \rangle = (N-1) \int dy \frac{zxy}{1+zxy} \sigma(y). \tag{B.18}$$

The behaviour of $\langle k(x) \rangle$ is proportional to x for small values of x and then 'saturates' to the maximum value $N-1$ for large x, consistently with the requirement of no multiple or self-loops. Park and Newman [463] studied this effect for a power-law distribution $\sigma(x) \propto x^{-\tau}$ with various values of the exponent τ (see Fig. B.7). They found that this behaviour has two important consequences on the topology: firstly, the degree distribution $P(k)$ behaves as a power-law with the same exponent τ for small values of x, but then diplays a cut-off ensuring $k \leq N-1$ (see Fig. B.8a).

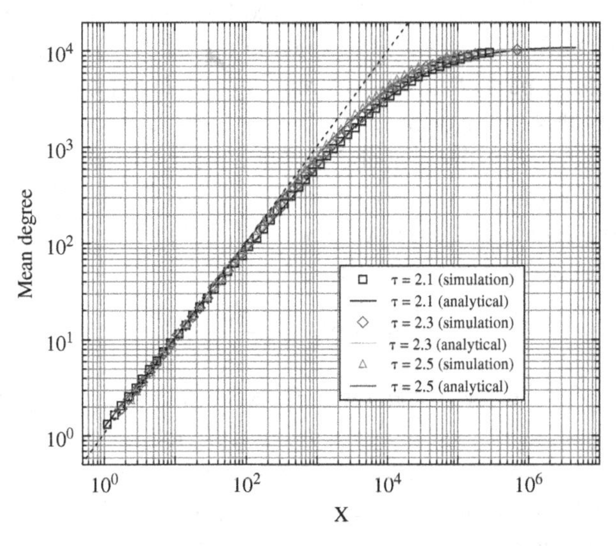

Fig. B.7. Average degree $\langle k(x) \rangle$ of vertices versus x corresponding to the choice $\sigma(x) \propto x^{-\tau}$ for three values of τ. The trend is initially linear and then saturates to the asymptotic value $k \to N-1$ (after ref. [463]).

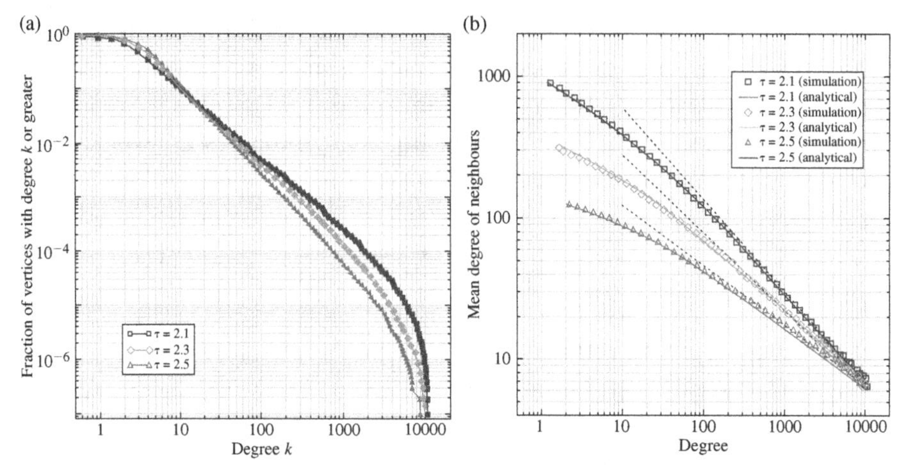

Fig. B.8. (a) Cumulative degree distribution $P_>(k)$ corresponding to the choice $\sigma(x) \propto x^{-\tau}$ for three values of τ. (b) Average nearest neighbour degree $\bar{k}^{nn}(k)$ for the same three choices of the exponent τ. Here isolated symbols correspond to numerical simulations, while solid lines are the analytical predictions (after ref. [463]).

Secondly, the average nearest neighbour degree turns out to be a decreasing function of the degree (see Fig. B.8b). As expected, the absence of multiple links generates an 'effective repulsion' between high-degree vertices, resulting in some spurious disassortativity. Moreover, the studied mechanism generates a $k^{nn}(k)$, which is not a power-law, even if it approaches a power-law behaviour asymptotically (see Fig. B.8b). These results allowed Park and Newman to confirm that, as suggested by Maslov *et al.* [405] by making use of the local rewiring algorithm (see Section B.5.1), part of the disassortativity displayed by the Internet can be accounted for by this mechanism.

For directed networks, the probability that a directed link from i to j is there is a function $p_{ij} = p(x_i, y_j)$ of *two* quantities x_i and y_j playing a role analogous to that of the expected out- and in-degrees $\tilde{k}_i^{out}$, $\tilde{k}_j^{in}$ in the directed version of the Chung–Lu model defined in Eq. (B.14):

$$p_{ij} = p(x_i, y_j) = \frac{z x_i y_j}{1 + z x_i y_j}. \tag{B.19}$$

Before turning to a different (but related) class of models, we briefly note that all the (directed) versions of the configuration model yield a trivial reciprocity structure. To see this, we first note that in the 'canonical' versions with a fixed number of links, the network is randomized in any respect once the in- and out-degree sequences are specified. This means that, in the randomized ensemble, two graphs with the same degree sequences but with different numbers of reciprocated links have the same statistical weight. As a result, reciprocated links occur completely by chance, with no tendency towards a non-random reciprocity structure. Therefore the ensemble average of the reciprocity is trivial. In the 'grand-canonical' versions, it is easy to obtain the same result since the probability p_{ij} that a link from i to j is there is independent on whether the reciprocal link from j to i is there. In other words, the two links do not affect the presence of each other, and the reciprocity is trivial.

B.6 Hidden-variable models

Hidden variable (HV in the following) models were first introduced by Caldarelli *et al.* [99] and by Söderberg [568] and later studied in deeper detail by Boguñá and Pastor-Satorras [73] and by Servedio *et al.* [555]. These models consider a static network with N vertices. Each vertex i is assigned a 'hidden variable' or 'fitness' x_i, drawn from a given statistical distribution $\sigma(x)$, which determines the probability that i connects to other vertices. In the simplest (and most studied) case of an undirected network, the probability q_{ij} that two vertices i and j are connected is a function $q(x_i, x_j)$ of their fitness values. Note that we require $q_{ij} = q_{ji}$, since there is no preferred ordering of i and j when drawing an undirected link between them. The ensemble of networks is constructed by keeping the fitness values $\{x_i\}_{i=1}^N$ fixed and repeating the random assignment of links. The expected topological properties of the model depend on the fitness distribution $\sigma(x)$, which we assume to be normalized so that $\int dx\, \sigma(x) = 1$, and on the functional form of $q(x_i, x_j)$. Clearly, if each vertex has the same fitness value $x_i = x$, corresponding to a delta-like $\sigma(x)$, the RG model is recovered. The same occurs for the special case of a constant connection probability $q(x_i, x_j) = q$ independently of $\sigma(x)$.

We note that hidden-variable models can be regarded as a generalization of the models described in Section B.5.2, where the form of the connection probability $q(x_i, x_j)$ is not restricted to Eq. (B.16), and the variable x_i plays the role of any 'physical' quantity assumed to determine the role of vertex i in the network. This makes hidden-variable models particularly suitable to detect the organizing mechanisms shaping the topology of real networks, since the fitness x can be in principle identified with an empirical quantity associated to each vertex. By contrast, in the configuration model no explicit assumption is made on the underlying mechanism determining network topology, since the degree sequence is taken as an input information without explanation. At the moment, the only other models making assumptions on the network formation process are those belonging to the class of evolving models described in Section B.4. While evolving models are particularly suitable to model networks whose future topology is likely to be determined essentially by their past one, without any additional 'external' information, hidden-variable models are better adapted to model static networks where the topological properties are essentially determined by some additional information of non-topological nature, which is, however, intrinsically related to the role played by each vertex in the network. Understanding which of the two mechanisms (if any) is the relevant one for a particular network sheds light on its organizing principles.

Many low-order expected properties of hidden variable models can be computed analytically and confirmed by numerical simulations. As a useful rule, note that the ensemble average of the generic element b_{ij} of the adjacency matrix of the network is

$$\langle b_{ij} \rangle = q(x_i, x_j) \tag{B.20}$$

and that, owing to the independence of different links, the ensemble average of products of adjacency matrix elements equals

$$\langle b_{ij} b_{kl} \rangle = \begin{cases} \langle b_{ij}^2 \rangle = \langle b_{ij} \rangle = q(x_i, x_j) & \text{if } i = k \text{ and } j = l \text{ or } i = l \text{ and } j = k \\ \langle b_{ij} \rangle \langle b_{kl} \rangle = q(x_i, x_j) q(x_k, x_l) & \text{else.} \end{cases} \tag{B.21}$$

Equations (B.20) and (B.21), along with their higher-order generalizations, allow to compute many expected topological properties. To start with, the expected number of links is obtained by making use of the definition (A.14) and Eq. (B.20):

$$\langle L^u \rangle = \sum_{i=1}^{N} \sum_{j<i} q(x_i, x_j) = \frac{N(N-1)}{2} \int dx \int dy \, q(x, y) \sigma(x) \sigma(y), \tag{B.22}$$

where we have replaced the discrete sums over all the $N(N-1)/2$ pairs of vertices with integrals over the fitness variable. The expected connectance is therefore

$$\langle c^u \rangle = 2 \frac{\langle L^u \rangle}{N(N-1)} = \int dx \int dy \, q(x, y) \sigma(x) \sigma(y). \tag{B.23}$$

Similarly, we can compute the expected degree of a vertex by means of Eqs. (A.6) and (B.20) as

$$\langle k_i \rangle = \sum_{j \neq i} q(x_i, x_j). \tag{B.24}$$

Note that the expected degree of a vertex with fitness x depends on x alone:

$$\langle k(x) \rangle = (N-1) \int dy \, q(x, y) \sigma(y). \tag{B.25}$$

The above result implies that, if $k(x)$ can be inverted to yield $x(k)$, then the degree distribution $P(k)$ can be expressed in terms of the fitness distribution $\sigma(x)$:

$$P(k) = \frac{dx(k)}{dk} \sigma[x(k)]. \tag{B.26}$$

The expected average nearest-neighbour degree can be computed as a generalization of Eq. (B.5) by making use of the property (B.21) and the definition (A.21):

$$\langle k_i^{nn} \rangle = \frac{\sum_{j \neq i} q(x_i, x_j) \langle k_j \rangle}{\langle k_i \rangle} = \frac{\sum_{j \neq i} \sum_{k \neq j} q(x_i, x_j) q(x_j, x_k)}{\sum_{j \neq i} q(x_i, x_j)} \tag{B.27}$$

and therefore the expected average nearest-neighbour degree of a vertex with fitness x depends on x alone:

$$\langle k^{nn}(x) \rangle = \frac{(N-1) \int dy \int dz \, q(x, y) q(y, z) \sigma(y) \sigma(z)}{\int dy \, q(x, y) \sigma(y)}. \tag{B.28}$$

Finally, by exploiting Eqs. (A.31) and (B.21) we can express the expected clustering coefficient of vertex i as:

$$\langle C_i \rangle = \frac{\sum_{j \neq i} \sum_{k \neq i, j} q(x_i, x_j) q(x_j, x_k) q(x_k, x_i)}{\left[\sum_{j \neq i} q(x_i, x_j) \right]^2} \tag{B.29}$$

where, for clarity of the notation, we assumed $\langle k_i \rangle^2 - \langle k_i \rangle \approx \langle k_i \rangle^2$ in the denominator (the full expression should be used for low-degree vertices). In this case too, the expected clustering coefficient of a vertex with fitness x depends on x alone:

$$\langle C(x) \rangle = \frac{\int dy \int dz \, q(x, y) q(y, z) q(z, x) \sigma(y) \sigma(z)}{\left[\int dy \, q(x, y) \sigma(y) \right]^2}. \tag{B.30}$$

Note that since the expected values $\langle C(x) \rangle$ and $\langle k^{nn}(x) \rangle$ depend only on the fitness x, they must be equal to their expected average values $\langle \bar{C}(x) \rangle$ and $\langle \bar{k}^{nn}(x) \rangle$ over all vertices with fitness x: $\langle C(x) \rangle = \langle \bar{C}(x) \rangle$ and $\langle k^{nn}(x) \rangle = \langle \bar{k}^{nn}(x) \rangle$. Therefore the knowledge of $x(k)$, inserted into Eqs. (B.28) and (B.30), allows to obtain the expected expressions for the $\langle \bar{k}^{nn}(k) \rangle$ and $\langle \bar{C}(k) \rangle$ curves.

Also the expected global properties can in principle be computed, even if with much more analytical effort. For instance, the cluster structure and the onset of the giant component in HV models has been analysed by Söderberg [568] through an extension of the technique used for random graphs and for the configuration model.

B.6.1 The threshold fitness model

An interesting choice proposed by Caldarelli *et al.* [99] is when the fitness distribution is exponential

$$\sigma(x) = e^{-x} \qquad x \in [0, +\infty) \tag{B.31}$$

and the connection probability is a step function with a threshold x_0:

$$q(x_i, x_j) = \Theta(x_i + x_j - x_0). \tag{B.32}$$

With this choice, two vertices are connected if the sum of their fitness values exceeds x_0. It is easy to show [73, 99], by making use of Eqs. (B.25) and (B.26), that the degree distribution is scale-free with exponent -2. This can also be confirmed by numerical simulations (see Fig. B.9). The behaviour of $\bar{k}^{nn}(k)$ and $\bar{C}(k)$ is interesting too, and the network can be shown [73, 99] to display the properties of disassortativity and clustering hierarchy introduced in Sections A.3 and A.4. The $\bar{k}^{nn}(k)$ and $\bar{C}(k)$ curves obtained by numerical simulations are also shown in Fig. B.9.

B.6.2 The directed case

The hidden variable model can be easily generalized to directed networks. The main difference is that in this case *two* quantities x_i and y_i can be associated to each vertex i, controlling the expected out-degree and in-degree respectively. The two fitness variables can have different statistical distributions $\sigma(x)$ and $\rho(y)$. The probability that a directed link from i to j is there is a function of both quantities:

$$p_{ij} = p(x_i, y_j). \tag{B.33}$$

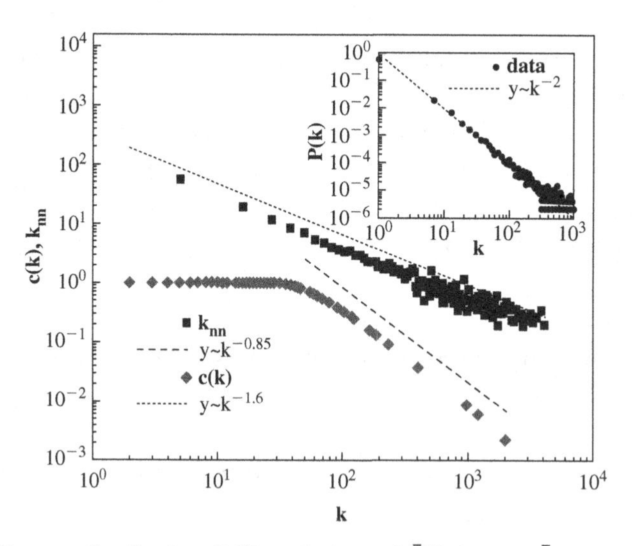

Fig. B.9. Degree distribution $P(k)$ and plots of $\bar{k}^{nn}(k)$ and $\bar{C}(k)$ in the fitness model with exponential fitness distribution and threshold connection probability (after ref. [99]).

Note that now it is possible to have $p_{ij} \neq p_{ji}$, differently from the undirected version of the model. The expressions for the expected properties are straightforward generalizations of Eqs. (B.25), (B.26), (B.28) and (B.30). For instance, we can write the expected out- and in-degree of a vertex with fitness values x and y as

$$\langle k^{out}(x)\rangle = (N-1)\int dy\; p(x,y)\rho(y) \tag{B.34}$$

$$\langle k^{in}(y)\rangle = (N-1)\int dx\; p(x,y)\sigma(x) \tag{B.35}$$

and, if the above expression are invertible, the expected out- and in-degree distributions read

$$P^{out}(k^{out}) = \frac{dx(k^{out})}{dk^{out}}\sigma[x(k^{out})] \tag{B.36}$$

$$P^{in}(k^{in}) = \frac{dy(k^{in})}{dk^{in}}\rho[y(k^{in})]. \tag{B.37}$$

In this case too, the configuration model defined by Eq. (B.19) can be recovered as a particular case. Using an argument similar to that used in Section B.5.2, we can conclude that the expected reciprocity of hidden-variable models is trivial.

B.7 Models with reciprocity

We showed that the directed version of all models described so far generate no reciprocity structure, in striking contrast with what empirically observed and discussed in Section A.3. In order to define local models that generate non-trivial reciprocity, one must first note that the reciprocity ρ defined in Eq. (A.30) aggregates the information about the connection properties of individual pairs of vertices. Let $p_{ij} \equiv p(i \to j)$ denote the probability that a link is drawn from vertex i to vertex j. In the general case, the probability $p_{ij}^{\leftrightarrow}$ of having a pair of mutual links between i and j is given by

$$p_{ij}^{\leftrightarrow} \equiv p(i \to j \cap j \to i) = r_{ij}p_{ji} = r_{ji}p_{ij}, \tag{B.38}$$

where r_{ij} is the *conditional probability* of having a link from i to j *given that* the mutual link from j to i is there:

$$r_{ij} \equiv p(i \to j | j \to i). \tag{B.39}$$

Note that $\overline{r}_{ij} \equiv \sum_{i\neq j} r_{ij}/N(N-1) = r$, motivating the choice of the symbol. The expected value of ρ reads

$$\langle\rho\rangle = \frac{\sum_{i\neq j} p_{ij}r_{ji} - (\sum_{i\neq j} p_{ij})^2/N(N-1)}{\sum_{i\neq j} p_{ij} - (\sum_{i\neq j} p_{ij})^2/N(N-1)}. \tag{B.40}$$

In the models discussed above the presence of the mutual link does not affect the connection probability, or in other words $r_{ij} = p_{ij}$ and $p_{ij}^{\leftrightarrow} = p_{ij}p_{ji}$. This yields $\langle\rho\rangle = 0$ in Eq. (B.40), confirming that such models generate areciprocal networks.

The only way to include reciprocity in the models is by considering a nontrivial form ($r_{ij} \neq p_{ij}$) of the conditional probability (hence the information required to generate the network is no longer specified by p_{ij} alone). This allows us to introduce, beyond $p_{ij}^{\leftrightarrow}$, the probability

$$p_{ij}^{\rightarrow} \equiv p(i \rightarrow j \cap j \nrightarrow i) = p_{ij} - r_{ij}p_{ji} \tag{B.41}$$

of having a *single* link from i to j (and no reciprocal link from j to i), and the probability $p_{ij}^{\nleftrightarrow}$ (fixed by the equality $p_{ij}^{\rightarrow} + p_{ji}^{\rightarrow} + p_{ij}^{\leftrightarrow} + p_{ij}^{\nleftrightarrow} = 1$) of having no link between i and j. The network can then be generated by drawing, *for each single vertex pair*, a link from i to j, a link from j to i, two mutual links or no link with the corresponding probabilities $p_{ij}^{\rightarrow}$, $p_{ji}^{\rightarrow}$, $p_{ij}^{\leftrightarrow}$ and $p_{ij}^{\nleftrightarrow}$ respectively.

The two probabilities (B.38) and (B.41) completely specify the reciprocity structure of the network, and also suggest related new quantities. For instance, if we define the *reciprocal degree* of a vertex i as the number $k_i^{\leftrightarrow}$ of reciprocated links of i

$$k_i^{\leftrightarrow} \equiv \sum_{j \neq i} a_{ij}a_{ji} \tag{B.42}$$

we can easily compute its expected value as

$$\langle k_i^{\leftrightarrow} \rangle = \sum_{j \neq i} r_{ij}p_{ji} = \sum_{j \neq i} p_{ij}^{\leftrightarrow}. \tag{B.43}$$

Similarly, we can define the numbers $k_i^{\rightarrow}$ and $k_i^{\leftarrow}$ of non-reciprocated out-going and in-coming links of a vertex i

$$k_i^{\rightarrow} \equiv \sum_{j \neq i} a_{ij}(1 - a_{ji}) = k_i^{in} - k_i^{\leftrightarrow} \tag{B.44}$$

$$k_i^{\leftarrow} \equiv \sum_{j \neq i} a_{ji}(1 - a_{ij}) = k_i^{out} - k_i^{\leftrightarrow} \tag{B.45}$$

and express their expected values as

$$\langle k_i^{\rightarrow} \rangle = \sum_{j \neq i} (p_{ij} - r_{ij}p_{ji}) = \sum_{j \neq i} p_{ij}^{\rightarrow} \tag{B.46}$$

$$\langle k_i^{\leftarrow} \rangle = \sum_{j \neq i} (p_{ji} - r_{ji}p_{ij}) = \sum_{j \neq i} p_{ji}^{\rightarrow}. \tag{B.47}$$

In terms of the above quantities one can also obtain the explicit expressions for the quantities computed on the undirected version of a directed network through the mapping (A.5). For instance, we have for the 'undirected' degree

$$\begin{aligned} k_i &= \sum_{j \neq i} b_{ij} = \sum_{j \neq i} (a_{ij} + a_{ji} - a_{ij} a_{ji}) \\ &= k_i^{\rightarrow} + k_i^{\leftarrow} + k_i^{\leftrightarrow} = k_i^{in} + k_i^{out} - k_i^{\leftrightarrow}. \end{aligned} \quad \text{(B.48)}$$

Finally, we note that the number of reciprocated links $L^{\leftrightarrow}$ can be obtained in terms of the reciprocal degrees as

$$L^{\leftrightarrow} = \sum_{i=1}^{N} k_i^{\leftrightarrow} \quad \text{(B.49)}$$

and similarly, the number $L^{\rightarrow}$ of non-reciprocated links can be written as

$$L^{\rightarrow} \equiv L - L^{\leftrightarrow} = \sum_{i=1}^{N} \sum_{j \neq i} a_{ij}(1 - a_{ji}) = \sum_{i=1}^{N} k_i^{\rightarrow} = \sum_{i=1}^{N} k_i^{\leftarrow}. \quad \text{(B.50)}$$

Appendix C

Higher-order topological properties

S. AHNERT, T. FINK AND G. CALDARELLI

C.1 Introduction

This appendix addresses network properties that are not captured by the local (vertex to vertex) structure alone. As one moves beyond the first neighbours of a vertex, various higher-order topological properties can be considered. Beyond the correlations between pairs of vertices, the next levels of complexity involve subgraphs of three, four, etc., vertices.

Non-random subgraphs consisting of a few (usually three to five) vertices are called motifs. On the contrary, larger and dense subgraphs (of no specified size) are called communities. We shall describe motifs, and an example of network dynamics involving them, in Section C.2. In Section C.3 we then turn to the description of communities and some algorithms to detect them.

C.2 Network motifs and boolean dynamics

Network motifs are small subgraphs in a larger network, typically of three to five nodes in size (see Fig. C.1). Alon and co-workers [391, 417] showed that some motifs occur much more often than would be expected by chance in many real-world networks. Moreover they showed that the relative frequencies of these motifs can be used to divide real-world networks into 'superfamilies' of networks.

The feed-forward network has been identified as an important ingredient in several biological processes [59, 391]. An example is a pulse generator in synthetic biology [59] which can be modelled by with a switch (left node), an inhibitor (top node) and an output (right node). The pulse generation itself is represented by the following sequence of active (1) and inactive (0) states of switch/inhibitor/output: $000 \Rightarrow 100 \rightarrow 111 \rightarrow 110$, where $\Rightarrow$ denotes switching on the generator (i.e., changing the constant rule on the switch node from 0 to 1). Another example using is a sign-sensitive delay element in transcription

Networks in Cell Biology, ed. M. Buchanan, G. Caldarelli, P. De Los Rios, F. Rao and M. Vendruscolo.
Published by Cambridge University Press.

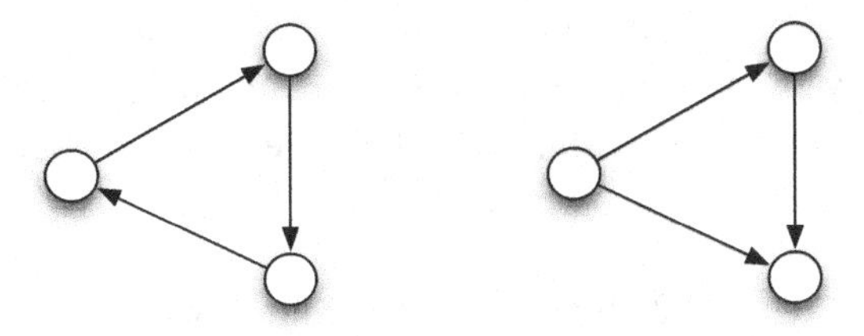

Fig. C.1. Two well-studied 3-node, directed network motifs: the feedback loop (left), and the feed-forward loop (right).

networks [391]. This is a circuit that responds slowly to the off–on and rapidly to the on–off switching of a signal.

Feedback is also an essential part of various biological mechanisms. Examples include (i) a cellular digital clock, which generates series of pulses of protein expression, and (ii) negative autoregulation. The first can be modelled using the 2-node feedback network (⇄) while the second requires an additional self-interaction (→). The dynamics are represented by ⚲ and ⌇.

C.2.1 Network dynamics: boolean networks

Dynamical networks are found in the brain [325], genetic regulation and transcription [236], the World Wide Web [84], ecological food webs [211]. In a dynamical network, associated with each node is a state which can change over time. Typically the value of a state at a given time depends upon the earlier states of other nodes connected by edges to the node in question.

One of the simplest models of network dynamics is a *boolean network*. In this model, the states can only take on two discrete values, 0 or 1, and all of the states are updated simultaneously.

A boolean network has two parts: a directed network of N nodes and from 0 to N^2 edges; and a set of local update rules for the nodes. An update rule assigns an output of 0 or 1 for each of the possible combinations of inputs. The state of the network is the vector of values of the N nodes, and thus there are 2^N possible network states. The detailed manner in which network states flow into other network states is called the network dynamic.

This section is divided into four parts. In the first subsection, we describe directed networks, on which boolean dynamics take place; in the second, we summarize boolean functions. We consider dynamics for small numbers of nodes in the third subsection, and dynamics for large numbers of nodes in the fourth (Kauffman networks).

It is worth noting that the notion of a boolean network is an extremely general one, and many simpler types of dynamical systems can be cast in the framework of boolean networks. For example, one-dimensional cellular automata, studied extensively by Wolfram and others, are boolean networks on a type of one-dimensional (1-d) lattice with identical update rules for each node.

C.2.2 Motifs in directed networks

In a boolean network, links between nodes need not be symmetric: just because the state of node i depends on the state of node j does not mean that the reverse is true. Furthermore, it is important to distinguish between nodes with and without self-interactions. At any given time step, a node with a self-interaction has access to its previous value, which it may incorporate into its update rule in conjunction with any other inputs. A node with no interactions whatsoever must remain fixed throughout all time at 0 or 1.

The number of networks grows with N as 2, 10, 104, 3044, . . .; this is the number of structures of finite relations (Sloane A000595, The On-Line Encyclopedia of Integer Sequences). In Fig. C.2, we show explicitly the 10 2-node networks and the 104 3-node networks. When all possible N^2 edges are present, the network is called fully connected.

C.2.3 Boolean functions

The state of a boolean network is updated discretely at fixed time points: at time $t + 1$ the value of each node is a boolean function of the nodes connected to it (as

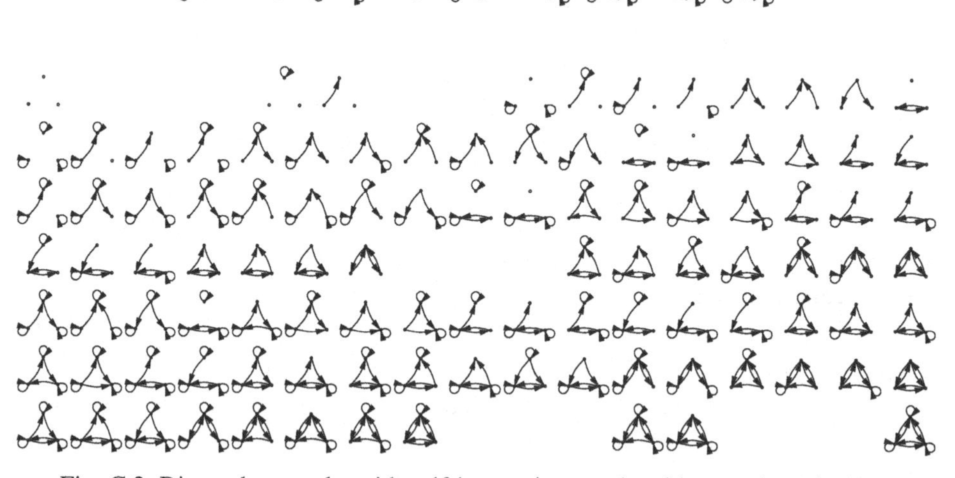

Fig. C.2. Directed networks with self-interactions, ordered by number of edges. There are 10 distinct networks on 2-nodes and 104 networks on 3-nodes. For 10-nodes, the number of networks is of the order of Avogadro's number.

Table C.1. *The 2^{2^1} boolean functions for 1 input (I_1). Rule R_1 maps all the states to 0; R_2 maps all the states to 1; R_3 preserves the states; and R_4 inverts them*

I_1	R_1	R_2	R_3	R_4
0	0	1	0	1
1	0	1	1	0

Table C.2. *The 2^{2^2} boolean functions for 2 inputs (I_1, I_2)*

I_1	I_2	R_1	R_2	R_3	R_4	R_5	R_6	R_7	R_8
0	0	1	0	0	1	0	1	1	0
0	1	1	0	0	1	1	0	0	1
1	0	1	0	1	0	0	1	0	0
1	1	1	0	1	0	1	0	0	0
I_1	I_2	R_9	R_{10}	R_{11}	R_{12}	R_{13}	R_{14}	R_{15}	R_{16}
0	0	0	0	0	1	1	1	1	0
0	1	0	0	1	0	1	1	0	1
1	0	1	0	1	1	0	1	0	1
1	1	0	1	1	1	1	0	1	0

inputs) at time t. A boolean function is just a technical name for a specification of the outputs for all possible combination of inputs; it is essentially a lookup table. If there are k inputs to a given node, the number of combinations of input states is 2^k. The number of possible boolean functions is 2 to this number, namely 2^{2^k}. We explicitly list the possible boolean functions for a node with 1 and 2 inputs in Tables C.1 and C.2.

It is sometimes useful to refer to the joint boolean function for a network, which is the ordered set of local boolean functions. The number of joint boolean functions is the product of the number of local functions. Thus if a network has nodes $1, 2, 3, \ldots$ with in-degree $k_1, k_2, k_3, \ldots$, then the number of joint boolean functions is $2^{2^{k_1}+2^{k_2}+2^{k_3}+\cdots}$. For a fully connected N-node network, there are $(2^N)^{2^N}$ joint boolean functions.

C.2.4 Boolean dynamics

In accordance with the boolean function, each state in a network is followed by another new state until the dynamics enters into an endless loop, or cycle. The

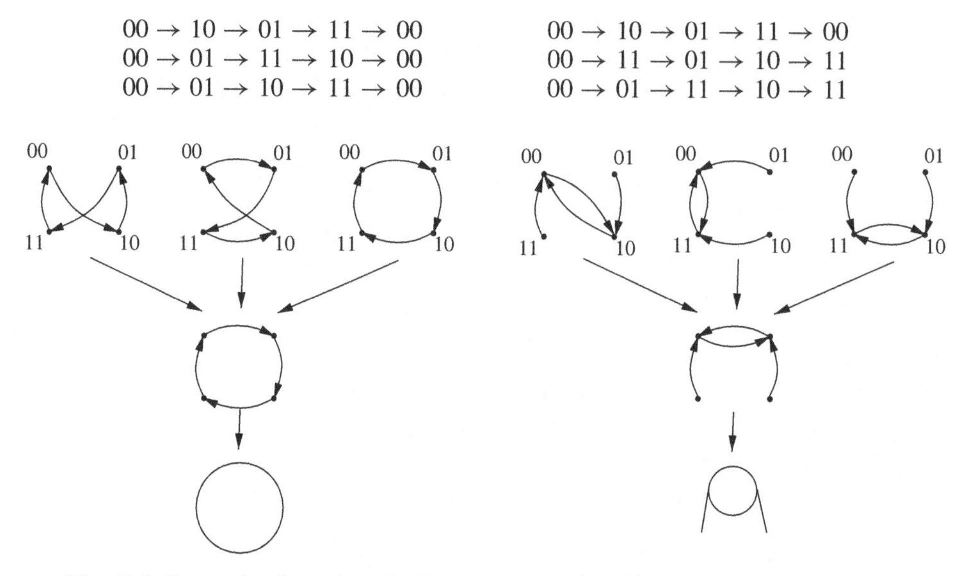

Fig. C.3. Dynamics for a 2-node (4-state) network. Different dynamics have the same graph structure, or type. The bottom-most symbols are a shorthand for the dynamic topology.

2-node network, for example, has 4 possible states: 00, 01, 10, 11. The behaviour of a boolean function is best appreciated by drawing its graph. When we leave the labels on the graph, we call the graph a dynamic, but when we drop the labels, we call the graph a *type of dynamic*. We show some possible dynamics on these states in Fig. C.3.

There is a one-to-one correspondence between boolean functions and dynamics, but the number of *types* of dynamics is much smaller, since multiple labelled graphs may have the same unlabelled graph. For 2 nodes, there are 4 network states and 256 dynamics but only 19 types of dynamics, namely:

where it is understood that all graphs contain 4-nodes and the circles ∘ ∘ ○ ◯ are cycles of length $1, 2, 3, 4$.

For $N = 3$, there are 104 networks (Fig. C.2 bottom) and $8^8 \simeq 1.7 \times 10^7$ boolean functions, which generate 951 types of dynamics. However, not all of the 104 networks can yield all possible types. While the fully connected network can exhibit the full range of behaviour, the less connected networks exhibit a surprisingly limited range of dynamics. For example, the distribution of types of dynamics given a random boolean function on the feedback (△) and feed-forward (△) networks are

Table C.3. *The number of networks, boolean functions and dynamics as a function of network size N. The last two columns apply to the fully connected network. How do these quantities diminish with the cutting of network bonds?*

Nodes	Networks	Boolean functions	Unlabelled dynamics
1	2	2^2	3
2	10	4^4	19
3	104	8^8	951
4	3044	16^{16}	3799624
5	291968	32^{32}	9.06×10^{13}
10	10^{23}	10^{3000}	?

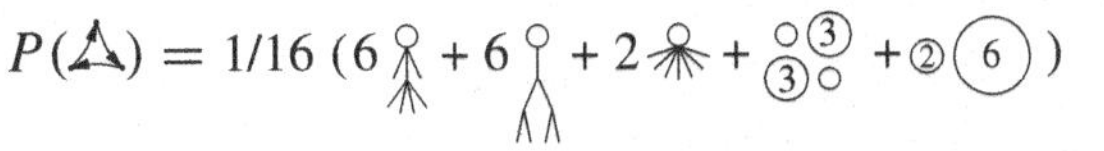

$$P(\triangle) = 1/16\,(6\,\cdot + 6\,\cdot + 2\,\cdot + \cdot_{(3)\,(3)} + (2)\,(6)) \quad \text{(C.1)}$$

$$P(\triangle) = 1/16\,(5\,\cdot + 2\,\cdot + 3\,\cdot + 2\,\cdot + 2\,\cdot + \cdot + \cdot). \quad \text{(C.2)}$$

The feedback network exhibits a broad range of cycle lengths (1, 2, 3, 6), whereas the feed-forward shows a variety of basins of attraction but only around 1-cycles. See also Table A3.3.

C.2.5 *Kauffman networks*

So far we have examined boolean networks with small numbers of nodes. What can be said about boolean dynamics on complex networks?

One well-studied model is called a Kauffman network. In this model there are exactly k randomly chosen inputs to every node, and the boolean function at each node is randomly and independently chosen. Kauffman networks exhibit three forms of behaviour, as parameterized by k:

$k < 2$: frozen dynamics (short cycles);
$k > 2$: chaotic dynamics (exponentially long cycles);
$k_c = 2$: marginal dynamics (power law cycles).

C.3 Community detection

Despite the rapid progress made in this field in recent years, there is still no strict consensus on how to define communities in a network. In principle one can use measures of clustering – that is how many vertices within a subset of the graph are connected to each other – or cluster together vertices that point to the same neighbours (even if they are not connected to each other) and therefore cluster vertices through their similarity. Finally when available one could use extra information related to the function of the vertices. The best choice for detecting the presence of communities is to determine them on a case by case basis. Here we will present some methods applying clustering measures to spot the community structure from the topology alone.

C.3.1 Edges deletion algorithms

Newman and Girvan [222, 436] introduced an approach for measuring community structure in unweighted, undirected graphs. The algorithm employs the previously defined *edge betweenness* in the following manner:

- compute the betweenness in the system,
- remove the edge with the largest betweenness,
- repeat from first step until no edges remain.

The idea is that in the presence of a clustered graph, most of the edges will be within the communities, with only a few edges to bridge them. In such a situation these edges will result in a large value of betweenness (Newman has also presented a generalization of this community detection algorithm to weighted networks [437]). Such a generalization can be fruitfully represented by a network of resistors such that the betweenness of a vertex or edge can be represented as the electric current flowing through it when a difference of potential is applied to two different vertices as shown in Fig. C.4. A similar approach computes betweenness as a random walk [692, 693]. In both cases the computation is more costly.

Actually, one of the drawbacks of this procedure is that the computation of the betweenness is particularly time consuming. Different methods have been

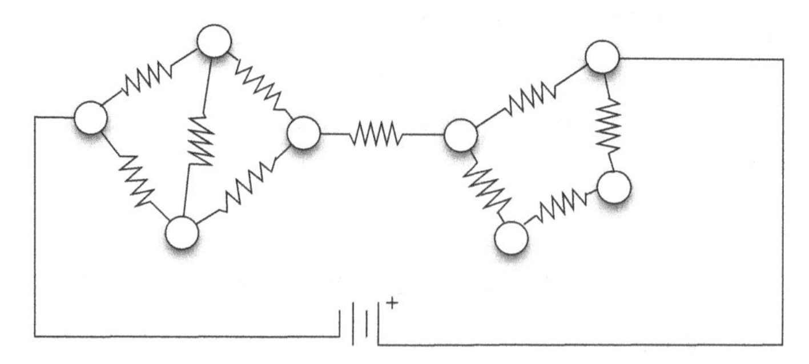

Fig. C.4. A network of resistors, betweenness is computed by means of current flowing in the various sites.

presented in order to compute this quantity approximately, for example by computing the electric current iteratively, to a given precision [664].

C.3.2 Spectral properties

Another powerful method for extracting the community structure from the topology of the graph is given by the so called 'spectral properties'. The starting point is given by the adjacency matrix of the graph combined with another matrix $\mathbf{K}$ whose entries are all zero apart from on the diagonal where the element k_{ii} is given by the degree k_i of the ith vertex. Once these two matrices are introduced we consider two of their possible combinations:

$$\mathbf{L} = \mathbf{K} - \mathbf{A}$$
$$\mathbf{N} = \mathbf{A}\mathbf{K}^{-1}. \tag{C.3}$$

The first matrix is called the *Laplacian* matrix and its elements l_{ij} are equal to -1 if there is an edge between i and j, and l_{ii} equal to the degree of vertex i on the diagonal. If we consider a vector ϕ containing values of a quantity defined on the vertices, the vector $\phi' = \mathbf{L}\phi$ is simply the Laplacian operator on this quantity. The second matrix is called the *normal* matrix and its elements are $n_{ij} = -1/k_i$ if there is an edge between i and j. The n_{ij} give the probability for a random walker to pass from one vertex to its neighbours. For both matrices the structure of the eigenvectors reflects the structure of the communities. In the case of the Laplacian matrix, a disconnected graph with m separate components will result in a block diagonal matrix. This means that there are m degenerate eigenvectors corresponding to the smallest eigenvalue, which is zero. If few edges are bridging the various components, a strong community structure still exists and the degeneration of the eigenvector is removed. Therefore, by looking at the eigenvector related to eigenvalues around zero, we can recover the block structure of the matrix, and thereby, the community structure of the network.

C.3.3 MCL clustering

The community structure of a graph can be investigated numerically by using a random walker algorithm such as the MCL algorithm [175, 611] which is particularly suited to networks with directed weighted edges and even loops (self-edges). The algorithm works as follows:

(i) start with the transition matrix T of the network and normalize each column of the matrix to obtain a stochastic matrix S;
(ii) compute S^2;

(iii) take the pth power ($p > 1$) of every element of S^2 and normalize each column to one;

(iv) go back to step (ii).

After several iterations MCL converges to a matrix $S_{MCL(p)}$ invariant under transformations (ii) and (iii). Only a few lines of $S_{MCL(p)}$ have some non-zero entries that give the clusters as separated basins (in general there is exactly one non-zero entry per column). Step (iii) reinforces the high-probability walks at short time scale at the expense of the low-probability walks. The parameter p tunes the granularity of the clustering. If p is large, the effect of step (iii) becomes stronger and the random walks are likely to end up in small 'basins of attraction' of the network, resulting in several small clusters. On the other hand, a small p produces larger clusters. In the limit of $p = 1$, only one cluster is detected.

Appendix D
Elementary mathematical concepts

A. GABRIELLI AND G. CALDARELLI

D.1 A primer on statistics and probability theory

This section provides an overview of some of the statistical tools and concepts which are useful for data analysis and the study of complex networks. Our emphasis will be on the practical application of probability theory, rather than its mathematical foundations, which is why we will confine ourselves to self-consistent definitions of the basic ingredients of applied statistics, rather than their derivation from first principles. For those who desire a more rigorous and more detailed treatment of the material, a celebrated introduction to probability theory can be found in [190], which discusses the contents of this chapter in much greater detail.

D.2 Events and probabilities

Tossing a coin – with an outcome of 'heads' or 'tails' – is one of the simplest examples of a probabilistic *event*. More complicated examples could be to obtain 'five' and 'two' when throwing a pair of dice, the ball landing on a red number in a game of roulette, or the spreading of an infection from an infected individual to a healthy one. In all these cases the set of all possible outcomes of an experiment is the *sample space*. An event can be defined as any member (or subset) of the sample space.

Technical part In set theory we can write that very simply: Ω is the sample space and any set $A \subset \Omega$ is an event in the following sense. If ω is the outcome of the experiment, A is verified if $\omega \in A$. By definition if A and B are two events, their union $C = A \cup B$ and their intersection $C = A \cap B$ also have to be events of the same experiment. In other words the set of events for a given experiment has to form a *closed system* under the operations of intersection and union. This property implies the existence of a null event, or empty set $\emptyset$ among the probabilistic events, because two sets can have an empty intersection.

Networks in Cell Biology, ed. M. Buchanan, G. Caldarelli, P. De Los Rios, F. Rao and M. Vendruscolo. Published by Cambridge University Press.

The probability $P(A)$ as a measure of the likelihood of a given event A is an intuitive concept, ubiquitous in everyday experience. Attempts to formulate this intuitive concept in a rigorous way, however, have given rise to differing interpretations and approaches. The most widely adopted interpretation identifies $P(A)$ as the frequency of the outcome A in an infinite number of independent repetitions of the same experiment.

Technical part Another definition of probability was introduced by Kolmogorov [331], who formulated the theory of probability in an axiomatic form for which the probability measure $P(A)$ is assigned *a priori* to every set A in the space of events. In order to verify these probabilities, one has to construct a suitable experiment from which they can be deduced. We define the probability (or statistical weight) $P(A)$ of a set A, as a function of A which satisfies the following axioms:

$P(A) \geq 0$ for all sets A;
$P(\Omega) = 1$;

if A_i $(i = 1, 2, 3, ...)$ is a countable (finite or infinite) sequence of non-overlapping sets, i.e. $A_i \cap A_J = \emptyset$ for all $i \neq j$, meaning that if A_i is verified A_j is not, then $[P\left(\bigcup_i A_i\right) = \sum_i P(A_i)]$, where $\bigcup_i A_i$ means the union of all the set A_i. The collection of sets $\{A_i\}$ is also called a collection of *mutually exclusive events*. The above axioms are all that is needed to construct the complete theory of probability. For instance one can show that the axioms imply:

$$P(\emptyset) = 0;$$

if $\overline{A}$ is the complementary set of A, i.e. the set of all events contained in Ω but not in A, we have $[P(\overline{A}) = 1 - P(A)]$.

A pictorial view of the meaning of probability is given in Fig. D.1. Once the area (probability) of the sample space Ω is fixed to one, the probability $P(A)$ is given by the area of the set/event A. From the same figure it is evident that two different events can overlap (this is the case of A and B), or not (A and C). This means that in the first case some results of the probabilistic experiment satisfy both conditions A and B, while in the second case no such result exists, and the events A and C are mutually exclusive.

D.3 Joint and conditional probabilities

The third axiom in the previous section rests upon the notion of mutually exclusive events, for which the intersection is empty. When the intersection $C = A \cap B$ between two sets A and B is non-empty, we can define the *joint probability*, given

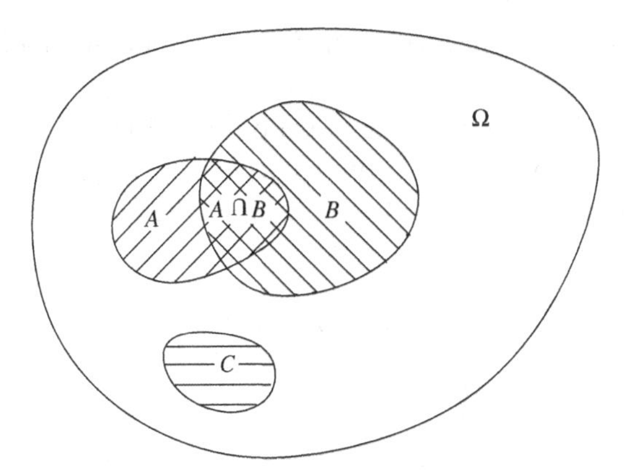

Fig. D.1. Pictorial representation of the sample space Ω and the events A, B and C. The sets A and B overlap and therefore can by verified by the same experimental outcome, while C is disjoint with respect to both of them.

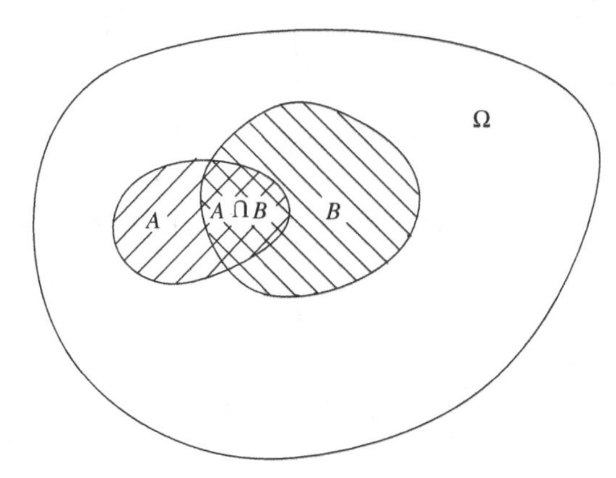

Fig. D.2. Pictorial representation of the sample space Ω and the intersection of the events A and B.

by $P(C) = P(A \cap B)$, that both events A and B are verified in the experiment, i.e. that the outcome $\omega \in A \cap B$. This is a fundamental concept underlying many of the probabilistic relations used in this book. Pictorially (see Fig. D.2) we can say that, for an area $P(\Omega)$ of Ω which is fixed to one, the probability $P(A \cap B)$ is given by the area of the intersection $A \cap B$.

Given the same two sets A and B, another fundamental concept is that of the *conditional probability* $P(A|B)$ of obtaining A, given a previous outcome B. The conditional probability $P(A|B)$ is given by:

$$P(A|B) = \frac{P(A \cap B)}{P(B)}. \tag{D.1}$$

The explanation of Eq. (D.1) is as follows: as B has already been verified, we need to limit the sample space to all possible outcomes of the experiment which satisfy B. Thus we have to normalize the joint distribution $P(A \cap B)$ by dividing it by $P(B)$.

Let us now consider some important relations involving these two probabilities. We start by taking a collection of sets $\{A_i\}$ with $i = 1, 2, \ldots, n$, including the case $n \to \infty$. Suppose that this collection is mutually exclusive and complete, i.e. satisfying

$$A_i \cap A_j = \emptyset \text{ for } i \neq j,$$
$$\bigcup_{i=1}^{n} A_i = \Omega.$$

Because of these two relations, it is clear that any other set B can be decomposed into the union of its intersections with the sets A_i, i.e. in the language of set theory:

$$\bigcup_{i=1}^{n}(B \cap A_i) = B \cap \left(\bigcup_{i=1}^{n} A_i\right) = B \cap \Omega = B\,.$$

Using the third axiom of the previous section we can now write:

$$\sum_{i=1}^{n} P(B \cap A_i) = P\left[\bigcup_{i=1}^{n}(B \cap A_i)\right] = P(B)\,.$$

Consequently, using also Eq. (D.1), we have

$$\sum_{i=1}^{n} P(B|A_i)P(A_i) = P(B).$$

Using this formula and the fact that

$$P(B \cap A_j) = P(B|A_j)P(A_j) = P(A_j|B)P(B)$$

we can now write the well-known *Bayes' theorem* which is commonly used in statistical inference:

$$P(A_j|B) = \frac{P(A_j)P(B|A_j)}{\sum_{i=1}^{n} P(A_i)P(B|A_i)}\,. \tag{D.2}$$

This formula gives the probability that the cause of an event B is a particular A_j among a mutually exclusive and complete collection of possible causes ($\{A_i\}$).

Finally, we introduce the concept of *independent events*. We can say that two events A and B are independent if the occurrence of B has no influence on the occurence of A and vice versa. Formally this can be expressed as:

$$\begin{cases} P(A|B) = P(A) \\ P(B|A) = P(B) \end{cases}\,. \tag{D.3}$$

This implies that

$$P(A \cap B) = P(A)P(B). \tag{D.4}$$

In other words the probability of both A and B occurring is given by the product of the simple probabilities of the two events. Similarly, n events $A_1, A_2, \ldots A_n$ are considered to be independent if for any subset $A_{i_1}, A_{i_2}, \ldots A_{i_k}$ with $k \leq n$ we have

$$P\left(A_{i_1} \cap A_{i_2} \ldots \cap A_{i_k}\right) = P\left(A_{i_1}\right) P\left(A_{i_2}\right) \ldots P\left(A_{i_k}\right). \tag{D.5}$$

Note that in general this is a much stronger condition than the binary mutual independence for which

$$P(A_i \cap A_j) = P(A_i)P(A_j), \tag{D.6}$$

for all pairs of events A_i, A_j with $i, j = 1, 2, \ldots, n$ and $i \neq j$. In fact there are cases in which Eq. (D.5) is satisfied but Eq. (D.6) is not.

D.4 Random variables and probability distributions

In many probabilistic experiments, the outcome takes the form of a number x[1] as for example the number of individuals affected by a certain disease. In general the numerical sample space Ω can either be discrete or continuous. In both cases we speak of a *random* or *stochastic* variable X which runs over all the possible values x in the sample space Ω.

For simplicity let us start with a single random variable X which can take only a discrete set of values x_i with $i = 1, 2, \ldots, N$ where N can be finite or infinite. All statistical properties of X can be derived from the set of probabilities P_i that the variable X takes the values x_i where $i = 1, 2, \ldots, N$. The probabilities have to satisfy the *normalization* property

$$\sum_{i=1}^{N} P_i = 1, \tag{D.7}$$

stating that one of all the possible outcomes will certainly happen.

In the case of continuous random variables, i.e. the height of a person, it is meaningless to talk about the probability that the variable X takes a particular, single value. This corresponds to saying that in the above example we want to know the probability that a person is 1.75 m tall with infinite precision. Actually, in this case the probability of any given specific value is vanishingly small. It makes more sense to define probabilities that the same measure is above or below a certain value (i.e. the probability that the height is below 2 m).

[1] More generally, of an n-dimensional vector $\mathbf{x} \equiv (x^{(1)}, x^{(2)}, \ldots, x^{(n)})$.

In all these cases the statistical properties of X are expressed by the *probability distribution* $P(x) = Prob(X < x)$ which gives the probability that the variable X takes a value smaller than x. Its characteristic properties are: (i) it is non-negative and non-decreasing at all x, (ii) $P(-\infty) = 0$, (iii) $P(+\infty) = 1$. If $P(x)$ is differentiable in x, the probability that X takes a value in the infinitesimal interval $[x, x + dx)$ can be written as

$$Prob(x \leq X < x + dx) = p(x)dx \text{ with } x \in \Omega \tag{D.8}$$

with

$$p(x) = \frac{dP(x)}{dx}, \tag{D.9}$$

which is called the *probability density function* (PDF) of the random variable X. It is simple to verify that $p(x)$ is non-negative at all x.

Technical part Note that if the variable X takes values only in some subset $\mathcal{I} \subset \mathcal{R}$, one can consider $p(x)$ as defined $\forall x \in \mathcal{R}$ (e.g. a closed interval $[a, b]$) and vanishing identically for $x \notin \mathcal{I}$. This implies that $P(x) = 0$ for all x smaller than the left extremal of $\mathcal{I}$ and $P(x) = 1$ for all x larger than its right extremal.

Note also that the discrete case can be included in the continuous description by using the Dirac delta function $\delta(x)$.[2]

Technical part If the variable X can take the values x_i with respective probabilities P_i with $i = 1, 2, \dots N$, one can write its PDF as

$$p(x) = \textstyle\sum_{i=1}^{N} P_i \delta(x - x_i).$$

By extending this equation we can write a *mixed* PDF with outcomes drawn from the continuous interval of $\mathcal{R}$ as well as the discrete set of x_i with $i = 1, 2, \dots N$. We obtain:

$$p(x) = p_0(x) + \textstyle\sum_{i=1}^{N} P_i \delta(x - x_i),$$

where $p_0(x)$ is a smooth non-negative function and P_i is the finite probability of having the value x_i, with the obvious normalization constraint $\int_{-\infty}^{+\infty} dx\, p(x) = 1$. Note that integration between $-\infty$ and x in one of the two previous equations gives a probability distribution $P(x)$ with a discontinuous step of height P_i at each x_i.

[2] The Dirac delta function $\delta(x)$ can be defined as the function (or more technically the *distribution*) given by $\delta(x) = 0$ for $x \neq 0$ and $\delta(0) = +\infty$ such that $\int_{x_0}^{x_1} dx \delta(x) = 1$ for any $x_0 < 0$ and $x_1 > 0$. A simple way to define it is to consider the family of normalized Gaussian functions $(\lambda/\pi)^{1/2} \exp(-\lambda x^2)$ in the limit of $\lambda \to +\infty$.

D.5 Mean values, standard deviations and fluctuations

Once we collect the results of our analysis, we are in most of the cases interested in the definition of a *mean value* as well as the indetermination attached to it.

The mean value $\langle \ldots \rangle$ of any function $F = f(\mathbf{X})$ is defined as

$$\langle F \rangle = \int_{-\infty}^{+\infty} dx_1 \ldots \int_{-\infty}^{+\infty} dx_n \, p_n(\mathbf{x}) f(\mathbf{x}) \,. \tag{D.10}$$

In the case of a discrete numerical sample space $\{\mathbf{x}_i\}$, with respective probabilities $\{P_i\}$, the above formula reduces to

$$\langle F \rangle = \sum_{i=1}^{N} P_i f(\mathbf{x}_i) \,. \tag{D.11}$$

This means that if we operate M independent trials of a probabilistic experiment with a random variable X which has outcomes $\mathbf{x}$ occurring with probabilities $p_n(\mathbf{x})$, then

$$\langle F \rangle = \lim_{M \to \infty} \frac{1}{M} \sum_{l=1}^{M} f[\mathbf{x}(l)] \,, \tag{D.12}$$

where $\mathbf{x}(l)$ is the value taken by $\mathbf{X}$ in the lth trial.

Technical part From Eqs. (D.10) and (D.11) it is also evident that the average of a linear superposition $af(\mathbf{X}) + bg(\mathbf{X})$ of two functions $F = f(\mathbf{X})$ and $G = g(\mathbf{X})$ of $\mathbf{X}$, with $a, b \in \mathcal{R}$, is simply the linear superposition of the averages of F and G with the same coefficients: $\langle a\, f(\mathbf{X}) + b\, g(\mathbf{X}) \rangle = a \langle f(\mathbf{X}) \rangle + b \langle g(\mathbf{X}) \rangle$.

In other words the 'average' operator is *linear*. Note, however, that in general the same does not hold for multiplication: $[\langle f(\mathbf{X}) g(\mathbf{X}) \rangle \neq \langle f(\mathbf{X}) \rangle \langle g(\mathbf{X}) \rangle]$. The difference between these two quantities leads to the concept of *statistical correlation*, to which we shall return later.

The mean value $\langle X \rangle$, as defined in Eqs. (D.10) and (D.11), is the most important *characteristic* value of a random variable X.

Technical part The mean value can be easily generalized to random vectors $\mathbf{X}$, such that $\langle \mathbf{X} \rangle = \left(\langle X^{(1)} \rangle, \langle X^{(2)} \rangle, \ldots \langle X^{(n)} \rangle \right)$. In general the mean value differs from the *most probable* value of the same variable, which is given by the value x_M at which the PDF $p(x)$ is maximal – or in the discrete case, by the value x_i which corresponds to the largest member of the set $\{P_i\}$. The value x_M indicates which value of X is typically observed in a *single* measurement, while the mean value gives its average over many measurements.

Another important characteristic quantity of a random variable X is the *squared average value* $\langle X^2 \rangle$ that gives a measure of the confidence we can have in taking the mean as a characteristic outcome of our experiment. Its main application is in the definition of the *variance* σ^2:

$$\sigma^2 = \langle (X - \langle X \rangle)^2 \rangle = \langle X^2 \rangle - \langle X \rangle^2. \quad \text{(D.13)}$$

The variance indicates how a PDF $p(x)$ is spread around the mean value $\langle X \rangle$. It thus measures the degree of uncertainty in a single measurement of the variable X, and the *dispersion* of the measured values of X in a large number of independent trials. The quantity σ, which is the square root of the variance, is usually called the *standard deviation* and gives the typical fluctuation between two independent measures of the variable X.

Mean and variance The *mean*, or *average*, or *expected value* for a certain event is given by summing up all the possible values of an event with their probability. For a fair die whose faces are equiprobable the mean value $\langle x \rangle$ in a roll is given by

$$\langle x \rangle = 1/6(1 + 2 + 3 + 4 + 5 + 6),$$

that is 3.5. Note that the mean does not need to be an event in itself (no roll will give value 3.5, but the sum of 1000 rolls will be a value very near to 3500). This example also explains in which sense the mean can be considered as the best guess for the outcome of an event. It is immediate to check that if we sum the distance from mean for any possible event this gives 0. On the other hand if we consider the distance independently from its sign (i.e. if we get 1 in the roll we are 2.5 far away from the mean, exactly as it happens when we get 6). Therefore a good recipe to have all these distance as positive numbers is to compute the *variance* σ^2 as the sum of the square values of the distance from the mean for all the possible events (weighted with their probability). In the above example of the roll of a die this would give

$$\sigma^2 = 1/6[(1-3.5)^2 + (2-3.5)^2 + (3-3.5)^2 + (4-3.5)^2 + (5-3.5)^2 + (6-3.5)^2],$$

that is 2.91, corresponding to a distance σ 1.708. If we take again 1000 different rolls (the central limit theorem applies) this would bring to a variance $\sigma_{1000} = 2910$ corresponding to an expected distance σ 53.944 from the mean of 3500.

D.6 Moments and correlations

All the above derivation can be recovered formally in the framework of the distribution moments. The quantity

$$\langle X^m \rangle = \int_{-\infty}^{+\infty} dx\, x^m p(x) \text{ for } m \in \mathcal{N} \quad \text{(D.14)}$$

is called the *integer moment* of order m of the random variable X. Using these moments one can in turn construct the *centred moments* $\langle (X - \langle X \rangle)^m \rangle$ of order m. As the order increases, the sequence of moments reveals more and more details about the shape of the PDF $p(x)$. The quantity $\langle (X - \langle X \rangle)^3 \rangle$, for instance, measures the degree of asymmetry of the PDF around $\langle X \rangle$. Furthermore, the larger m is, the more information the moment contains about the large $|x|$ tails of $p(x)$. It is possible to show [190] that, if the integer moments $\langle X^m \rangle$ for all $m \in \mathcal{N}$ are finite, knowledge of them is equivalent to knowing the PDF $p(x)$. This situation implies that $p(x)$ decreases at large x faster than any negative power of x (e.g. as $\exp(-ax^b)$ with $a, b > 0$ or $p(x)$ with limited support).

When $p(x)$ has an integrable power law tail [e.g. $p(x) \sim 1/(a^\alpha + |x|^\alpha)$ with $a, \alpha > 0$], moments of order $m \geq m_0$ diverge. If $p(x) \sim |x|^{-\alpha}$ with $\alpha > 1$ at large x, then $m_0 = [\alpha]$, with $[\alpha]$ the integer part of α, if α is non-integer and $m_0 = \alpha - 1$ if α is integer. These kinds of PDF are commonly encountered when studying the physics of scale-invariant phenomena (e.g. Lévy flights [367], second-order phase transition [267], non-equilibrium critical phenomena [257], self-organized critical systems [291]), and refer to systems that show large or diverging fluctuations in at least some of the observables.

We now discuss the statistical meaning of the mixed moments of different random variables and introduce the concept of *statistical correlation*. Let us consider a pair of random variables (X, Y) with a JPDF $p_2(x, y)$. Using this function it is possible to evaluate not only the moments of the single variables X and Y introduced above, but also the *mixed* moments of these two variables. Such mixed moments play a fundamental role in statistical analysis of stochastic phenomena. The simplest mixed moment is the *correlation coefficient*, defined as:

$$G_{lm} = \langle X^l Y^m \rangle \text{ with } (l, m) \in \mathcal{N}^{\in} . \tag{D.15}$$

If $p_2(x, y)$ factorizes into the product of two functions $p(x)$ and $q(y)$

$$p_2(x, y) = p(x) q(y) , \tag{D.16}$$

the two variables are statistically independent. This means that the knowledge of the value of one of the two variables does not influence the probability distribution of the other variable. In fact we can define the *conditional* PDF of X, given a value y of Y, is in general

$$p_c(x|y) = \frac{p_2(x, y)}{q(y)} ,$$

where, in this case,

$$q(y) = \int_{-\infty}^{+\infty} dx \, p_2(x, y) .$$

Consequently, if Eq. (D.16) is valid, we have $p_c(x|y) = p(x)$ where $p(x) = \int_{-\infty}^{+\infty} dy\, p_2(x, y)$, and thus

$$G_{lm} = \langle X^l \rangle \langle Y^m \rangle \; \forall (l, m) \in \mathcal{N}^{\in} .$$

In order to define the degree of statistical dependence (or correlation) of a pair of variables, it makes sense to define the *centred* correlation coefficients

$$C_{lm} = \langle X^l Y^m \rangle - \langle X^l \rangle \langle Y^m \rangle . \tag{D.17}$$

The larger the modulus of these coefficients, the more correlated are the two variables X and Y.

The most important of these coefficients (and usually the most accessible in an experiment) is

$$C_{11} = \langle XY \rangle - \langle X \rangle \langle Y \rangle = \langle (X - \langle X \rangle)(Y - \langle Y \rangle) \rangle ,$$

which is called the *covariance* of the two variables X and Y. If $C_{11} > 0$, then a positive fluctuation $x > \langle X \rangle$ ($x < \langle X \rangle$) indicates a higher probability that Y will also exhibit a positive fluctuation $y > \langle Y \rangle$ ($y < \langle Y \rangle$) in the same experiment. By contrast, if $C_{11} < 0$, then $x > \langle X \rangle$ means a higher probability that $y < \langle Y \rangle$.

Let us now generalize this to the case of n variables $(X^{(1)}, X^{(2)}, \dots X^{(n)})$. The general correlation coefficient is

$$G_{l_1 l_2 \dots l_n} = \langle [X^{(1)}]^{l_1} [X^{(2)}]^{l_2} \dots [X^{(n)}]^{l_n} \rangle .$$

It is possible to show that, if $G_{l_1 l_2 \dots l_n}$ is finite for all $(l_1, l_2, \dots l_n) \in \mathcal{N}^{\backslash}$, the knowledge of all these coefficients is equivalent to the knowledge of the JPDF $p_n(x^{(1)}, x^{(2)}, \dots x^{(n)})$. As in the case of $n = 2$, a particularly important role in statistical applications is played by the $n \times n$ symmetric pair *correlation matrix* $g(i, j) = \langle X^{(i)} X^{(j)} \rangle$ and/or the *covariance matrix*

$$c(i, j) = \left[g_{ij} - \langle X^{(i)} \rangle \langle X^{(j)} \rangle \right] . \tag{D.18}$$

The statistical interpretation of these two matrices is the same as in the two-variable case. Note that any correlation or covariance matrix has to be positive definite.

One of the most important applications of these statistical tools is a random variable $X^{(l)}$ representing an observable X at a point l in time or space. Under these circumstances the set of random variables $\{X^{(l)}\}$ with $-\infty < l < +\infty$ is called a *stochastic process*. Consequently the two matrices $g(i, j)$, $c(i, j)$ become the spatial or temporal correlation and covariance matrices of the observable X. A particularly useful case is the *stationary* stochastic process, which occurs when the statistical features of the stochastic process are translationally invariant, such that for $g_m(i_1, i_2, \dots i_m) = \langle X^{(i_1)} X^{(i_2)} \dots X^{(i_m)} \rangle$, we have

$g_m(i_1, i_2, \ldots i_m) = g_m(i_1 + i_0, i_2 + i_0, \ldots i_m + i_0)$ for any integer i_0 and set of indices $(i_1, i_2, \ldots i_m)$. As a result, any stationary stochastic process has a mean $m = \langle X^{(i)} \rangle$ and variance $\sigma^2 = \left[\langle [X^{(i)}]^2 \rangle - \langle X^{(i)} \rangle^2\right]$ which are independent of i. Furthermore the matrix elements $g(i, j) = g(i - j)$ and $c(i, j) = c(i - j)$ only depend on the distances $(i - j)$.

The statistical theory of a stochastic process based only on the first (mean values) and second moments (variances and correlation matrix) of the stochastic variables is usually called *correlation theory.*

D.7 Common distributions

In this section we will discuss some commonly encountered probability distributions of both discrete and continuous random variables. A detailed description of these distributions and others can be found in [190].

- *Binomial distribution* Consider a probabilistic experiment with two possible outcomes, for example 'yes' and 'no', with respective probabilities p and $q = 1 - p$. Let us now assume to perform N independent trials of this experiment. The random variable we want to study is the number of times we obtain 'yes' out of N trials. The probability of having exactly n 'yes' over N trials is given by the *binomial distribution*:

$$P_b(n; p, N) = \binom{N}{n} p^n q^{N-n} \text{ with } n = 0, 1, 2, \ldots N\,, \tag{D.19}$$

where $p^n q^{N-n}$ is the probability of obtaining a particular combination of n 'yes' and $N - n$ 'no', and $\binom{N}{n}$ gives the number of such combinations. The mean and variance of the binomial distribution are given by:

$$\begin{cases} \langle n \rangle = Np \\ \sigma_n^2 = \langle n^2 \rangle - \langle n \rangle^2 = Npq\,. \end{cases} \tag{D.20}$$

We can calculate $\langle n \rangle = \sum_{n=1^N} n P_b(n; p, N)$ pulling Np out of the sum, cancelling the n with the $n!$ in the denominator of $\binom{N}{n}$ and shifting the sum index n by one to $n' = n-1$. The sum is then visibly equal to 1, so that the mean is equal to Np. The derivation of the variance follows similar lines.

Note that the average number of successes grows proportionally to the number of trials N, but the size of a typical fluctuation from this value σ_n grows only as $\sqrt{N}$.

- *Poisson distribution* Let us now think of particles that occupy points on a segment $[0, L)$. The probability p_1 of a particle being located inside an infinitesimal segment $[x, x + dx)$ is λdx where λ is a non-zero, positive constant. Correspondingly, the probability p_0 of having no particle is $(1 - \lambda dx)$. We assume that the occupation states of any two disjoint segments are independent probabilistic events. It is then possible to show

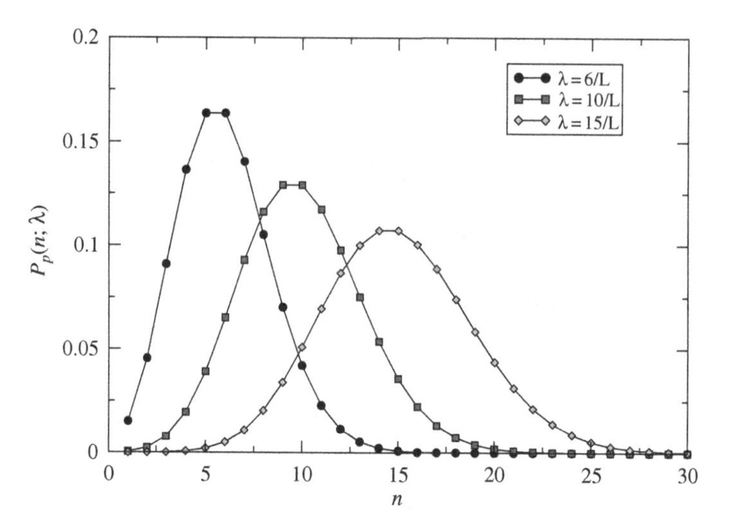

Fig. D.3. The shape of a Poisson distribution for various choices of the parameters. Note that the maximum is around the mean value, and also that as this maximum (the mean) increases by varying the parameters, so does the variance.

that the probability distribution of the total number n of particles in the segment is a Poisson distribution (see also Fig. D.3), given by:

$$P_p(n;\lambda) = \frac{(\lambda L)^n}{n!} e^{-\lambda L} \quad \text{with } n = 0, 1, 2, \ldots + \infty\,. \tag{D.21}$$

Note that the same distribution is obtained if we consider the statistics of random events occurring in a given time interval L with a time rate λ, and without any correlation between events. An example of such a process is the nuclear decay of a radioactive substance. The mean and variance of the distribution in Eq. (D.21) are:

$$\begin{cases} \langle n \rangle = \lambda L \\ \sigma_n^2 \equiv \langle n^2 \rangle - \langle n \rangle^2 = \lambda L\,. \end{cases} \tag{D.22}$$

As for the binomial distribution in Eq. (D.20), we see that $\langle n \rangle$ grows proportionally to the size of the system, now given by L, with average density λ. Again the fluctuations σ_n grow more slowly, as $\sqrt{L}$. In other words, as L and N become larger, their distributions become more sharply peaked.

The Poisson distribution can be derived from Eq. (D.19) in the limit $N \to \infty$ and $p \to 0$ with fixed $Np = \lambda L$.

Note that Eq. (D.21) is obtained also if multiple occupation of an infinitesimal segment is permitted but with a probability of order higher than dx, so that $p_1 = \lambda dx$ and $p_{>1} = O(dx^2)$, and $p_0 = 1 - p_1 - p_{>1}$, where $p_{>1}$ is the probability of having more than one particle in the segment $[x, x + dx)$ and $O(dx^2)$ indicates an infinitesimal of higher order than dx.

- *Gaussian distribution* A continuous random variable X with values $x \in (-\infty, +\infty)$ is Gaussian distributed if its PDF is

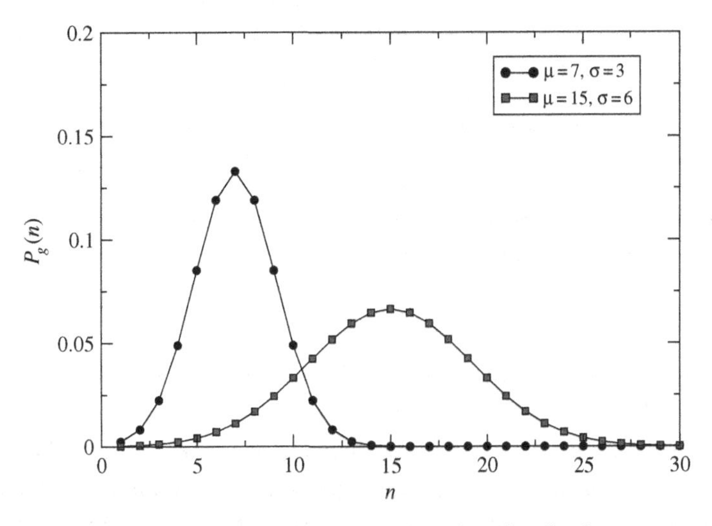

Fig. D.4. The bell-shape of a Gaussian distribution.

$$p_g(x) = \frac{1}{\sqrt{2\pi\sigma^2}} \exp\left[-\frac{(x-m)^2}{2\sigma^2}\right]. \tag{D.23}$$

It is simple to verify that

$$\begin{cases} \langle X \rangle = m \\ \langle X^2 \rangle - \langle X \rangle^2 = \sigma^2 . \end{cases} \tag{D.24}$$

In the case in which $m = 0$ and $\sigma = 1$, Eq. (D.23) is called the *standard* Gaussian PDF.

The Gaussian distribution (see Fig. D.4) is the most well-known probability distribution of continuous variables, for two main reasons: (1) its role in the *central limit theorem* (see below), and (2) its property that, for a random variable X of which only the mean and variance are known, the Gaussian distribution is the probability distribution which maximizes the *Shannon entropy* and therefore has to be considered the most likely distribution for X.

It is possible to show that for large N and arbitrary $p \in (0, 1)$, the binomial distribution Eq. (D.19) converges to the Gaussian PDF Eq. (D.23) in the following sense:

(i) let us define the variable $X = \frac{n-Np}{\sqrt{Npq}}$;

(ii) in the limit $N \to \infty$ the variable X can be seen as a continuous variable with values in $(-\infty, +\infty)$ and with PDF given by Eq. (D.23), where $m = 0$ and $\sigma = 1$.

- *Multivariate Gaussian distribution* A set of n arbitrarily correlated continuous random variables $(X_1, X_2, \ldots X_n)$, each with values in $(-\infty, +\infty)$, is said to be Gaussian distributed if its JPDF is

$$p_g(x_1, x_2, \ldots x_n) = \sqrt{\frac{\|\hat{A}\|}{(2\pi)^n}} \exp\left[-\frac{1}{2}\sum_{i,j=1}^{N}(x_i - m_i)A_{ij}(x_j - m_j)\right], \tag{D.25}$$

where $\hat{A} = ((A_{ij}))$ is a positive definite matrix[3] and $\|\hat{A}\| > 0$ its determinant. One can show [98] that

$$\begin{cases} \langle X_i \rangle = m_i & \text{for } i = 1, 2, \dots n \\ \langle (X_i - m_i)(X_j - m_j) \rangle = [\hat{A}^{-1}]_{ij} & \text{for } i, j = 1, 2, \dots n \,. \end{cases} \tag{D.26}$$

Furthermore it is important to note that the covariance matrix $\hat{c} = ((c_{ij}))$, defined in Eq. (D.18), is the inverse of the matrix $\hat{A}$:

$$\hat{c} = [\hat{A}]^{-1} \,.$$

An important feature of both Eqs. (D.21) and (D.23) is that all the moments, respectively $\langle n^m \rangle$ and $\langle X^m \rangle$, are finite for all $m \geq 0$. This is because both of these probability distributions decrease faster than any negative power for large values of their respective arguments. As discussed earlier, negative power law distributions occur in systems with large fluctuations of observables, and correspondingly have slowly decreasing tails. An example would be a random variable X with a PDF $p(x)$ decaying as $x^{-5/2}$ at large x. The mean value $\langle X \rangle$ is finite, but the variance diverges, which means that in a single measurement of X we are likely to find a value which is very far away from the mean. This would be very unlikely in all the distributions with finite moments.

D.8 Law of large numbers

Let us now consider N random variables $X^{(i)}$ with $i = 1, 2, \dots N$, which all share the same mean value $m = \langle X^{(i)} \rangle$. We define a new random variable Y_N as

$$Y_N = \frac{1}{N} \sum_{i=1}^{N} X^{(i)} \,,$$

which is the arithmetic average of the variables $\{X^{(i)}\}$. The *law of large numbers* states that if the covariance $c(i, j)$ defined in Eq. (D.18) vanishes sufficiently rapidly for $|i - j| \to \infty$ then

$$\lim_{N \to \infty} Y_N = \frac{1}{N} \sum_{i=1}^{N} \langle X^{(i)} \rangle = m \,. \tag{D.27}$$

This means that in the large N limit the quantity Y_N is no longer random but instead is exactly equal to m. In other words the PDF of the variable Y_N becomes the Dirac delta function $\delta(y - m)$.

[3] Note that if the matrix $\hat{A}$ is not positive definite, the PDF is not normalizable.

In order to derive this result, we write

$$\langle Y_N \rangle = \frac{1}{N} \sum_{i=1}^{N} \langle X^{(i)} \rangle = m\,. \tag{D.28}$$

In the second step we evaluate its variance

$$\sigma^2(Y_N) \equiv \langle Y_N^2 \rangle - \langle Y_N \rangle^2 = \frac{1}{N^2} \sum_{i,j=1}^{N} c(i,j)\,. \tag{D.29}$$

Therefore if $\sum_{i,j=1}^{N} c(i,j)$ grows slower than N^2 for $N \to \infty$, then $\sigma^2(Y_N)$ vanishes in this limit, and Y_N has no more fluctuations around its mean value m. This gives rise to an infinitely narrow distribution – in other words, the Dirac delta distribution. If $c(i,j)$ vanishes for $|i-j| \to \infty$ *at least* as rapidly as a negative power law $|i-j|^{-a}$ with $a > 0$, then the condition of the law of large numbers is fulfilled. This means that the decay of $c(i,j)$ can be very slow, and the law of large numbers can therefore be invalid even for very large $|i-j|$ even if it holds in the limit.

D.9 Central limit theorem

In this section we state without proof one of the most important theorems of probability theory, which has many important consequences in the applications of statistics. The central limit theorem describes the collective effect of fluctuations in a stochastic system. We will first formulate the central limit theorem in its more basic form which nevertheless is widely applicable:

Let $X^{(i)}$ with $i = 1, 2, \ldots N$ be a set of *independent* random variables with zero mean $\langle X^{(i)} \rangle = 0$, and a fixed variance $\langle [X^{(i)}]^2 \rangle - \langle X^{(i)} \rangle^2 = \sigma^2$ which is independent of i. Now define a variable S_N as

$$S_N = \frac{1}{\sqrt{N}} \sum_{i=1}^{N} X^{(i)}\,. \tag{D.30}$$

In the limit of $N \to \infty$, the PDF of the variable S_N becomes a Gaussian $p(s)$ with zero mean $m = 0$ and variance σ^2:

$$p(s) = \frac{1}{\sqrt{2\pi\sigma^2}} \exp\left(-\frac{s^2}{2\sigma^2}\right). \tag{D.31}$$

If $\langle X^{(i)} \rangle = m^{(i)} \neq 0$, the variable S_N satisfying the central limit theorem has to be redefined as

$$S_N = \frac{1}{\sqrt{N}} \sum_{i=1}^{N} \left(X^{(i)} - m^{(i)}\right). \tag{D.32}$$

Note that differently from the law of large numbers, where the sum in Y_N is divided by N, the sum in S_N is divided by $\sqrt{N}$.

A more general version of the central limit theorem, in which the variances $\sigma^{(i)}$ can depend on i, is given by defining a variable U_N such that:

$$U_N = \frac{1}{z_N} \sum_{i=1}^{N} \left(X^{(i)} - m^{(i)} \right) ,$$

where

$$z_N^2 = \sum_{i=1}^{N} [\sigma^{(i)}]^2 .$$

One can show that, if the *Lindeberg condition*

$$\lim_{N \to \infty} \left[\frac{1}{z_N^2} \sum_{i=1}^{N} \int_{|x| > t z_N} dx\, x^2 p^{(i)}(x) \right]$$

is satisfied for any fixed $t > 0$, then in the limit $N \to \infty$ the PDF of U_N is

$$p(u) = \frac{1}{\sqrt{2\pi}} \exp\left(-\frac{u^2}{2} \right) .$$

In other words, the central limit theorem states that for an observable $\mathcal{O}$ which is subject to random fluctuations or uncertainty, and which can be expressed as a sum of a large number N of other, independent variables, then the fluctuations of $\mathcal{O}$ are expected to be Gaussian. Furthermore, if the size of all fluctuations of the contributions is bounded, then the typical fluctuation (or standard deviation) of $\mathcal{O}$ is expected to be of order $\sqrt{N}$.

Finally, if the variables $\{X^{(i)}\}$ are not statistically independent, but the covariance $c(i, j)$ defined in Eq. (D.18) is sufficiently short ranged (rapidly decreasing) in $|i - j|$, we can again expect U_N to become Gaussian for large N. Note however that the conditions on $c(i, j)$ for the validity of the central limit theorem are more restrictive than those for the law of large numbers [98]. In fact it is difficult to construct a stationary stochastic process that does not satisfy the law of large numbers, but it is easy to find one that violates the central limit theorem.

References

[1] Bigg database, http://bigg.ucsd.edu/
[2] Biocyc home, http://biocyc.org/
[3] Kegg: Kyoto encyclopedia of genes and genomes, http://www.genome.jp/kegg/
[4] Omim home.
[5] Makita Y., Nakao M., Ogasawara N., Nakai K. DBTBS: database of transcriptional regulation in bacillus subtilis and its contribution to comparative genomics. *Nucleic Acids Research* **32**, D75–77 (2004).
[6] Grainger D. C., Hurd D., Harrisou M., Holdstock J., Busby S. J. Studies of the distribution of Escherichia coli CAMP-receptor protein and RNA polymerase along the E. coli chromosome. *Proceedings of the National Academy of Sciences USA* **102**, 17693–17698 (2005).
[7] Abdi H. (2003). Partial least squares regression (pls-regression). In *Encyclopedia for Research Methods for the Social Sciences*. Sage, Thousand Oaks (CA).
[8] Adams M. *et al* (2000). The genome sequence of *Drosophila melanogaster*. *Science* **287**, 5461, 2185–2195.
[9] Aebersold R., Mann M. (2003). Mass spectrometry-based proteomics. *Nature* **422**, 6928, 198–207.
[10] Alarid E. (2006). Lives and times of nuclear receptors. *Molecular Endocrinology* **20**, 1972–1981.
[11] Alber T. (1992). Structure of the leucine zipper. *Currents Opinions in Genetics and Development* **2**, 205–210.
[12] Albert R., Barabási A.-L. (2002). Statistical mechanics of complex networks. *Reviews of Modern Physics* **74**, 47–97.
[13] Albert R., Jeong H., Barabási A.-L. (1999). Diameter of the World Wide Web. *Nature* **401**, 130–131.
[14] Albert R., Jeong H., Barabási A.-L. (2000). Error and attack tolerance of complex networks. *Nature* **406**, 378–382.
[15] Albert R., Jeong H., Barabási A.-L. (2001). Errata: Error and attack tolerance of complex networks. *Nature* **409**, 542.
[16] Alberts B., Johnson A., Lewis J., Raff M., Roberts K., Walter P. (2002). *The Molecular Biology of the Cell*. Garland Publishing, New York.
[17] Aldridge B., Burke J., Lauffenburger D., Sorger P. (2006). Physicochemical modelling of cell signalling pathways. *Nature Cell Biology* **8**, 1195–1203.
[18] Allison S., Milner J. (2004). Remodelling chromatin on a global scale: a novel protective function of p53. *Carcinogenesis* **25**, 1551–1557.

[19] Alon U. (2006). *An Introduction to Systems Biology: Design Principles of Biological Circuits*, 1st edn. Chapman & Hall/CRC.
[20] Alon U. (2007). Network motifs: theory and experimental approaches. *Nature Reviews Genetics* **8**, 450–461.
[21] Aloy P., Boettcher B., Ceulemans H. *et al.* (2004). Structure-based assembly of protein complexes in yeast. *Science* **303**, 2026–2029.
[22] Aloy P., Ceulemans H., Stark A., Russell R. B. (2003). The relationship between sequence and interaction divergence in proteins. *Journal of Molecular Biology* **332**, 989–998.
[23] Aloy P., Russell R. (2002a). The third dimension for protein interactions and complexes. *Trends in Biochemical Sciences* **27**, 633–638.
[24] Aloy P., Russell R. (2004). Ten thousand interactions for the molecular biologist. *Nature Biotechnology* **22**, 1317–1321.
[25] Aloy P., Russell R. (2006). Structural systems biology: modelling protein interactions. *Nature Reviews Molecular Cell Biology* **7**, 188–197.
[26] Aloy P., Russell R. B. (2002b). Potential artefacts in protein-interaction networks. *FEBS Letters* **530**, 253–254.
[27] Altschul S., Madden T., Schaffer A. *et al.* (1997). Gapped blast and psi-blast: a new generation of protein database search programs. *Nucleic Acids Research* **25**, 3389–3402.
[28] Ambesi-Impiombato A., Bansal M., Lió P., di Bernardo D. (2006). Computational framework for the prediction of transcription factor binding sites by multiple data integration. *BMC Neuroscience* **7 Suppl 1**, S1–S8.
[29] Ambros V. (2004). The functions of animal microRNAs. *Nature* **431**, 350–355.
[30] Anand G., Law D., Mandell J. *et al.* (2003). Identification of the protein kinase a regulatory rialpha-catalytic subunit interface by amide h/2h exchange and protein docking. *Proceedings of the National Academy of Sciences of the United States of America* **100**, 13264–13269.
[31] Andreadis A. (2005). Tau gene alternative splicing: expression patterns, regulation and modulation of function in normal brain and neurodegenerative diseases. *Biochimica Biophysica Acta* **1739**, 91–103.
[32] Apic G., Gough J., Teichmann S. (2001). Domain combinations in archaeal, eubacterial and eukaryotic proteomes. *Journal of Molecular Biology* **310**, 311–325.
[33] Aravind L., Anantharaman V., Balaji S., Babu M. M., Iyer L. (2005). The many faces of the helix-turn-helix domain: transcription regulation and beyond. *FEMS Microbiology Review* **29**, 231–262.
[34] Aravind L., Koonin E. (1999). DNA-binding proteins and evolution of transcription regulation in the archaea. *Nucleic Acids Research* **27**, 4658–4670.
[35] Arifuzzaman M., Maeda M., Itoh A. *et al.* (2006). Large-scale identification of protein–protein interaction of *Escherichia coli* k-12. *Genome Research* **16**, 5, 686–691.
[36] Arnone M., Davidson E. (1997). The hardwiring of development: organization and function of genomic regulatory systems. *Development* **124**, 1851–1864.
[37] Arnone M., Martin E., Davidson E. (1998). Cis-regulation downstream of cell type specification: a single compact element controls the complex expression of the cyiia gene in sea urchin embryos. *Development* **125**, 1381–1395.
[38] Aronheim A. (1997). Improved efficiency sos recruitment system: expression of the mammalian gap reduces isolation of ras gtpase false positives. *Nucleic Acids Research* **25**, 16, 3373–3374.

[39] Audic Y., Hartley R. (2004). Post-transcriptional regulation in cancer. *Biology of the Cell* **96**, 479–498.
[40] Babich V., Aksenov N., Alexeenko V., Oei S., Buchlow G., Tomilin N. (1999). Association of some potential hormone response elements in human genes with the alu family repeats. *Gene* **239**, 341–349.
[41] Babu M. M., Teichmann S. (2003). Evolution of transcription factors and the gene regulatory network in *Escherichia coli. Nucleic Acids Research* **31**, 1234–1244.
[42] Babu M. M., Luscombe N. M., Aravind L., Gerstein M., Teichmann S. A. Structure and evolution of transcriptional regulatory networks. *Curr. Opin. Struct. Biol.* **14**, 283–291 (2004).
[43] Babu M. M., Teichmann S., Aravind L. (2006). Evolutionary dynamics of prokaryotic transcriptional regulatory networks. *Journal of Molecular Biology* **358**, 614–633.
[44] Bader G., Heilbut A., Andrews B., Tyers M., Hughes T., Boone C. (2003). Functional genomics and proteomics: charting a multidimensional map of the yeast cell. *Trends in Cell Biology* **13**, 7, 344–356.
[45] Bader G. D., Donaldson I., Wolting C., Ouellette B. F., Pawson T., Hogue C. W. (2001). Bind – the biomolecular interaction network database. *Nucleic Acids Research* **29**, 1 (January), 242–245.
[46] Bailey T., Elkan C. (1994). Fitting a mixture model by expectation maximization to discover motifs in biopolymers. *Proceedings International Conference Intelligent Systems Molecular Biology* **2**, 28–36.
[47] Balaji S., Iyer L., Aravind L., Babu M. M. (2006). Uncovering a hidden distributed architecture behind scale-free transcriptional regulatory networks. *Journal of Molecular Biology* **360**, 204–212.
[48] Balazsi G., Barabasi A., Oltvai Z. (2005). Topological units of environmental signal processing in the transcriptional regulatory network of *Escherichia coli. Proceedings of the National Academy of Sciences USA* **102**, 7841–7846.
[49] Ban N., Nissen P., Hansen J., Moore P. B., Steitz T. A. (2000). The complete atomic structure of the large ribosomal subunit at 2.4 Å resolution. *Science* **289**, 5481, 905–920.
[50] Bao Q., Shi Y. (2007). Apoptosome: a platform for the activation of initiator caspases. *Cell Death Differentiation* **14**, 56–65.
[51] Bar-Joseph Z., Gerber G., Lee T. *et al.* (2003). Computational discovery of gene modules and regulatory networks. *Nature Biotechnology* **21**, 1337–1342.
[52] Barabási A.-L. (2002). *Linked: the New Science of Networks.* Perseus, New York.
[53] Barabási A.-L., Albert R., Jeong H. (1999). Mean-field theory for scale-free random networks. *Physica A* **272**, 173–187.
[54] Barabási A.-L., Jeong H., Néda Z., Ravasz E., Schubert A., Vicsek T. (2002). Evolution of the social network of scientific collaborations. *Physica A* **311**, 590.
[55] Barábasi A.-L., Oltvai Z. (2004). Network biology: understanding the cell's functional organization. *Nature Genetics* **5**, 101–114.
[56] Barrat A., Barthélemy M., Pastor-Satorras R., Vespignani A. (2004). The architecture of complex weighted networks. *Proceedings of the National Academy of Sciences of the United States of America* **101**, 3747–3752.
[57] Barrat A., Weigt M. (2000). On the properties of small-world networks. *The European Physical Journal B* **13**, 547–560.
[58] Bassler B. (2002). Small talk. Cell-to-cell communication in bacteria. *Cell* **109**, 421–424.

[59] Basu S., Mehreja R., Thiberge S., Chen M., Weiss R. (2004). Spatiotemporal control of gene expression with pulse-generating networks. *Proceedings of the National Academy of Sciences of the United States of America* **101**, 6355–6360.

[60] Baumbach J., Wittkop T., Rademacher K., Rahmann S., Brinkrolf K., Tauch A. (2007). Coryneregnet 3.0 – an interactive systems biology platform for the analysis of gene regulatory networks in corynebacteria and escherichia coli. *Journal of Biotechnology* **129**, 279–289.

[61] Becskei A., Seraphin B., Serrano L. (2001). Positive feedback in eukaryotic gene networks: cell differentiation by graded to binary response conversion. *Embo J.* **20**, 2528–2535.

[62] Becskei A., Serrano L. (2000). Engineering stability in gene networks by autoregulation. *Nature* **405**, 590–593.

[63] Ben-Naim E., Toroczkai Z. (2005). *Complex Networks (Lecture Notes in Physics)*. Secaucus, NJ, USA: Springer-Verlag New York, Inc.

[64] Benecke A., Gaudon C., Gronemeyer H. (2001). Transcriptional integration of hormone and metabolic signals by nuclear receptors. In *Transcription Factors*, J. Locker, ed. Academic Press, San Diego, CA, 167–214.

[65] Berg J., Tymoczko J., Stryer L. (2006). *Biochemistry*, 6th edn. W. H. Freeman.

[66] Bevilacqua A., Ceriani M., Capaccioli S., Nicolin A. (2003). Post-transcriptional regulation of gene expression by degradation of messenger RNAs. *Journal of Cell Physiology* **195**, 356–372.

[67] Beyer A., Bandyopadhyay S., Ideker T. (2007). Integrating physical and genetic maps: from genomes to interaction networks. *Nature Review Genetics* **8**, 9, 699–710.

[68] Bhalla U., Iyengar R. (1999). Emergent properties of networks of biological signaling pathways. *Science* **283**, 381–387.

[69] Bharathan G., Janssen B.-J., Kellogg E., Sinha N. (1997). Did homeodomain proteins duplicate before the origin of angiosperms, fungi, and metazoa? *Proceedings of the National Academy of Sciences of the United States of America* **94**, 13749–13753.

[70] Biggins J., Koh J. (2007). Chemical biology of steroid and nuclear hormone receptors. *Current Opinion in Chemical Biology* **11**, 99–110.

[71] Blume-Jensen P., Hunter T. (2001). Oncogenic kinase signalling. *Nature* **411**, 355–365.

[72] Boccaletti S., Latora V., Moreno Y., Chavez M., Hwang D. (2006). Complex networks: Structure and dynamics. *Physics Reports* **424**, 175–308.

[73] Boguñá M., Pastor-Satorras R. (2003). Class of correlated random networks with hidden variables. *Physical Review E* **68**, 036112.

[74] Bollobás B. (1985). *Random Graphs*. Academic Press, London.

[75] Bonifer C., Lefevre P., Tagoh H. (2006). The regulation of chromatin and DNA-methylation patterns in blood cell development. *Current Topics in Microbiological and Immunology* **310**, 1–12.

[76] Bornholdt S., Schuster H. (2002). *Handbook of Graphs and Networks: From the Genome to the Internet*. Wiley VCH, Berlin.

[77] Bouwmeester T., Bauch A., Ruffner H. *et al.* (2004). A physical and functional map of the human tnf-alpha/nf-kappa b signal transduction pathway. *Nature Cell Biology* **6**, 2, 97–105.

[78] Brant-Zawadzki P., Schmid D., Jiang H., Weyrich A., Zimmerman G., Kraiss L. (2007). Translational control in endothelial cells. *Journal of Vascular Surgery* **45 Suppl A**, A8–A14.

[79] Bredesen D., Rao R., Mehelen P. (2006). Cell death in the nervous system. *Nature* **443**, 796–802.
[80] Breitkreutz D., Braiman-Wiksman L., Daum N., Denning M., Tennenbaum T. (2007). Protein kinase C family: on the crossroads of cell signalling in skin and tumor epithelium. *Journal of Cancer Research Clinical Oncology* **133**, 793–808.
[81] Brennan R. (1993). The winged-helix DNA-binding motif: another helix-turn-helix takeoff. *Cell* **74**, 773–776.
[82] Brennan R., Matthews B. (1989). The helix-turn-helix DNA binding motif. *Journal of Biological Chemistry* **264**, 1903–1906.
[83] Brilli M., Fani R., Lió P. (2007). Motifscorer: using a compendium of microarrays to identify regulatory motifs. *Bioinformatics* **23**, 493–495.
[84] Broder A., Kumar R., Maghoul F. *et al.* (2000). Graph structure in the web. *Computer Network* **33**, 309–320.
[85] Broida A., Claffy K. (2001). Internet topology: Connectivity of IP graphs. In *Scalability and Traffic Control in IP Networks,* in *Proceedings SPIE*, S. Fahmy and K. Park, eds. vol. 4526. International Society for Optical Engineering, Bellingham, WA, pp. 172–187.
[86] Brown J., Stoyanov J., Kidd S., Hobman J. (2003). The merr family of transcriptional regulators. *FEMS Microbiology Review* **27**, 145–163.
[87] Browning D., Busby S. (2004). The regulation of bacterial transcription initiation. *Nature Review Microbiology* **2**, 57–65.
[88] Buchanan M. (2002a). *Nexus: Small Worlds and the Ground-breaking Science of Networks*. Norton, New York.
[89] Buchanan M. (2002b). *Small World: Uncovering Nature's Hidden Network.* Weidenfeld and Nicolson, London.
[90] Buck M., Lieb J. (2004). Chip-chip: considerations for the design, analysis, and application of genome-wide chromatin immunoprecipitation experiments. *Genomics* **83**, 349–360.
[91] Bulyk M., McGuire A., Masuda N., Church G. (2004). A motif co-occurrence approach for genome-wide prediction of transcription-factor-binding sites in escherichia coli. *Genome Research* **14**, 201–208.
[92] Burdge G., Hanson M., Slater-Jefferies J., Lillycrop L. (2007). Epigenetic regulation of transcription: a mechanism for inducing variations in phenotype (fetal programming) by differences in nutrition during early life? *British Journal of Nutrition* **97**, 1036–1046.
[93] Buryanov Y., Shevchuk T. (2005). DNA methyltransferases and structural-functional specificity of eukaryotic dna modification. *Biochemistry (Mosc)* **70**, 730–742.
[94] Bussemaker H., Li H., Siggia E. (2001). Regulatory element detection using correlation with expression. *Nature Genetics* **27**, 167–171.
[95] Butland G., Peregrín-Alvarez J., Li J. *et al.* (2005). Interaction network containing conversed and essential protein complexes in *Escherichia coli. Nature* **433**, 531–537.
[96] Buttinelli M., Panetta G., Rhodes D., Travers A. (1999). The role of histone h1 in chromatin condensation and transcriptional repression. *Genetica* **106**, 117–124.
[97] Cagney G., Uetz P., Fields S. (2000). High-throughput screening for protein–protein interactions using two-hybrid assay. *Methods in Enzymology* **328**, 3–14.
[98] Caldarelli G. (2007). *Scale-Free Networks*. Oxford University Press, Oxford.

[99] Caldarelli G., Capocci A., De Los Rios P., Muñoz M. (2002). Scale free networks from varying vertex intrinsic fitness. *Physical Review Letters* **89**, 258702.

[100] Camas F., Blazquez J., Poyatos J. (2006). Autogenous and nonautogenous control of response in a genetic network. *Proceedings of the National Academy of Sciences USA* **103**, 12718–12723.

[101] Cani P., Delzenne N. (2007). Gut microflora as a target for energy and metabolic homeostasis. *Current Opinion in Clinical Nutrition and Metabolic Care* **10**, 729–734.

[102] Carafoli E. (2003). The calcium-signalling saga: tap water and protein crystals. *Nature Reviews Molecular Cell Biology* **4**, 326–332.

[103] Carey M., Smale S. (2000). *Transcriptional Regulation in Eukaryotes: Concepts, Strategies, and Techniques*. Cold Spring Harbor Laboratory Press, Cold Spring Harbor, New York.

[104] Cawley S., Bekiranov S., Ng H. *et al.* (2004). Unbiased mapping of transcription factor binding sites along human chromosomes 21 and 22 points to widespread regulation of noncoding RNAs. *Cell* **116**, 499–509.

[105] Cech T., Atkins J. (2005). *The RNA World (Cold Spring Harbor Monograph Series)*, 3rd edn. Cold Spring Harbor Laboratory Press.

[106] Chan S.-L., Yu V. (2004). Proteins of the Bcl-2 family in apoptosis signalling: from mechanistic insights to therapeutic opportunities. *Clinical Experimental Pharmacology and Physiology* **31**, 119–128.

[107] Chang L., Karin M. (2001). Mammalian map kinase signalling cascades. *Nature* **410**, 37–40.

[108] Chang M., Jaehning J. (1997). A multiplicity of mediators: alternative forms of transcription complexes communicate with transcriptional regulators. *Nucleic Acids Research* **25**, 4861–4865.

[109] Cheadle C., Fan J., Cho-Chung Y.-S. *et al.* (2005). Stability regulation of mrna and the control of gene expression. *Annals of the New York Academy of Sciences* **1058**, 196–204.

[110] Chen Q., Chang H., Govindan R., Jamin S., Shenker S., Willinger W. (2002). The origin of power laws in internet topologies revisited. In *Proceedings of the 1st Annual Joint Conference of the IEEE Computer and Communications Societies*. IEEE Computer Society.

[111] Chen Z., Bhoj V., Seth R. (2006). Ubiquitin, TAK1 and IKK: is there a connection? *Cell Death Differentiation* **13**, 687–692.

[112] Chua G., Morris Q., Sopko R. *et al.* (2006). Identifying transcription factor functions and targets by phenotypic activation. *Proceedings of the National Academy of Sciences USA* **103**, 12045–12050.

[113] Chung F., Lu L. (2002). Connected components in random graphs with given expected degre sequences. *Annals of Combinatorics* **6**, 125ñ–145.

[114] Citri A., Yarden Y. (2004). EGF-ERBB signalling: towards the systems level. *Nature Reviews Molecular Cell Biology* **7**, 505–516.

[115] Clark K., Halay E., Lai E., Burley S. (1993). Co-crystal structure of the HNF-3/fork head DNA-recognition motif resembles histone H5. *Nature* **364**, 412–420.

[116] Cliften P., Sudarsanam P., Desikan A. *et al.* (2003). Finding functional features in saccharomyces genomes by phylogenetic footprinting. *Science* **301**, 71–76.

[117] Cobb M., Goldsmith E. (1995). How map kinases are regulated. *Journal Biological Chemistry* **270**, 14843–14846.

[118] Cohen P. (1989). The structure and regulation of protein phosphatases. *Annual Review Biochemistry* **58**, 453–508.
[119] Cohen P., Frame S. (2001). The renaissance of GSK3. *Nature Reviews Molecular Cell Biology* **2**, 769–775.
[120] Cohen R., Erez K., ben-Avraham D., Havlin S. (2000). Resilience of the Internet to random breakdowns. *Physical Review Letters* **85**, 4626–4628.
[121] Cohen R., Erez K., ben-Avraham D., Havlin S. (2001). Breakdown of the Internet under intentional attack. *Physical Review Letters* **86**, 3682.
[122] Conant G., Wagner A. (2003). Convergent evolution of gene circuits. *Nature Genetics* **34**, 264–266.
[123] Conlon E., Liu X., Lieb J., Liu J. (2003). Integrating regulatory motif discovery and genome-wide expression analysis. *Proceedings of the National Academy of Sciences of the United States of America* **100**, 3339–3344.
[124] Conradi C., Flockerzi D., Raisch J., Stelling J. (2007). Subnetwork analysis reveals dynamic features of complex (bio)chemical networks. *Proceedings of the National Academy of Sciences of the United States of America* **104**, 19175–19180.
[125] Conti M., Nemoz G., Sette C., Vicini E. (1995). Recent progress in understanding the hormonal regulation of phosphodiesterases. *Endocrinology Review* **16**, 370–389.
[126] Cooper T. (2005). Alternative splicing regulation impacts heart development. *Cell* **120**, 1–2.
[127] Cordell H. (2002). Epistasis: what it means, what it doesn't mean, and statistical methods to detect it in humans. *Human Molecular Genetics* **11**, 2463–2468.
[128] Costanzo M., Crawford M., Hirschman J. *et al.* (2001). YPD^{TM}, PombePD^{TM} and WormPD^{TM}: model organism volumes of the BioKnowledge^{TM} Library, an integrated resource for protein information. *Nucleic Acids Research* **29**, 75–79.
[129] Coulson R., Enright A., Ouzounis C. (2001). Transcription-associated protein families are primarily taxon-specific. *Bioinformatics* **17**, 95–97.
[130] Covert M., Palsson B. (2002). Transcriptional regulation in constraints-based metabolic models of *Escherichia coli*. *The Journal of Biological Chemistry* **277**, 28058–28064.
[131] Crick F. (1958). On protein synthesis. *The Symposia of the Society for Experimental Biology* **12**, 138–163.
[132] Crick F. (1970). Central dogma of molecular biology. *Nature* **227**, 561–563.
[133] Dailey L., Basilico C. (2001). Coevolution of hmg domains and homeodomains and the generation of transcriptional regulation by sox/pou complexes. *Journal of Cell Physiology* **186**, 315–328.
[134] Danon L., Duch J., Arenas A., Díaz-Guilera A. (2007). Community structure identification. In *Large Scale Structure and Dynamics of Complex Networks*, G. Caldarelli and A. Vespignani, eds. World Scientific, pp. 93–114.
[135] Davidson E. (2001). *Genomic Regulatory Systems: Development and Evolution*. Academic Press, San Diego, CA.
[136] Davis R. (2000). Signal transduction by the JNK group of MAP kinases. *Cell* **103**, 239–252.
[137] de Solla Price D. (1965). Networks of scientific papers. *Science* **149**, 510–515.
[138] Deane C., Salwiński A., Xenarios I., Eisenberg D. (2002). Protein interactions: two methods for assessment of the reliability of high throughput observations. *Molecular & Cellular Proteomics* **1**, 349–356.

[139] Deeds E., Ashenberg O., Shakhnovich E. (2006). A simple physical model for scaling in protein–protein interaction networks. *Proceedings of the National Academy of Sciences of the United States of America* **103**, 311–316.
[140] Dennis E., Rhee S., Billah M., Hannun Y. (1991). Role of phospholipases in generating lipid second messengers in signal transduction. *FASEB Journal* **5**, 2068–2077.
[141] Deppmann C., Alvania R., Taparowsky E. (2006). Cross-species annotation of basic leucine zipper factor interactions: Insight into the evolution of closed interaction networks. *Molecular Biology and Evolution* **23**, 1480–1492.
[142] Hu Z., Killion P. J., Iyer V. R. Genetic reconstruction of a functional transcriptional regulatory network. *Nat. Genet.* **39**, 683–687 (2007).
[143] Devlin T. (2002). *Textbook of Biochemistry*. Wiley-Liss.
[144] D'Haeseleer P., Church G. (2004). Estimating and improving protein interaction error rates. *Proceedings of the IEEE Computational Systems Bioinformatics Conference*, pp. 216–223.
[145] Ding B., Gentleman R. (2005). Classification using generalized partial least squares. *Journal of Computational and Graphical Statistics* **14**, 280–298.
[146] Dobrin R., Beg Q., Barabasi A.-L., Oltvai Z. (2004). Aggregation of topological motifs in the *Escherichia coli* transcriptional regulatory network. *BMC Bioinformatics* **5**, 10.
[147] Dobrodumov A., Gronenborn A. (2003). Filtering and selection of structural models: combining docking and NMR. *Proteins* **53**, 18–32.
[148] Dodd I., Egan J. (1990). Improved detection of helix-turn-helix DNA-binding motifs in protein sequences. *Nucleic Acids Research* **18**, 5019–5026.
[149] Dodd I., Egan J. (1987). Systematic method for the detection of potential lambda cro-like DNA-binding regions in proteins. *Journal of Molecular Biology* **194**, 557–564.
[150] Dominguez C., Boelens R., Bonvin A. (2003). HADDOCK: a protein–protein docking approach based on biochemical or biophysical information. *Journal of the American Chemical Society* **125**, 1731–1737.
[151] Dorogovtsev S., Goltsev A., Mendes J. (2002). Pseudofractal scale-free web. *Physical Review E* **65**, 066122.
[152] Dorogovtsev S., Mendes J. (2001). Language as an evolving word web. *Proceedings of The Royal Society London B* **268**, 2603–2606.
[153] Dorogovtsev S., Mendes J. (2002). Evolution in networks. *Advances in Physics* **51**, 1079–1187.
[154] Dorogovtsev S., Mendes J. (2003). *Evolution of Networks From Biological Nets to the Internet and the WWW*. Oxford University Press, Oxford.
[155] Duarte N., Becker S., Jamshidi N. *et al.* (2007). Global reconstruction of the human metabolic network based on genomic and bibliomic data. *Proceedings of the National Academy of Sciences of the United States of America* **104**, 1777–1782.
[156] Duboule D. (1994). *Guidebook to the Homeobox Genes*. Oxford University Press.
[157] Dyson F. (1999). *Origins of Life*, 2nd edn. Cambridge University Press.
[158] Ebenhöh O., Handorf T., Heinrich R. (2004). Structural analysis of expanding metabolic networks. *Genome Informatics. International Conference on Genome Informatics* **15**, 35–45.
[159] Ebenhöh O., Heinrich R. (2001). Evolutionary optimization of metabolic pathways. Theoretical reconstruction of the stoichiometry of ATP and NADH producing systems. *Bulletin of Mathematical Biology* **63**, 21–55. PMID: 11146883.

[160] Eddy S. (1996). Hidden Markov models. *Current Opinions in Structure Biology* **6**, 361–365.

[161] Eddy S. (2001). Non-coding RNA genes and the modern RNA world. *Nature Reviews Genetics* **2**, 919–929.

[162] Edwards A., Kus B., Jansen R., Greenbaum D., Greenblatt J., Gerstein M. (2002). Bridging structural biology and genomics: assessing protein interaction data with known complexes. *Trends in Genetics* **18**, 10, 529–536.

[163] Edwards J., Palsson B. (1999). Systems properties of the *Haemophilus influenzae* rd metabolic genotype. *The Journal of Biological Chemistry* **274**, 17410–17416.

[164] Edwards J., Palsson B. (2000). The *Escherichia coli* mg1655 in silico metabolic genotype: its definition, characteristics, and capabilities. *Proceedings of the National Academy of Sciences of the United States of America* **97**, 5528–5533.

[165] Eigen M. (1979). *The Hypercycle: A Principle of Natural Self Organization.* Springer-Verlag.

[166] Eisen M., Spellman P., Brown P., Botstein D. (1998). Cluster analysis and display of genome-wide expression patterns. *Proceedings of the National Academy of Sciences of the United States of America* **95**, 14863–14868.

[167] Eisenberg D., McLachlan A. (1986). Solvation energy in protein folding and binding. *Nature* **319**, 6050, 199–203.

[168] Eisenberg E., Levanon E. (2003). Preferential attachment in the protein network evolution. *Physical Review Letters* **91**, 138701.

[169] Elena S., Lenski R. (1997). Test of synergistic interactions among deleterious mutations in bacteria. *Nature* **390**, 395–398.

[170] Elliott D., Grellscheid S. (2006). Alternative RNA splicing regulation in the testis. *Reproduction* **132**, 811–819.

[171] Elowitz M., Levine A., Siggia E., Swain P. (2002). Stochastic gene expression in a single cell. *Science* **297**, 1183–1186.

[172] Emmerling M., Dauner M., Ponti A. *et al.* (2002). Metabolic flux responses to pyruvate kinase knockout in *Escherichia coli*. *Journal of Bacteriology* **184**, 152–164.

[173] ENCODE Project Consortium. (2007). Identification and analysis of functional elements in 1% of the human genome by the encode pilot project. *Nature* **447**, 799–816.

[174] Enright A., Iliopoulos I., Kyrpides N., Ouzounis C. (1999). Protein interaction maps for complete genomes based on gene fusion events. *Nature* **402**, 6757, 86–90.

[175] Enright A., Van Dongen S., Ouzounis C. (2002). An efficient algorithm for large-scale detection of protein families. *Nucleic Acids Research* **30**, 1575–1584.

[176] Erdős P., Rényi A. (1959). On random graphs. *Publicationes Mathematicae Debrecen* **6**, 290–297.

[177] Erdős P., Rényi A. (1960). On the evolution of random graphs. *Publications of the Mathematical Institute of the Hungarian Academy of Sciences* **5**, 17–61.

[178] Eskin E., Pevzner P. (2002). Finding composite regulatory patterns in DNA sequences. *Bioinformatics* **18**, S354–S363.

[179] Euskirchen G., Royce T., Bertone P. *et al.* (2004). Creb binds to multiple loci on human chromosome 22. *Molecular and Cellular Biology* **24**, 3804–3814.

[180] Eyckerman S., Verhee A., der Heyden J. *et al.* (2001 Dec). Design and application of a cytokine-receptor-based interaction trap. *Nature Cell Biology* **3**, 12, 1114–1119.

[181] Fabre J. (1925). *Souvenirs entomologique.* Delagrave.

[182] Fairall L., Schwabe J. (2001). DNA binding by transcription factors. In *Transcription Factors*, J. Locker, ed. Academic Press, San Diego, California, 65–84.
[183] Faith J., Hayete B., Thaden J. *et al.* (2007). Large-scale mapping and validation of escherichia coli transcriptional regulation from a compendium of expression profiles. *PLoS Biology* **5**, e8.
[184] Falkowski P. (2006). Evolution. tracing oxygen's imprint on earth's metabolic evolution. *Science* **311**, 1724–1725.
[185] Faloutsos M., Faloutsos P., Faloutsos C. (1999). On power-law relationships of the internet topology. *Proceedings ACM SIGCOMM, Computer Communication Review* **29**, 251–262.
[186] Famili I., Palsson B. (2003). The convex basis of the left null space of the stoichiometric matrix leads to the definition of metabolically meaningful pools. *Biophysical Journal* **85**, 16–26. PMID: 12829460.
[187] Fassler J., Landsman D., Acharya A., Moll J., Bonovich M., Vinson C. (2002). B-ZIP proteins encoded by the *Drosophila* genome: evaluation of potential dimerization partners. *Genome Research* **12**, 1190–1200.
[188] Feist A., Henry C., Reed J. *et al.* (2007). A genome-scale metabolic reconstruction for *Escherichia coli* k-12 mg1655 that accounts for 1260 orfs and thermodynamic information. *Molecular Systems Biology* **3**, 121. PMID: 17593909.
[189] Fell D., Wagner A. (2000). The small world of metabolism. *Nature Biotechnology* **189**, 1121–1122.
[190] Feller W. (1968). *An Introduction to Probability Theory and its Applications.* Wiley & Sons, New York.
[191] Feng J., Fouse S., Fan G. (2007). Epigenetic regulation of neural gene expression and neuronal function. *Pediatric Research* **61**, 58R–63R.
[192] Ferrer i Cancho R., Solé R. (2001). The small world of human language. *Proceedings of the Royal Society London B* **268**, 2261–2265.
[193] Ferrigno O., Virolle T., Djabari Z., Ortonne J., White R., Aberdam D. (2001). Transposable B2 SINE elements can provide mobile RNA polymerase II promoters. *Nature Genetics* **28**, 77–81.
[194] Fields S. (2007). Molecular biology. Site-seeing by sequencing. *Science* **316**, 1441–1442.
[195] Fields S., Song O. (1989). A novel genetic system to detect protein–protein interactions. *Nature* **340**, 6230, 245–246.
[196] Fields S., Sternglanz R. (1994). The two-hybrid system: an assay for protein-protein interactions. *Trends in Genetics* **10**, 286–292.
[197] Fiers W., Contreras R., Duerinck F. *et al.* (1976). Complete nucleotide sequence of bacteriophage MS2 RNA: primary and secondary structure of the replicase gene. *Nature* **260**, 500–507.
[198] Flajolet M., Rotondo G., Daviet L. *et al.* (2000). A genomic approach to the *Hepatitis C* virus. *Gene* **242**, 369–379.
[199] Flake G., Lawrence S., Giles C. (2000). Efficient identification of web communities. In *Proceedings of the Sixth International Conference on Knowledge Discovery and Data Mining*. Boston: ACM, pp. 150–160.
[200] Fleischmann R., Adams M., White O. *et al.* (1995). Whole-genome random sequencing and assembly of *Haemophilus influenzae*. *Science* **269**, 496–512.
[201] Förster J., Famili I., Fu P., Palsson B., Nielsen J. (2003). Genome-scale reconstruction of the *Saccharomyces cerevisiae* metabolic network. *Genome Research* **13**, 244–253.

[202] Frampton J., Leutz A., Gibson T., Graf T. (1989). DNA-binding domain ancestry. *Nature* **342**, 134.
[203] Freeman B., Yamamoto K. (2001). Continuous recycling: a mechanism for modulatory signal transduction. *Trends in Biochemical Science* **26**, 285–290.
[204] Freeman L. (1977). A set of measures of centrality based upon betweenness. *Sociometry* **40**, 35–41.
[205] Fry I. (2000). *The Emergence of Life on Earth: A Historical and Scientific Overview*, 1st edn. Rutgers University Press.
[206] Fujibuchi W., Anderson J., Landsman D. (2001). Prospect improves cis-acting regulatory element prediction by integrating expression profile data with consensus pattern searches. *Nucleic Acids Research* **29**, 3988–3996.
[207] Gagliardi D., Stepien P., Temperley R., Lightowlers R., Chrzanowska-Lightowlers Z. (2004). Messenger RNA stability in mitochondria: different means to an end. *Trends in Genetics* **21**, 36.
[208] Gajiwala K., Burley S. (2000). Winged helix proteins. *Current Opinion in Structural Biology* **10**, 110–116.
[209] Gallegos M., Schleif R., Bairoch A., Hofmann K., Ramos J. (1997). Arac/xyls family of transcriptional regulators. *Microbiology and Molecular Biological Reviews* **61**, 393–410.
[210] Garavelli J. (2004). The resid database of protein modifications as a resource and annotation tool. *Proteomics* **4**, 6, 1527–1533.
[211] Garlaschelli D., Caldarelli G., Pietronero L. (2003). Universal scaling relations in food webs. *Nature* **423**, 165–168.
[212] Garlaschelli D., Loffredo M. (2004). Patterns of link reciprocity in directed networks. *Physical Review Letters* **93**, 268701.
[213] Gasch A., Huang M., Metzner S., Botstein D., Elledge S., Brown P. (2001). Genomic expression responses to DNA-damaging agents and the regulatory role of the yeast atr homolog mec1p. *Molecular Biology of the Cell* **12**, 2987–3003.
[214] Gavin A.-C., Aloy P., Grandi P. *et al.* (2006). Proteome survey reveals modularity of the yeast cell machinery. *Nature* **440**, 631–636.
[215] Gavin A.-C., Bösche M., Krause R. *et al.* (2002). Functional organization of the yeast proteome by systematic analysis of protein complexes. *Nature* **415**, 141–147.
[216] Ge H., Liu Z., Church G., Vidal M. (2001). Correlation between transcriptome and interactome mapping data from *saccharomyces cerevisiae*. *Nature Genetics* **29**, 482–486.
[217] Gehring W., Affolter M., Burglin T. (1994). Homeodomain proteins. *Annual Review of Biochemistry* **63**, 487–526.
[218] Gertz J., Elfond G., Shustrova A. *et al.* (2003). Inferring protein interactions from phylogenetic distance matrices. *Bioinformatics* **19**, 16, 2039–2045.
[219] Gill S., Pop M., Deboy R. *et al.* (2006). Metagenomic analysis of the human distal gut microbiome. *Science* **312**, 1355–1359.
[220] Ginalski K. (2006). Comparative modeling for protein structure prediction. *Current Opinion in Structural Biology* **16**, 172–177.
[221] Giot L., Bader J., Brouwer C. *et al.* (2003). A Protein Interaction Map of *Drosophila melanogaster*. *Science* **302**, 1727–1736.
[222] Girvan M., Newman M. (2002). Community structure in social and biological networks. *Proceedings of the National Academy of Sciences of the United States of America* **99**, 7821–7826.
[223] Glass C., Rose D., Rosenfeld M. (1997). Nuclear receptor coactivators. *Current Opinion in Cell Biology* **9**, 222–232.

[224] Glover J., Harrison S. (1995). Crystal structure of the heterodimeric bZIP transcription factor c-Fos–c-Jun bound to DNA. *Nature* **373**, 257–261.

[225] Goh C., Cohen F. (2002). Co-evolutionary analysis reveals insights into protein–protein interactions. *Journal of Molecular Biology* **324**, 177–192.

[226] Golemis E., Khazak V. (1997). Alternative yeast two-hybrid systems. The interaction trap and interaction mating. *Methods in Molecular Biology* **63**, 197–218.

[227] Goll J., Uetz P. (2006). The elusive yeast interactome. *Genome Biology 7,* 6, 223.

[228] Goodyer C., Zogopolos C., Schwartzbauer G., Zheng H., Hendy G., Menon R. (2001). Organization and evolution of the human growth hormone receptor 5′-flanking region. *Endocrinology* **142**, 1923–1934.

[229] Gopisetty G., Ramachandran K., Singal R. (2006). DNA methylation and apoptosis. *Molecular Immunology* **43**, 1729–1740.

[230] Gottesman S. (1984). Bacterial regulation: global regulatory networks. *Annual Review Genetics* **18**, 415–441.

[231] Govindan R., Tangmunarunkit H. (2000). Heuristic for internet map discovery. *Proceeding of the 19th annual joint conference of IEEE Computer and Communications Societies* **3**, 1371–1380.

[232] Gray J. (2006). High-resolution protein–protein docking. *Current Opinion in Structural Biology* **16**, 183–193.

[233] Greil F., Moorman C., van Steensel B. (2006). Damid: mapping of in vivo protein-genome interactions using tethered dna adenine methyltransferase. *Methods Enzymology* **410**, 342–359.

[234] Grigoriev A. (2001). A relationship between gene expression and protein interactions on the proteome scale: analysis of the bacteriophage t7 and the yeast saccharomyces cerevisiae. *Nucleic Acids Research* **29**, 17, 3513–3519.

[235] Griner E., Kazanietz M. (2007). Protein kinase C and other diacylglycerol effectors in cancer. *Nature Review on Cancer* **7**, 281–294.

[236] Guelzim N., Bottani S., Bourgine P., Kepes F. (2002). Topological and causal structure of the yeast transcriptional regulatory network. *Nature Genetics* **31**, 60–63.

[237] Guet C., Elowitz M., Hsing W., Leibler S. (2002). Combinatorial synthesis of genetic networks. *Science* **296**, 1466–1470.

[238] Guhaniyogi J., Brewer G. (2001). Regulation of mRNA stability in mammalian cells. *Gene* **265**, 11–23.

[239] Gygi S., Rist B., Gerber S., Turecek F., Gelb M., Aebersold R. (1999). Quantitative analysis of complex protein mixtures using isotope-coded affinity tags. *Nature Biotechnology* **17**, 10, 994–999.

[240] Hallikas O., Palin K., Sinjushina N. *et al.* (2006). Genome-wide prediction of mammalian enhancers based on analysis of transcription-factor binding affinity. *Cell* **124**, 47–59.

[241] Hamm H. (1998). The many faces of G protein signaling. *Journal of Biological Chemistry* **273**, 669–672.

[242] Han J.-D., Bertin N., Hao T. *et al.* (2004). Evidence for dynamically organized modularity in the yeast protein–protein interaction network. *Nature* **430**, 88–93.

[243] Hancock J. (2003). RAS proteins: different signals from different locations. *Nature Reviews Molecular Cell Biology* **4**, 373–384.

[244] Handorf T., Ebenhöh O., Heinrich R. (2005). Expanding metabolic networks: scopes of compounds, robustness, and evolution. *Journal of Molecular Evolution* **61**, 498–512.

[245] Harbison C., Gordon D., Lee T. *et al.* (2004). Transcriptional regulatory code of a eukaryotic genome. *Nature* **431**, 99–104.
[246] Harrison S. (1991). A structural taxonomy of DNA-binding domains. *Nature* **353**, 715–719.
[247] Hart G., Ramani A., Marcotte E. (2006). How complete are current yeast and human protein-interaction networks? *Genome Biology 7,* 11, 120.
[248] Hartwell L., Hopfield J., Leibler S., Murray A. (1999). From molecular to modular cell biology. *Nature* **402**, C47–C52.
[249] Hasty J., McMillen D., Isaacs F., Collins J. (2001). Computational studies of gene regulatory networks: in numero molecular biology. *Nature Review Genetics* **2**, 268.
[250] Hazzalin C., Mahadevan L. (2002). MAPK-regulated transcription: a continuously variable gene switch? *Nature Review in Molecular Cell Biology* **3**, 30–40.
[251] Heinrich R., Schuster S. (1996). *The Regulation Of Cellular Systems*, 1st edn. Springer.
[252] Heinrich R., Schuster S. (1998). The modelling of metabolic systems. structure, control and optimality. *Bio Systems* **47**, 61–77.
[253] Heintzman N., Ren B. (2007). The gateway to transcription: identifying, characterizing and understanding promoters in the eukaryotic genome. *Cellular and Molecular Life Sciences* **64**, 386–400.
[254] Henikoff S., Greene E., Pietrokovski S., Bork P., Attwood T., Hood L. (1997). Gene families: the taxonomy of protein paralogs and chimeras. *Science* **278**, 5338, 609–614.
[255] Hillenmeyer M., Fung E., Wildenhain J. *et al.* (2008). The chemical genomic portrait of yeast: uncovering a phenotype for all genes. *Science* **320**, 5874, 362–365.
[256] Hinnebusch A. (2005). Translational regulation of GCN4 and the general amino acid control of yeast. *Annual Review of Microbiology* **59**, 407–50.
[257] Hinrichsen H. (2000). Nonequilibrium critical phenomena and phase-transitions into absorbing states. *Advances in Physics* **49**, 815–958.
[258] Hipfner D., Cohen S. (2004). Connecting proliferation and apoptosis in development and disease. *Nature Reviews Molecular Cell Biology* **5**, 805–815.
[259] Ho Y., Gruhler A., Heilbut A. *et al.* (2002). Systematic identification of protein complexes in *Saccharomyces cerevisiae* by mass spectrometry. *Nature* **415**, 180–183.
[260] Hoeflich K., Ikura M. (2002). Calmodulin in action: diversity in target recognition and activation mechanisms. *Cell* **108**, 739–742.
[261] Hollams E., Giles K., Thomson A., Leedman P. (2002). mRNA stability and the control of gene expression: implications for human disease. *Neurochemical Research* **27**, 957–980.
[262] Holme P., Huss M., Jeong H. (2003). Subnetwork hierarchies in biochemical pathways. *Bioinformatics* **19**, 532–538.
[263] Holmes R., Soloway P. (2006). Regulation of imprinted DNA methylation. *Cytogenetics Genome Research* **113**, 122–129.
[264] Holter N., Maritan A., Cieplak M., Fedoroff N., Banavar J. (2001). Dynamic modeling of gene expression data. *Proceedings of the National Academy of Sciences of the United States of America* **98**, 1693.
[265] Horak C., Luscombe N., Qian J. *et al.* (2002). Complex transcriptional circuitry at the g1/s transition in saccharomyces cerevisiae. *Genes Development* **16**, 3017–3033.

[266] Hu C.-D., Kerppola T. (2003). Simultaneous visualization of multiple protein interactions in living cells using multicolor fluorescence complementation analysis. *Nature Biotechnology* **21**, 5, 539–545.

[267] Huang K. (1987). *Statistical Mechanics*. J. Wiley, New York.

[268] Hubbard S., Miller W. (2007). Receptor tyrosine kinases: mechanisms of activation and signaling. *Current Opinion in Cell Biology* **19**, 117–123.

[269] Hudson M., Snyder M. (2006). High-throughput methods of regulatory element discovery. *Biotechniques* **41**, 673–681.

[270] Huffman J., Brennan R. (2002). Prokaryotic transcription regulators: more than just the helix-turn-helix motif. *Current Opinion in Structural Biology* **12**, 98–106.

[271] Hughes J., Estep P., Tavazoie S., Church G. (2000). Computational identification of cis-regulatory elements associated with groups of functionally related genes in Saccharomyces cerevisiae. *Journal of Molecular Biology* **296**, 1205–1214.

[272] Hunter A., LaCasse E., Korneluk R. (2007). The inhibitors of apoptosis (IAPs) as cancer targets. *Apoptosis* **12**, 1543–1568.

[273] Hurley J. (1999). Structure, mechanism, and regulation of mammalian adenylyl cyclase. *Journal of Biological Chemistry* **274**, 7599–7602.

[274] Hurst H. (1994). Transcription factors. 1: bZIP proteins. *Protein Profile* **1**, 123–168.

[275] Huynen M., Bork P. (1998). Measuring genome evolution. *Proceedings of the National Academy of Sciences of the United States of America* **95**, 5849–5856.

[276] Ibarra R., Edwards J., Palsson B. (2002). *Escherichia coli* k-12 undergoes adaptive evolution to achieve in silico predicted optimal growth. *Nature* **420**, 186–189.

[277] Ihmels J., Bergmann S., Barkai N. (2004). Defining transcription modules using large-scale gene expression data. *Bioinformatics* **20**, 1993–2003.

[278] Ihmels J., Friedlander G., Bergmann S., Sarig O., Ziv Y., Barkai N. (2002). Revealing modular organization in the yeast transcriptional network. *Nature Genetics* **31**, 370–377.

[279] Imhof A., Bonaldi T. (2005). "Chromatomics" the analysis of the chromatome. *Molecular Biosystems* **1**, 112–116.

[280] Imielinski M., Belta C., Rubin H., Halász A. (2006). Systematic analysis of conservation relations in *Escherichia coli* genome-scale metabolic network reveals novel growth media. *Biophysical Journal* **90**, 2659–2672.

[281] Irvine R., Schell M. (2001). Back in the water: the return of the inositol phosphates. *Nature Reviews Molecular Cell Biology* **2**, 327–338.

[282] Isalan M., Lemerle C., Michalodimitrakis K. *et al.* (2008). Evolvability and hierarchy in rewired bacterial gene networks. *Nature* **452**, 840–845.

[283] Ito T., Chiba T., Ozawa R., Yoshida M., Hattori M., Sakaki Y. (2001). A comprehensive two-hybrid analysis to explore the yeast protein interactome. *Proceedings of the National Academy of Sciences of the United States of America* **98**, 4569–4574.

[284] Ito T., Ota K., Kubota H. *et al.* (2002). Roles for the two-hybrid system in exploration of the yeast protein interactome. *Molecular & Cellular Proteomics* **1**, 8, 561–566.

[285] Ito T., Tashiro K., Muta S. *et al.* (2000). Toward a protein-protein interaction map of the budding yeast: a comprehensive system to examine two-hybrid interactions in all possible combinations between the yeast proteins. *Proc. Natl. Acad. Sci. U.S.A.* **97**, 1143–1147.

[286] Jackson-Fisher A., Chitikila C., Mitra M., Pugh B. (1999). A role for tbp dimerization in preventing unregulated gene. *Molecular Cell* **3**, 717–727.
[287] Janes K., Yaffe M. (2006). Data-driven modelling of signal-transduction networks. *Nature Reviews Molecluar Cell Biology* **7**, 820–828.
[288] Jansen R., Lan N., Qian J., Gerstein M. (2002). Integration of genomic datasets to predict protein complexes in yeast. *Journal of Structural and Functional Genomics* **2**, 2, 71–81.
[289] Jansen R., Yu H., Greenbaum D. *et al.* (2003). A bayesian networks approach for predicting protein–protein interactions from genomic data. *Science* **302**, 5644, 449–453.
[290] Gralla J. D., Collado-Vides J. (1996). In *Cellular and Molecular Biology: Escherichia coli and Salmonella (2nd edn)*, vol. 79. American Society for Microbiology Washington, D.C. USA, 1232–1245.
[291] Jensen H. (1998). *Self-Organized Criticality*. Cambridge University Press, Cambridge.
[292] Jeong H., Mason S., Barabási A.-L., Oltvai Z. (2001). Lethality and centrality in protein networks. *Nature* **411**, 41–42.
[293] Jeong H., Néda Z., Barabási A.-L. (2003). Measuring preferential attachment for evolving networks. *Europhysics Letters* **61**, 567.
[294] Jeong H., Tombor B., Albert R., Oltvai Z., Barabási A.-L. (2000). The large-scale organization of metabolic networks. *Nature* **407**, 651–654.
[295] Jin F., Avramova L., Huang J., Hazbun T. (2007). A yeast two-hybrid smart-pool-array system for protein-interaction mapping. *Nature Methods 4,* 5, 405–407.
[296] Johnson D., Mortazavi A., Myers R., Wold B. (2007). Genome-wide mapping of in vivo protein–DNA interactions. *Science* **316**, 1497–1502.
[297] Johnsson N., Varshavsky A. (1994). Split ubiquitin as a sensor of protein interactions in vivo. *Proceedings of the National Academy of Sciences of the United States of America* **91**, 22, 10340–10344.
[298] Jones R., Gordus A., Krall J., MacBeath G. (2006). A quantitative protein interaction network for the erbb receptors using protein microarrays. *Nature* **439**, 7073, 168–174.
[299] Jordan J., Landau E., Iyengar R. (2000). Signaling networks: the origins of cellular multitasking. *Cell* **103**, 193–200.
[300] Joung J., Ramm E., Pabo C. (2000). A bacterial two-hybrid selection system for studying protein–DNA and protein–protein interactions. *Proceedings of the National Academy of Sciences of the United States of America* **97**, 13, 7382–7387.
[301] Joyce A., Palsson B. (2006). The model organism as a system: integrating "omics" data sets. *Nature Reviews Molecular Cell Biology* **7**, 198–210.
[302] Jung S., Kim S., Kahng B. (2002). A geometric fractal growth model for scale free networks. *Physical Review E* **65**, 056101.
[303] Kalir S., Mangan S., Alon U. (2005). A coherent feed-forward loop with a sum input function prolongs flagella expression in *Escherichia coli. Molecular System Biology* **1**, 2005.0006.
[304] Kalir S., McClure J., Pabbaraju K. *et al.* (2001). Ordering genes in a flagella pathway by analysis of expression kinetics from living bacteria. *Science* **292**, 2080–2083.
[305] Kanehisa M., Goto S. (2000). KEGG: Kyoto encyclopedia of genes and genomes. *Nucleic Acids Research* **28**, 27–30.

[306] Karimova G., Pidoux J., Ullmann A., Ladant D. (1998). A bacterial two-hybrid system based on a reconstituted signal transduction pathway. *Proceedings of the National Academy of Sciences of the United States of America* **95**, 10, 5752–5756.

[307] Karp P., Riley M., Saier M., Paulsen I., Paley S., Pellegrini-Toole A. (2000). The EcoCyc and MetaCyc databases. *Nucleic Acids Research* **28**, 56–59.

[308] Kathiresan S., Melander O., Guiducci C. *et al.* (2008). Six new loci associated with blood low-density lipoprotein cholesterol, high-density lipoprotein cholesterol or triglycerides in humans. *Nature Genetics* **40**, 189–197.

[309] Katz M., Amnit I., Yarden Y. (2007). Regulation of MAPKs by growth factors and receptor tyrosine kinases. *Biochimica et Biophysica Acta* **1773**, 1161–1176.

[310] Kauffman K., Prakash P., Edwards J. (2003). Advances in flux balance analysis. *Current Opinion in Biotechnology* **14**, 491–496.

[311] Kauffman S. (1969). Metabolic stability and epigenesis in randomly constructed genetic nets. *Journal of Theoretical Biology* **22**, 434–467.

[312] Kauffman S. (1971). Gene regulation networks: a theory for their structure and global behaviour. In *Current Topics in Developmental Biology 6*, A. Moscana and A. Monroy, eds. Academic Press, New York, pp. 145–182.

[313] Kauffman S. (1993). *The Origins of Order.* Oxford University Press, Oxford.

[314] Kel A., Gossling E., Reuter I., Cheremushkin E., Kel-Margoulis O., Wingender E. (2003). Match: A tool for searching transcription factor binding sites in DNA sequences. *Nucleic Acids Research* **31**, 3576–3579.

[315] Kellis M., Patterson N., Endrizzi M., Birren B., Lander E. (2003). Sequencing and comparison of yeast species to identify genes and regulatory elements. *Nature* **423**, 241–254.

[316] Kholodenko B. (2006). Cell-signalling dynamics in time and space. *Nature Reviews Molecular Cell Biology* **7**, 165–176.

[317] Kholodenko B., Kiyatkin A., Bruggeman F., Sontag E., Westerhoff H., Hoek J. (2002). Untangling the wires: a strategy to trace functional interactions in signaling and gene networks. *Proceedings of the National Academy of Sciences of the United States of America* **99**, 12841–12846.

[318] Kitano H. (2002). Systems biology: a brief overview. *Science* **295**, 1662–1664.

[319] Klamt S., Stelling J. (2003). Two approaches for metabolic pathway analysis? *Trends in Biotechnology* **21**, 64–69.

[320] Klamt S., Stelling J., Ginkel M., Gilles E. (2003). Fluxanalyzer: exploring structure, pathways, and flux distributions in metabolic networks on interactive flux maps. *Bioinformatics* **19**, 261–269.

[321] Kleidon A., Lorenz R. (2004). *Non-equilibrium Thermodynamics and the Production of Entropy: Life, Earth, and Beyond*, 1st edn. Springer.

[322] Kleinberg J. (2000). Navigation in a small world. *Nature* **406**, 6798, 845.

[323] Kleinberg J., Kumar S., Raghavan P., Rajagopalan S., Tomkins A. (1999). The web as a graph: measurements, models and methods. In *Proc. of the Int. Conf. on Combinatorics and Computing, COCOON'99*. Berlin: Springer-Verlag, 1. Tokyo.

[324] Klemm K., Eguíluz V. (2002). Growing scale-free networks with small-world behaviour. *Physical Review E* **65**, 057102.

[325] Koch M., Norris D., Hund-Georgiadis M. (2002). An investigation of functional and anatomical connectivity using magnetic resonance imaging. *Neuroimage* **16**, 241–250.

[326] Kochen, M., Ed. (1989). *The Small World.* Ablex, Norwood, NJ.

[327] Kodandapani R., Pio F., Ni C. *et al.* (1996). A new pattern for helix-turn-helix recognition revealed by the PU.1 ETS-domain-DNA complex. *Nature* **380**, 456–460.

[328] Kohout T., Lefkowitz R. (2003). Regulation of G protein-coupled receptor kinases and arrestins during receptor desensitization. *Molecular Pharmacology* **63**, 9–18.

[329] Kolch W. (2005). Coordinating ERK/MAPK signalling through scaffolds and inhibitors. *Nature Reviews Molecular Cell Biology* **6**, 827–837.

[330] Koleske A., Young R. (1995). The RNA polymerase ii holoenzyme and its implications for gene regulation. *Trends in Biochemical Sciences* **20**, 113–116.

[331] Kolmogorov N. (1956). *Foundation of the Theory of Probability (English edition from the original of 1933).* Chelsea Publishing Company, New York.

[332] Komarova N., Zou X., Nie Q., Bardwell L. (2005). A theoretical framework for specificity in cell signaling. *Molecular System Biology* **1**, 2005.0023.

[333] Korner H., Sofia H., Zumft W. (2003). Phylogeny of the bacterial superfamily of crp-fnr transcription regulators: exploiting the metabolic spectrum by controlling alternative gene programs. *FEMS Microbiological Review* **27**, 559–592.

[334] Kortemme T., Joachimiak L. A., Bullock A. N., Schuler A. D., Stoddard B. L., Baker D. (2004). Computational redesign of protein–protein interaction specificity. *Nature Structural and Molecular Biology* **11**, 371–379.

[335] Kozma S., Thomas G. (2002). Regulation of cell size in growth, development and human disease. *Bioessays* **24**, 65–71.

[336] Krogan N., Cagney G., Yu H. *et al.* (2006). Global landscape of protein complexes in the yeast saccharomyces cerevisiae. *Nature* **440**, 7084, 637–643.

[337] Kulesh D., Clive D., Zarlenga D., Greene J. (1987). Identification of interferon-modulated proliferation-related cDNA sequences. *Proceedings of the National Academy of Sciences of the United States of America* **84**, 8453–8457.

[338] Kumar S. (2007). Caspase function in programmed cell death. *Cell Death Differentiation* **14**, 32–43.

[339] Kumar S., Saradhi M., Chaturvedi N., Tyagi R. (2006). Intracellular localization and nucleocytoplasmic trafficking of steroid receptors: an overview. *Molecular Cell Endocrinology* **246**, 147–156.

[340] Kummerfeld S. K., Teichmann S. A. DBD: a transcription factor prediction database. *Nucleic Acids Res.* **34**, D74–81 (2006).

[341] Kuras L., Struhl K. (1999). Binding of tbp to promoters in vivo is stimulated by activators and requires pol ii holoenzyme. *Nature* **399**, 609–613.

[342] Lancet D., Sadovsky E., Seidemann E. (1993). Probability model for molecular recognition in biological receptor repertoires: significance to the olfactory system. *Proceedings of the National Academy of Sciences of the United States of America* **90**, 3715–3719. PMID: 8475121.

[343] Lander E., Linton L., Birren B. *et al.* (2001). Initial sequencing and analysis of the human genome. *Nature* **409**, 6822, 860–921.

[344] Lanning N., Carter-Su C. (2006). Recent advances in growth hormone signaling. *Reviews in Endocrine and Metabolic Disorders* **7**, 225–235.

[345] Lashkari D., DeRisi J., McCusker J. *et al.* (1997). Yeast microarrays for genome wide parallel genetic and gene expression analysis. *Proceedings of the National Academy of Sciences of the United States of America* **94**, 24, 13057–13062.

[346] Latchman D. (1998). *Eukaryotic Transcription Factors.* Academic Press, San Diego, CA.

[347] Lau N., Seto A., Kim J. *et al.* (2006). Characterization of the piRNA complex from rat testes. *Science* **313**, 363.

[348] Lauffenburger D. (2000). Cell signaling pathways as control modules: Complexity for simplicity. *Proceedings of the National Academy of Sciences of the United States of America* **97**, 5031–5033.
[349] Lawrence J. (1999). Selfish operons: the evolutionary impact of gene clustering in prokaryotes and eukaryotes. *Current Opinion in Genetics Development* **9**, 642–648.
[350] Lawrence S., Giles C. (1998). Searching the world wide web. *Science* **280**, 98–100.
[351] Lawrence S., Giles C. (1999). Accessibility of information on the web. *Nature* **400**, 107–109.
[352] Lazebnik Y. (2002). Can a biologist fix a radio? – or, what I learned while studying apoptosis. *Cancer Cell* **2**, 179–182.
[353] Le Breton M., Cormier P., Belle R., Mulner-Lorillon O., Morales J. (2005). Translational control during mitosis. *Biochimie* **87**, 805–811.
[354] Ledent V., Vervoort M. (2001). The basic helix-loop-helix protein family: comparative genomics and phylogenetic analysis. *Genome Research* **11**, 754–770.
[355] Lee T., Johnstone S., Young R. (2006). Chromatin immunoprecipitation and microarray-based analysis of protein location. *Nature Protocols* **1**, 729–748.
[356] Lee T., Rinaldi N., Robert F. *et al.* (2002). Transcriptional regulatory networks in *Saccharomyces cerevisiae. Science* **298**, 799–804.
[357] Lee T., Young R. (2000). Transcription of eukaryotic protein-coding genes. *Annual Review of Genetics* **34**, 77–137.
[358] Lefkowitz R., Shenoy S. (2005). Transduction of receptor signals by beta-arrestins. *Science* **308**, 512–517.
[359] Lehár J., Zimmermann G., Krueger A. *et al.* (2007). Chemical combination effects predict connectivity in biological systems. *Molecular Systems Biology* **3**, 80.
[360] Lemon B., Tjian R. (2000). Orchestrated response: a symphony of transcription factors for gene control. *Genes and Development* **14**, 2551–2569.
[361] Lensink M., Méndez R., Wodak S. (2007). Docking and scoring protein complexes: Capri 3rd edition. *Proteins* **69**, 704–718.
[362] Leonhardt H., Cardoso M. (2000). Dna methylation, nuclear structure, gene expression and cancer. *Journal of Cell Biochemistry. Supplement* **35**, 78–83.
[363] Lespinet O., Wolf Y., Koonin E., Aravind L. (2002). The role of lineage-specific gene family expansion in the evolution of eukaryotes. *Genome Research* **12**, 1048–1059.
[364] Lettini A., Guidoboni M., Fonsatti E., Anzalone L., Cortini E., Maio M. (2007). Epigenetic remodelling of dna in cancer. *Histologic and Histopathology* **22**, 1413–1424.
[365] Levitzki A. (1988). From epinephrine to cyclic amp. *Science* **241**, 800–806.
[366] Levy D., Darnell J. J. (2002). STATs: transcriptional control and biological impact. *Nature Reviews in Molecular Cell Biology* **3**, 651–662.
[367] Lévy P. (1925). *Calcul des probabilités*. Gauthier-Villars, Paris.
[368] Lewin B. (2000). *Genes VII*. Oxford University Press, Oxford.
[369] Li H., Rhodius V., Gross C., Siggia E. (2002). Identification of the binding sites of regulatory proteins in bacterial genomes. *Proceedings of the National Academy of Sciences USA* **99**, 11772–11777.
[370] Li J., Chory J. (1997). A putative leucine-rich repeat receptor kinase involved in brassinosteroid signal transduction. *Cell* **90**, 929–938.
[371] Li S., Armstrong C., Bertin N. *et al.* (2004). A map of the interactome network of the metazoan *C. elegans*. *Science* **303**, 5657, 540–543.

[372] Liebmann C. (2001). Regulation of MAP kinase activity by peptide receptor signalling pathway: paradigms of multiplicity. *Cell Signalling* **13**, 777–785.

[373] Lim C., Cao X. (2006). Structure, function, and regulation of STAT proteins. *Molecular BioSystems* **2**, 536–550.

[374] Liu X., Brutlag D., Liu J. (2002). An algorithm for finding protein–DNA interaction sites with applications to chromatin immunoprecipitation microarray experiments. *Nature Biotechnology* **20**, 835–839.

[375] Locker J. (2001). *Transcription Factors.* Academic Press, San Diego, CA.

[376] Lösel R., Wehling M. (2003). Nongenomic actions of steroid hormones. *Nature Review Molecular Cell Biology* **4**, 46–56.

[377] Lozada-Chavez I., Janga S., Collado-Vides J. (2006). Bacterial regulatory networks are extremely flexible in evolution. *Nucleic Acids Research* **34**, 3434–3445.

[378] Lu L., Arakaki A., Lu H., Skolnick J. (2003). Multimeric threading-based prediction of protein–protein interactions on a genomic scale: application to the saccharomyces cerevisiae proteome. *Genome Research* **13**, 6A (June), 1146–1154.

[379] Lu L., Lu H., Skolnick J. (2002). Multiprospector: an algorithm for the prediction of protein–protein interactions by multimeric threading. *Proteins* **49**, 3, 350–364.

[380] Luo Y., Batalao A., Zhou H., Zhu L. (1997). Mammalian two-hybrid system: a complementary approach to the yeast two-hybrid system. *Biotechniques* **22**, 2, 350–352.

[381] Luscombe N., Babu M. M., Yu H., Snyder M., Teichmann S., Gerstein M. (2004). Genomic analysis of regulatory network dynamics reveals large topological changes. *Nature* **431**, 308–312.

[382] Ma H., Buer J., Zeng A. (2004). Hierarchical structure and modules in the *Escherichia coli* transcriptional regulatory network revealed by a new top-down approach. *BMC Bioinformatics* **5**, 199.

[383] Madan Babu M. M., Teichmann S. (2003). Functional determinants of transcription factors in *Escherichia coli*: protein families and binding sites. *Trends Genetics* **19**, 75–79.

[384] Maeda Y., Sano M. (2006). Regulatory dynamics of synthetic gene networks with positive feedback. *Journal of Molecular Biology* **359**, 1107–1124.

[385] Maier T., Leibundgut M., Ban N. (2008). The crystal structure of a mammalian fatty acid synthase. *Science* **321**, 1315–1322.

[386] Malay A., Allen K., Tolan D. (2005). Structure of the thermolabile mutant aldolase b, a149p: molecular basis of hereditary fructose intolerance. *Journal of Molecular Biology* **347**, 135–144.

[387] Maller J. (2001). The elusive progesterone receptor in Xenopus oocytes. *Proceedings of the National Academy of Sciences of the United States of America* **98**, 8–10.

[388] Malter J. (2001). Regulation of mRNA stability in the nervous system and beyond. *Journal of Neuroscience Research* **66**, 311–316.

[389] Mandelbrot B. (1982). *The Fractal Geometry of Nature.* San Fransisco: Freeman.

[390] Mandell J., Falick A., Komives E. (1998). Identification of protein-protein interfaces by decreased amide proton solvent accessibility. *Proceedings of the National Academy of Sciences of the United States of America* **95**, 25, 14705–14710.

[391] Mangan S., Alon U. (2003). Structure and function of the feed-forward loop network motif. *Proceedings of the National Academy of Sciences of the United States of America* **100**, 11980–11985.

[392] Mangan S., Itzkovitz S., Zaslaver A., Alon U. (2006). The incoherent feed-forward loop accelerates the response-time of the gal system of *Escherichia coli. Journal of Molecular Biology* **356**, 1073–1081.
[393] Mangan S., Zaslaver A., Alon U. (2003). The coherent feedforward loop serves as a sign-sensitive delay element in transcription networks. *Journal of Molecular Biology* **334**, 197–204.
[394] Mangelsdorf D., Thummel C., Beato M. *et al.* (1995). The nuclear receptor superfamily: the second decade. *Cell* **83**, 835–839.
[395] Manning B., Cantley L. (2007). AKT/PKB signaling: navigating downstream. *Cell* **129**, 1261–1274.
[396] Marcotte E., Pellegrini M., Ng H.-L., Rice D., Yeates T., Eisenberg D. (1999). Detecting protein function and protein–protein interactions from genome sequences. *Science* **285**, 5428, 751–753.
[397] Marcotte E., Pellegrini M., Thompson M., Yeates T., Eisenberg D. (1999). A combined algorithm for genome-wide prediction of protein function. *Nature* **402**, 6757, 83–86.
[398] Marshall C. (1995). Specificity of receptor tyrosine kinase signaling: transient versus sustained extracellular signal-regulated kinase activation. *Cell* **80**, 179–185.
[399] Marti F., Xu C., Selvakumar A., Brent R., Dupont B., King P. (1998). Lck-phosphorylated human killer cell-inhibitory receptors recruit and activate phosphatidylinositol 3-kinase. *Proceedings of the National Academy of Sciences of the United States of America* **95**, 20, 11810–11815.
[400] Marti-Renom M., Stuart A., Fiser A., Sanchez R., Melo F., Sali A. (2000). Comparative protein structure modeling of genes and genomes. *Annual Review of Biophysical and Biomolecular Structure* **29**, 291–325.
[401] Martinez-Antonio A., Collado-Vides J. (2003). Identifying global regulators in transcriptional regulatory networks in bacteria. *Current Opinion in Microbiology* **6**, 482–489.
[402] Martinez-Antonio A., Janga S., Salgado H., Collado-Vides J. (2006). Internal-sensing machinery directs the activity of the regulatory network in *Escherichia coli. Trends Microbiology* **14**, 22–27.
[403] Marx C., Van Dien S., Lidstrom M. (2005). Flux analysis uncovers key role of functional redundancy in formaldehyde metabolism. *PLoS Biology* **3**, e16.
[404] Maslov S., Sneppen K., Eriksen K., Yan K. (2004). Upstream plasticity and downstream robustness in evolution of molecular networks. *BMC Evol Biol* **4**, 9.
[405] Maslov S., Sneppen K., Zaliznyak A. (2004). Detection of topological patterns in complex networks: correlation profile of the Internet. *Physica A* **333**, 529–540.
[406] Massagué J., Gomis R. (2006). The logic of TGF-beta signaling. *FEBS Letters* **580**, 2811–2820.
[407] Mata J., Marguerat S., Bahler J. (2005). Post-transcriptional control of gene expression: a genome-wide perspective. *Trends in Biochemical Sciences* **30**, 506–514.
[408] Matthews B., Ohlendorf D., Anderson W., Takeda Y. (1982). Structure of the DNA-binding region of lac repressor inferred from its homology with cro repressor. *Proceedings of the National Academy of Sciences of the United States of America* **79**, 1428–1432.
[409] Matys V., Kel-Margoulis O., Fricke E. *et al.* (2006). Transfac and its module transcompel: transcriptional gene regulation in eukaryotes. *Nucleic Acids Research* **34**, D108–110.

[410] Mayr B., Montminy M. (2001). Transcriptional regulation by the phosphorylation-dependent factor creb. *Nature Reviews Molecular Cell Biology* **2**, 599–609.

[411] McAdams H., Srinivasan B., Arkin A. (2004). The evolution of genetic regulatory systems in bacteria. *Nature Review Genetics* **5**, 169–178.

[412] McCue L., Thompson W., Carmack C. *et al.* (2001). Phylogenetic footprinting of transcription factor binding sites in proteobacterial genomes. *Nucleic Acids Research* **29**, 774–782.

[413] McGraith S., Holtzman T., Moss B., Fields S. (2000). Genome-wide analysis of vaccinia virus protein–protein interactions. *Proceedings of the National Academy of Sciences of the United States of America* **97**, 4879–4884.

[414] McGuire A., Hughes J., Church G. (2000). Conservation of DNA regulatory motifs and discovery of new motifs in microbial genomes. *Genome Research* **10**, 744–757.

[415] McKenna N., O'Malley B. (2002). Combinatorial control of gene expression by nuclear receptors and coregulators. *Cell* **108**, 465–474.

[416] Michell R. (2008). Inositol derivatives: evolution and functions. *Nature Reviews Molecular Cell Biology* **9**, 151–161.

[417] Milo R., Shen-Orr S., Itzkovitz S., Kashtan N., Chklovskii D., Alon U. (2002). Network motifs: Simple building blocks of complex networks. *Science* **298**, 824–827.

[418] Berridge M. J., Bootman M. D., Roderick H. L. (2003). Calcium signalling: dynamics, homeostasis and remodelling. *Nature Reviews Molecular Cell Biology* **4**, 517–529.

[419] Molloy M., Reed B. (1995). A critical point for random graphs with a given degree sequence. *Random Structures and Algorithms* **6**, 161–180.

[420] Moore J., Williams S. (2005). Traversing the conceptual divide between biological and statistical epistasis: systems biology and a more modern synthesis. *BioEssays* **27**, 637–646.

[421] Moreno-Campuzano S., Janga S., Perez-Rueda E. (2006). Identification and analysis of dna-binding transcription factors in bacillus subtilis and other firmicutes – a genomic approach. *BMC Genomics* **7**, 147.

[422] Morowitz H. (2004). *Beginnings of Cellular Life: Metabolism Recapitulates Biogenesis*. Yale University Press.

[423] Morowitz H., Kostelnik J., Yang J., Cody G. (2000). The origin of intermediary metabolism. *Proceedings of the National Academy of Sciences of the United States of America* **97**, 7704–7708.

[424] Moss T., Wallrath L. (2007). Connections between epigenetic gene silencing and human disease. *Mutation Research* **618**, 163–174.

[425] Murray P. (2007). The JAK-STAT signaling pathway: input and output integration. *Journal of Immunology* **178**, 2623–2629.

[426] Mwangi M., Siggia E. (2003). Genome wide identification of regulatory motifs in *Bacillus subtilis*. *BMC Bioinformatics* **4**, 18.

[427] Myers C. (2003). Software systems as complex networks: Structure, function, and evolvability of software collaboration graphs. *Physical Review E* **68**, 046116.

[428] Natarajan M., Lin K.-M., Hsueh R., Sternweis P., Ranganathan R. (2006). A global analysis of cross-talk in a mammalian cellular signalling network. *Nature Cell Biology* **8**, 571–580.

[429] Naveh-Many T., Nechama M. (2007). Regulation of parathyroid hormone mrna stability by calcium, phosphate and uremia. *Current Opinion in Nephrology and Hypertension* **16**, 305–310.

[430] Nelson D., Cox M. (2004). *Lehninger Principles of Biochemistry, Fourth Edition*, 4th edn. W. H. Freeman.
[431] Newman M. (2001a). Clustering and preferential attachment in growing networks. *Physical Review E* **64**, 025102(R).
[432] Newman M. (2001b). Scientific collaboration networks. I. Network construction and fundamental results. *Physical Review E* **64**, 016131.
[433] Newman M. (2001c). The structure of scientific collaboration networks. *Proceedings of the National Academy of Sciences of the United States of America* **98**, 404–409.
[434] Newman M. (2002). Assortative mixing in networks. *Physical Review Lett.* **89**, 208701.
[435] Newman M. (2003a). Mixing patterns in networks. *Physical Review E* **67**, 026126.
[436] Newman M. (2003b). The structure and function of complex networks. *The SIAM Review* **45**, 167–256.
[437] Newman M. (2004). Coauthorship networks and patterns of scientific collaboration. *Proceedings of the National Academy of Sciences of the United States of America* **101**, 5200–5205.
[438] Newman M. (2005). Power laws Pareto distributions and Zipf's laws. *Contemporary Physics* **46**, 323–351.
[439] Newman, M., Barabási, A.-L. and Watts, D., eds. (2003). *The Structure and Dynamics of Complex Networks*. Princeton University Press, Princeton.
[440] Newman M., Moore C., Watts D. (2000). Mean field solution of the small-world network model. *Physical Review Letters* **84**, 3201–3204.
[441] Newman M., Strogatz S., Watts D. (2001). Random graphs with arbitrary degree distributions and their applications. *Physical Review E* **64**, 026118.
[442] Newman M., Watts D. (1999). Scaling and percolation in the small-world network model. *Physical Review E* **60**, 7332–7342.
[443] Newton A. (1995). Protein kinase c: structure, function, and regulation. *Journal of Biological Chemistry* **270**, 28495–28498.
[444] Noble D. (2002). The rise of computational biology. *Nature Reviews Molecular Cell Biology* **3**, 460–463.
[445] Norman A., Mizwicki M., Norman D. (2004). Steroid-hormone rapid actions, membrane receptors and a conformational ensemble model. *Nature Review Drug Discovery* **3**, 27–41.
[446] Oberhardt M., Puchalka J., Fryer K., Martins dos Santos V., Papin J. (2008). Genome-scale metabolic network analysis of the opportunistic pathogen *Pseudomonas aeruginosa* pao1. *Journal of Bacteriology* **190**, 2790–2803.
[447] Ogata K., Morikawa S., Nakamura H. et al. (1994). Solution structure of a specific DNA complex of the Myb DNA-binding domain with cooperative recognition helices. *Cell* **79**, 639–648.
[448] Ohlendorf D., Anderson W., Fisher R., Takeda Y., Matthews B. (1982). The molecular basis of DNA–protein recognition inferred from the structure of cro repressor. *Nature* **298**, 718–723.
[449] Ohlendorf D., Anderson W., Matthews B. (1983). Many gene-regulatory proteins appear to have a similar α-helical fold that binds DNA and evolved from a common precursor. *Journal of Molecular Evolution* **19**, 109–114.
[450] Ong S.-E., Blagoev B., Kratchmarova I. *et al.* (2002). Stable isotope labeling by amino acids in cell culture, silac, as a simple and accurate approach to expression proteomics. *Molecular & Cellular Proteomics* **1**, 5, 376–386.
[451] Oparin A. (2003). *Origin of Life*, 2nd edn. Dover Publications.

[452] Orgel L. (2004). Prebiotic chemistry and the origin of the RNA world. *Critical Reviews in Biochemistry and Molecular Biology* **39**, 99–123.
[453] Orphanides G., Lagrange T., Reinberg D. (1998). The general transcription factors of RNA polymerase ii. *Genes and Development* **10**, 2657–2683.
[454] Ostareck-Lederer A., Ostareck D. (2004). Control of mRNA translation and stability in haematopoietic cells: the function of hnrnps k and e1/e2. *Biology of the Cell* **96**, 407–411.
[455] Otting G., Qian Y., Muller M., Affolter M., Gehring W., Wuthrich K. (1988). Secondary structure determination for the antennapedia homeodomain by nuclear magnetic resonance and evidence for a helix-turn-helix motif. *EMBO Journal* **7**, 4305–4309.
[456] Overbeek R., Larsen N., Pusch G. *et al.* (2000). WIT: integrated system for high-throughput genome sequence analysis and metabolic reconstruction. *Science* **28**, 123–125.
[457] Palin K., Ukkonen E., Brazma A., Vilo J. (2002). Correlating gene promoters and expression in gene disruption experiments. *Bioinformatics* **18**, S172–S180.
[458] Palsson B. (2006). *Systems Biology: Properties of Reconstructed Networks*, 1st edn. Cambridge University Press.
[459] Papa S., Bubici C., Zazzeroni F. *et al.* (2006). The NF-kB-mediated control of the JNK cascade in the antagonism of programmed cell death in health and disease. *Cell Death Differentiation* **13**, 712–729.
[460] Papin J., Hunter T., Palsson B., Subramaniam S. (2005). Reconstruction of cellular signalling networks and analysis of their properties. *Nature Reviews Molecular Cell Biology* **6**, 99–111.
[461] Papin J., Price N., Palsson B. (2002). Extreme pathway lengths and reaction participation in genome-scale metabolic networks. *Genome Research* **12**, 1889–1900.
[462] Papp B., Pal C., Hurst L. (2003). Evolution of cis-regulatory elements in duplicated genes of yeast. *Trends in Genetics* **19**, 417–422.
[463] Park J., Newman M. (2003). Origin of degree correlations in the internet and other networks. *Physical Review E* **68**, 026112.
[464] Pastor-Satorras R., Vázquez A., Vespignani A. (2001). Dynamical and correlation properties of the Internet. *Physical Review Letters* **87**, 258701.
[465] Pastor-Satorras R., Vespignani A. (2001). Epidemic spreading in scale-free networks. *Physical Review Letters* **86**, 3200–3203.
[466] Pastor-Satorras R., Vespignani A. (2004). *Evolution and Structure of Internet: A Statistical Physics Approach.* Cambridge University Press, Cambridge.
[467] Patil A., Nakamura H. (2005). Filtering high-throughput protein-protein interaction data using a combination of genomic features. *BMC Bioinformatics* **6**, 100.
[468] Pawson T., Nash P. (2003). Assembly of cell regulatory systems through protein interaction domains. *Science* **300**, 445–452.
[469] Pellegrini M., Marcotte E., Thompson M., Eisenberg D., Yeates T. (1999). Assigning protein functions by comparative genome analysis: protein phylogenetic profiles. *Proceedings of the National Academy of Sciences of the United States of America* **96**, 8, 4285–4288.
[470] Pennacchio L., Rubin E. (2001). Genomic strategies to identify mammalian regulatory sequences. *Nature Reviews Genetics* **2**, 100–109.
[471] Perez-Rueda E., Collado-Vides J. (2000). The repertoire of dna-binding transcriptional regulators in escherichia coli k-12. *Nucleic Acids Research* **28**, 1838–1847.

[472] Peter M., Krammer P. (2003). The CD95(APO-1/Fas) DISC and beyond. *Cell Death Differentiation* **10**, 26–35.
[473] Pharkya P., Burgard A., Maranas C. (2004). Optstrain: a computational framework for redesign of microbial production systems. *Genome Research* **14**, 2367–2376.
[474] Phizicki E., Field S. (1995). Protein–protein interactions: Methods for detection and analysis. *Microbiological Review* **58**, 94–123.
[475] Pierce K., Premont R., Lefkowitz R. (2002). Seven-transmembrane receptors. *Nature Reviews Molecular Cell Biology* **3**, 639–650.
[476] Pilpel Y., Sudarsanam P., Church G. (2001). Identifying regulatory networks by combinatorial analysis of promoter elements. *Nature Genetics* **29**, 153–159.
[477] Pirkkala L., Nykanen P., Sistonen L. (2001). Roles of the heat shock transcription factors in regulation of the heat shock response and beyond. *FASEB Journal* **15**, 1118–1131.
[478] Porath J., Carlsson J., Olsson I., Belfrage G. (1975 Dec 18). Metal chelate affinity chromatography, a new approach to protein fractionation. *Nature* **258**, 5536, 598–599.
[479] Price N., Papin J., Schilling C., Palsson B. (2003). Genome-scale microbial in silico models: the constraints-based approach. *Trends in Biotechnology* **21**, 162–169.
[480] Pritsker M., Liu Y., Beer M., Tavazoie S. (2004). Whole-genome discovery of transcription factor binding sites by network-level conservation. *Genome Research* **14**, 99–108.
[481] Ptacek J., Devgan G., Michaud G. *et al.* (2005). Global analysis of protein phosphorylation in yeast. *Nature* **438**, 7068, 679–684.
[482] Ptashne M. (2005). Regulation of transcription: from lambda to eukaryotes. *Trends Biochemical Science* **30**, 275–279.
[483] Pugh B. (1996). Mechanisms of transcription complex assembly. *Current Opinion in Cell Biology* **8**, 303–311.
[484] Pugh B. (2001). RNA polymerase ii transcription machinery. In *Transcription Factors*, J. Locker, ed. Academic Press, San Diego, CA, 1–16.
[485] Qian Z., Cai Y., Li Y. (2006). Automatic transcription factor classifier based on functional domain composition. *Biochemical Biophysical Research Communication* **347**, 141–144.
[486] Quina A., Buschbeck M., Di Croce L. (2006). Chromatin structure and epigenetics. *Biochemical Pharmacology* **72**, 1563–1569.
[487] Raff R. (1996). *The Shape of Life: Genes, Development, and the Evolution of Animal Form.* The University of Chicago Press, Chicago.
[488] Rain J., Selig L., De Reuse H. *et al.* (2001). The protein–protein interaction map of *Helicobacter pylori*. *Nature* **409**, 211–215.
[489] Rajewsky N., Socci N., Zapotocky M., Siggia E. (2002). The evolution of dna regulatory regions for proteo-gamma bacteria by interspecies comparisons. *Genome Research* **12**, 298–308.
[490] Ramakrishnan V., Finch J., Graziano V., Lee P., Sweet R. (1993). Crystal structure of globular domain of histone h5 and its implications for nucleosome binding. *Nature* **362**, 219–223.
[491] Ramani A., Marcotte E. (2003). Exploiting the co-evolution of interacting proteins to discover interaction specificity. *Journal of Molecular Biology* **327**, 1, 273–284.
[492] Rao C., Arkin A. (2000). Control motifs for intracellular regulatory networks. *Annual Review of Biomedical Engineering* **3**, 391–419.

[493] Ravasz E., Barabási A.-L. (2002). Hierarchical organization in complex networks. *Physical Review E* **67**, 026122.
[494] Ravasz E., Somera A., Mongru D., Oltvai Z., Barabási A.-L. (2002). Hierarchical organization of modularity in metabolic networks. *Science* **297**, 1551–1555.
[495] Raymond J., Segrè D. (2006). The effect of oxygen on biochemical networks and the evolution of complex life. *Science* **311**, 1764–1767.
[496] Razin A., Kantor B. (2005). DNA methylation in epigenetic control of gene expression. *Progress in Molecular and Subcellular Biology* **38**, 151–167.
[497] Redner S. (1998). How popular is your paper? An empirical study of the citation distribution. *European Physical Journal B* **4**, 131–135.
[498] Reik W. (2007). Stability and flexibility of epigenetic gene regulation in mammalian development. *Nature* **447**, 425–432.
[499] Reinberg, D. *et al.* (1998). The RNA polymerase ii general transcription factors: past, present, and future. *Cold Spring Harbor Symposia on Quantitative Biology* **63**, 83–103.
[500] Resendis-Antonio O., Freyre-Gonzalez J., Menchaca-Mendez R. *et al.* (2005). Modular analysis of the transcriptional regulatory network of *E. coli. Trends in Genetics* **21**, 16–20.
[501] Riechmann J., Heard J., Martin G. *et al.* (2000). Arabidopsis transcription factors: genome-wide comparative analysis among eukaryotes. *Science* **290**, 2105–2110.
[502] Riedl S., Shi Y. (2004). Molecular mechanisms of caspase regulation during apoptosis. *Nature Reviews Molecular Cell Biology* **5**, 897–907.
[503] Rigali S., Derouaux A., Giannotta F., Dusart J. (2002). Subdivision of the helix-turn-helix gntr family of bacterial regulators in the fadr, hutc, mocr, and ytra subfamilies. *Journal of Biological Chemistry* **277**, 12507–12515.
[504] Rigaut G., Shevchenko A., Rutz B., Wilm M., Séraphin B. (1999). A generic protein purification method for protein complex characterization and proteome exploration. *Nature Biotechnology* **17**, 1030–1032.
[505] Roach P. (1991). Multisite and hierarchical protein phosphorylation. *Journal of Biological Chemistry* **266**, 14139–14142.
[506] Robinson C., Sali A., Baumeister W. (2007). The molecular sociology of the cell. *Nature* **450**, 973–982.
[507] Ron D., Dressler H. (1992). Pgstag – a versatile bacterial expression plasmid for enzymatic labeling of recombinant proteins. *Biotechniques* **13**, 6, 866–869.
[508] Ronen M., Rosenberg R., Shraiman B. I., Alon U. (2002). Assigning numbers to the arrows: parameterizing a gene regulation network by using accurate expression kinetics. *Proceedings of the National Academy of Sciences USA* **99**, 10555–10560.
[509] Rosenfeld N., Elowitz M., Alon U. (2002). Negative autoregulation speeds the response times of transcription networks. *Journal of Molecular Biology* **323**, 785–793.
[510] Rosenwald S., Kafri R., Lancet D. (2002). Test of a statistical model for molecular recognition in biological repertoires. *Journal of Theoretical Biology* **216**, 327–336.
[511] Rual J., Venkatesan K., Hao T. *et al.* (2005). Towards a proteome-scale map of the human protein–protein interaction network. *Nature* **437**, 7062, 1173–1178.
[512] Russell R., Alber F., Aloy P. *et al.* (2004). A structural perspective on protein–protein interactions. *Current Opinion in Structural Biology* **14**, 3, 313–324.

[513] Ruvkun G. (2001). Glimpses of a tiny RNA world. *Science* **294**, 797–799.
[514] Ryu T., Jung J., Lee S. *et al.* (2007). bZIPdb: a database of regulatory information for human bZIP transcription factors. *BMC Genomics* **8**, 136.
[515] Salgado H., Gama-Castro S., Peralta-Gil M. *et al.* (2006). Regulondb (version 5.0): *Escherichia coli* K-12 transcriptional regulatory network, operon organization, and growth conditions. *Nucleic Acids Research* **34**, D394–397.
[516] Salgado H., Santos-Zavaleta A., Gama-Castro S. *et al.* (2001). RegulonDB (version 3.2): transcriptional regulation and operon organization in *Escherichia coli* K-12. *Nucleic Acids Research* **29**, 72–74.
[517] Sali A., Glaeser R., Earnest T., Baumeister W. (2003). From words to literature in structural proteomics. *Nature* **422**, 6928, 216–225.
[518] Saltiel A., Khan C. (2001). Insulin signalling and the regulation of glucose and lipid metabolism. *Nature* **414**, 799–806.
[519] Salvesen G., Cuckett C. (2002). IAP proteins: blocking the road to death's door. *Nature Reviews Molecular Cell Biology* **3**, 401–410.
[520] Sanger F., Air G., Barrell B. *et al.* (1977). Nucleotide sequence of bacteriophage *phi X174* DNA. *Nature* **265**, 5596, 687–695.
[521] Santos K., Mazzola T., Carvalho H. (2005). The prima donna of epigenetics: the regulation of gene expression by DNA methylation. *Brazilian Journal of Medical Biological Research* **38**, 1531–1541.
[522] Sassone-Corsi P. (1995). Transcription factors responsive to camp. *Annual Review of Cell and Developmental Biology* **11**, 355–377.
[523] Sauer R., Yocum R., Doolittle R., Lewis M., Pabo C. (1982). Homology among DNA-binding proteins suggests use of a conserved super-secondary structure. *Nature* **298**, 447–451.
[524] Sauer U. (2006). Metabolic networks in motion: 13C-based flux analysis. *Molecular Systems Biology* **2**, 62. PMID: 17102807.
[525] Saxon A., Diaz-Sanchez D., Zhang K. (1997). Regulation of the expression of distinct human secreted ige proteins produced by alternative RNA splicing. *Biochemical Society Transactions* **25**, 383–387.
[526] Schaefer U., Voloshanenko O., Willen D., Walczak H. (2007). Trail: a multifunctional cytokine. *Frontiers in Bioscience* **12**, 3813–3824.
[527] Scheid M., Woodgett J. (2001). PKB/AKT: functional insights from genetic models. *Nature Reviews Molecular Cell Biology* **2**, 760–768.
[528] Schena M., Shalon D., Davis R., Brown P. (1995). Quantitative monitoring of gene expression patterns with a complementary DNA microarray. *Science* **270**, 5235, 467–470.
[529] Schilling C., Letscher D., Palsson B. (2000). Theory for the systemic definition of metabolic pathways and their use in interpreting metabolic function from a pathway-oriented perspective. *Journal of Theoretical Biology* **203**, 229.
[530] Schindler C., Levy D., Decker T. (2007). JAK-STAT signaling: from interferons to cytokines. *Journal of Biological Chemistry* **282**, 20059–20063.
[531] Schlessinger J. (2000). Cell signaling by receptor tyrosine kinases. *Cell* **103**, 211–225.
[532] Schmierer B., Hill C. (2007). Tgf-beta-smad signal transduction: molecular specificity and functional flexibility. *Nature Reviews Molecular Cell Biology* **8**, 970–982.
[533] Schneider T., Stormo G., Gold L., Ehrenfeucht A. (1986). Information content of binding sites on nucleotide sequences. *Journal of Molecular Biology* **188**, 415–431.

[534] Schob H., Grossniklaus U. (2006). The first high-resolution DNA "methylome". *Cell* **126**, 1025–1028.
[535] Schuetz R., Kuepfer L., Sauer U. (2007). Systematic evaluation of objective functions for predicting intracellular fluxes in *Escherichia coli. Molecular Systems Biology* **3**, 119.
[536] Schultz S., Shields G., Steitz T. (1991). Crystal structure of a CAP-DNA complex: the DNA is bent by 90 degrees. *Science* **253**, 1001–1007.
[537] Schulze W., Deng L., Mann M. (2005). Phosphotyrosine interactome of the ERBb-receptor kinase family. *Molecular Systems Biology* **1**, 2005.0008.
[538] Schuster H. (2002). *Complex Adaptive Systems.* Scator Verlag, Saarbruskey, Germany.
[539] Schuster S., Dandekar T., Fell D. (1999). Detection of elementary flux modes in biochemical networks: a promising tool for pathway analysis and metabolic engineering. *Trends in Biotechnology* **17**, 53–60.
[540] Schuster S., Fell D., Dandekar T. (2000). A general definition of metabolic pathways useful for systematic organization and analysis of complex metabolic networks. *Nature Biotechnology* **18**, 326–332.
[541] Schwerk C., Schulze-Osthoff K. (2005). Regulation of apoptosis by alternative pre-mRNA splicing. *Molecular Cell* **19**, 1–13.
[542] Schwikowski B., Uetz P., Fields S. (2000). A network of protein-protein interactions in yeast. *Nature Biotechnology* **18**, 12, 1257–1261.
[543] Scott L., Mohlke K., Bonnycastle L. *et al.* (2007). A genome-wide association study of type 2 diabetes in Finns detects multiple susceptibility variants. *Science* **316**, 1341–1345.
[544] Segal E., Shapira M., Regev A. *et al.* (2003). Module networks: identifying regulatory modules and their condition-specific regulators from gene expression data. *Nature Genetics* **34**, 166–176.
[545] Segel L. (1992). *Biological Kinetics.* Cambridge University Press.
[546] Segrè D., Ben-Eli D., Deamer D., Lancet D. (2001). The lipid world. *Origins of Life and Evolution of the Biosphere : the Journal of the International Society for the Study of the Origin of Life* **31**, 119–145.
[547] Segrè D., Ben-Eli D., Lancet D. (2000). Compositional genomes: prebiotic information transfer in mutually catalytic noncovalent assemblies. *Proceedings of the National Academy of Sciences of the United States of America* **97**, 4112–4117.
[548] Segrè D., Deluna A., Church G., Kishony R. (2005). Modular epistasis in yeast metabolism. *Nature Genetics* **37**, 77–83.
[549] Segrè D., Lancet D. (2000). Composing life. *EMBO Reports* **1**, 217–222.
[550] Segrè D., Lancet D., Kedem O., Pilpel Y. (1998). Graded autocatalysis replication domain (gard): kinetic analysis of self-replication in mutually catalytic sets. *Origins of Life and Evolution of the Biosphere : the Journal of the International Society for the Study of the Origin of Life* **28**, 501–514.
[551] Segrè D., Shenhav B., Kafri R., Lancet D. (2001). The molecular roots of compositional inheritance. *Journal of Theoretical Biology* **213**, 481–491.
[552] Segrè D., Vitkup D., Church G. (2002). Analysis of optimality in natural and perturbed metabolic networks. *Proceedings of the National Academy of Sciences of the United States of America* **99**, 15112–15117.
[553] Serman A., Vlahovic M., Serman L., Bulic-Jakus B. (2006). DNA methylation as a regulatory mechanism for gene expression in mammals. *Collegium antroplogicum* **30**, 665–671.

[554] Serrano M., Boguñá M. (2003). Topology of the world trade web. *Physical Review E* **68**, 015101.
[555] Servedio V., Caldarelli G., Buttà P. (2004). Vertex intrinsic fitness: how to produce arbitrary scale-free networks. *Physical Review E* **70**, 056126.
[556] Shames D., Minna J., Gazdar A. (2007). DNA methylation in health, disease, and cancer. *Current Molecular Medicine* **7**, 85–102.
[557] Shastri A., Morgan J. (2005). Flux balance analysis of photoautotrophic metabolism. *Biotechnology Progress* **21**, 1617–1626.
[558] Shen-Orr S., Milo R., Mangan S., Alon U. (2002). Network motifs in the transcriptional regulation network of *E. coli*. *Nature Genetics* **31**, 64–68.
[559] Shi Y., Massagué J. (2003). Mechanisms of tgf-beta signalling from cell membrane to the nucleus. *Cell* **113**, 685–700.
[560] Shim J., Karin M. (2002). The control of mRNA stability in response to extracellular stimuli. *Molecular Cell* **14**, 323–331.
[561] Shlomi T., Berkman O., Ruppin E. (2005). Regulatory on/off minimization of metabolic flux changes after genetic perturbations. *Proceedings of the National Academy of Sciences of the United States of America* **102**, 7695–7700.
[562] Shlomi T., Eisenberg Y., Sharan R., Ruppin E. (2007). A genome-scale computational study of the interplay between transcriptional regulation and metabolism. *Molecular Systems Biology* **3**, 101.
[563] Shore P., Sharrocks A. (2001). Regulation of transcription by extracellular signals. In *Transcription factors*, J. Locker, ed. Academic Press, San Diego, CA, 113–135.
[564] Sigman M., Cecchi G. (2002). Global organization of the Wordnet lexicon. *Proceedings of the National Academy of Sciences of the United States of America* **99**, 1742–1747.
[565] Skrabanek L., Saini H., Bader G., Enright A. (2008). Computational prediction of protein–protein interactions. *Molecular Biotechnology* **38**, 1–17.
[566] Slonim N., Elemento O., Tavazoie S. (2006). Ab initio genotype–phenotype association reveals intrinsic modularity in genetic networks. *Molecular System Biology* **2**, 2006.0005.
[567] Smith G., Sternberg M. (2002). Prediction of protein–protein interactions by docking methods. *Current Opinion in Structural Biology* **12**, 1, 28–35.
[568] Söderberg B. (2002). General formalism for inhomogeneous random graphs. *Physical Review Letters* **66**, 066121.
[569] Sokal J. (1973). *Numerical Taxonomy*. Freeman, San Francisco.
[570] Solé R., Pastor-Satorras R., Smith E., Kepler T. (2002). A model of large-scale proteome evolution. *Advances in Complex Systems* **5**, 43–54.
[571] Song C., Havlin S., Makse H. (2005). Self-similarity of complex networks. *Nature* **433**, 7024, 392–395.
[572] Song C., Havlin S., Makse H. (2006). Origins of fractality in the growth of complex networks. *Nature Physics* **2**, 4, 275–281.
[573] Sprinzak E., Sattath S., Margalit H. (2003). How reliable are experimental protein-protein interaction data? *Journal of Molecular Biology* **327**, 5, 919–923.
[574] Stagljar I., Korostensky C., Johnsson N., te Heesen S. (1998). A genetic system based on split-ubiquitin for the analysis of interactions between membrane proteins in vivo. *Proceedings of the National Academy of Sciences of the United States of America* **95**, 9, 5187–5192.
[575] Stauber M., Prell A., Schmidt-Ott U. (2002). A single hox3 gene with composite bicoid and zerknult expression characteristics in non-cyclorrhaphan flies.

Proceedings of the National Academy of Sciences of the United States of America **99**, 274–279.
[576] Stegmaier P., Kel A., Wingender E. (2004). Systematic DNA-binding domain classification of transcription factors. *Genome Informatics* **15**, 276–286.
[577] Steitz T., Ohlendorf D., McKay D., Anderson W., Matthews B. (1982). Structural similarity in the DNA-binding domains of catabolite gene activator and cro repressor proteins. *Proceedings of the National Academy of Sciences of the United States of America* **79**, 3097–3100.
[578] Stelling J., Klamt S., Bettenbrock K., Schuster S., Gilles E. (2002). Metabolic network structure determines key aspects of functionality and regulation. *Nature* **420**, 190–193.
[579] Stelzl U., Worm U., Lalowski M. *et al.* (2005). A human protein–protein interaction network: a resource for annotating the proteome. *Cell* **122**, 6, 957–968.
[580] Stephanopoulos G. (2007). Challenges in engineering microbes for biofuels production. *Science* **315**, 801–804.
[581] Stephanopoulos G., Alper H., Moxley J. (2004). Exploiting biological complexity for strain improvement through systems biology. *Nature Biotechnology* **22**, 1261–1267.
[582] Stephanopoulos G., Aristidou A., Nielsen J. (1998). *Metabolic Engineering: Principles and Methodologies*, 1st edn. Academic Press.
[583] Strogatz S. (2001). Exploring complex networks. *Nature* **410**, 268–276.
[584] Stuart J., Segal E., Koller D., Kim S. (2003). A gene-coexpression network for global discovery of conserved genetic modules. *Science* **302**, 249–255.
[585] Sudarsanam P., Pilpel Y., Church G. (2002). Genome-wide co-occurrence of promoter elements reveals a cis-regulatory cassette of rRNA transcription motifs in *Saccharomyces cerevisiae*. *Genome Research* **12**, 1723–1731.
[586] Swindells M. (1995). Identification of a common fold in the replication terminator protein suggests a possible mode for DNA binding. *Trends in Biochemical Sciences* **20**, 300–302.
[587] Tadesse M., Vannucci M., Lió P. (2004). Identification of DNA regulatory motifs using bayesian variable selection. *Bioinformatics* **20**, 2553–2561.
[588] Taniguchi C., Emanuelli B., Kahn C. (2006). Critical nodes in signalling pathways: insight into insulin action. *Nature Reviews Molecular Cell Biology* **7**, 85–96.
[589] Tatusov R., Koonin E., Lipman D. (1997). A genomic perspective on protein families. *Science* **278**, 5338, 631–637.
[590] Taylor S. (1989). cAMP-dependent protein kinase. *Journal of Biological Chemistry* **264**, 8443–8446.
[591] Teichmann S., Babu M. M. (2004). Gene regulatory network growth by duplication. *Nature Genetics* **36**, 492–496.
[592] Terpe K. (2003). Overview of tag protein fusions: from molecular and biochemical fundamentals to commercial systems. *Applied Microbiology and Biotechnology* **60**, 5, 523–533.
[593] Thieffry D., Huerta A., Perez-Rueda E., Collado-Vides J. (1998). From specific gene regulation to genomic networks: a global analysis of transcriptional regulation in *Escherichia coli*. *Bioessays* **20**, 5, 433–440.
[594] Thiele I., Vo T., Price N., Palsson B. (2005). Expanded metabolic reconstruction of *Helicobacter pylori* (iit341 gsm/gpr): an in silico genome-scale characterization of single- and double-deletion mutants. *Journal of Bacteriology* **187**, 5818–5830.

[595] Thomas M., Chiang C. (2006). The general transcription machinery and general cofactors. *Critical Reviews in Biochemistry and Molecular Biology* **41**, 105–178.
[596] Titz B., Schlesner M., Uetz P. (2004). What do we learn from high-throughput protein interaction data? *Expert Reviews of Proteomics* **1**, 1, 111–121.
[597] Tong A., Evangelista M., Parsons A. *et al.* (2001). Systematic genetic analysis with ordered arrays of yeast deletion mutants. *Science* **294**, 2364–2368.
[598] Tong A., Lesage G., Bader G. *et al.* (2004). Global mapping of the yeast genetic interaction network. *Science* **303**, 808–813.
[599] Torres N., Voit E. (2002). *Pathway Analysis and Optimization in Metabolic Engineering*, 1st edn. Cambridge University Press.
[600] Tourriere H., Chebli K., Tazi J. (2002). mRNA degradation machines in eukaryotic cells. *Biochimie* **84**, 821–837.
[601] Tsai M.-J., O'Malley B. (1994). Molecular mechanisms of action of steroid/thyroid receptor superfamily members. *Annual Reviews of Biochemistry* **63**, 451–486.
[602] Turek-Plewa J., Jagodzinski P. (2005). The role of mammalian DNA methyltransferases in the regulation of gene expression. *Cellular and Molecular Biology Letters* **10**, 631–647.
[603] Tyson-Capper A. (2007). Alternative splicing: an important mechanism for myometrial gene regulation that can be manipulated to target specific genes associated with preterm labour. *BMC Pregnancy Childbirth* **7 Suppl. 1**, S13.
[604] Uetz P., Dong Y.-A., Zeretzke C. *et al.* (2006). Herpesviral protein networks and their interaction with the human proteome. *Science* **311**, 5758, 239–242.
[605] Uetz P., Fumagalli S., James D., Zeller R. (1996). Molecular interaction between limb deformity proteins (formins) and src family kinases. *Journal of Biological Chemistry* **271**, 52, 33525–33530.
[606] Uetz P., Stagljar I. (2006). The interactome of human EGF/ErbB receptors. *Molecular Systems Biology* **2**, 2006.0006.
[607] Uetz, P. *et al.* (2000). A comprehensive analysis of protein-protein interactions in Saccharomyces cerevisiae. *Nature* **403**, 623–627.
[608] Vagin V., Sigova A., Li C., Seitz H., Gvozdev V., Zamore P. (2006). A distinct small RNA pathway silences selfish genetic elements in the germline. *Science* **313**, 320.
[609] Valencia A., Pazos F. (2002). Computational methods for the prediction of protein interactions. *Current Opinion in Structural Biology* **12**, 3, 368–373.
[610] van den Beucken T., Koritzinsky M., Wouters B. (2006). Translational control of gene expression during hypoxia. *Cancer Biological Therapy* **5**, 749–55.
[611] van Dongen S. (2000). Ph.D. thesis.
[612] van Nimwegen E. (2003). Scaling laws in the functional content of genomes. *Trends Genetics* **19**, 479–484.
[613] Vanhaesebroeck B., Alessi D. (2000). The PI3K-PDK1 connection: more than just a road to PKB. *Biochemical Journal* **346**, 561–576.
[614] Vanyushin B. (2006). DNA methylation in plants. *Current Topics in Microbiological Immunology* **301**, 67–122.
[615] Varma A., Palsson B. (1994). Stoichiometric flux balance models quantitatively predict growth and metabolic by-product secretion in wild-type *Escherichia coli* w3110. *Applied and Environmental Microbiology* **60**, 3724–3731.
[616] Vázquez A., Flammini A., Maritan A., Vespignani A. (2003a). Global protein function prediction from protein–protein interaction networks. *Nature Biotechnology* **21**, 697–700.

[617] Vázquez A., Flammini A., Maritan A., Vespignani A. (2003b). Modelling of protein interaction networks. *ComPlexUs* **1**, 38–44.
[618] Vázquez A., Moreno Y., Boguñá M., Pastor-Satorras R., Vespignani A. (2003). Topology and correlations in structured scale-free networks. *Physical Review E* **67**, 046111.
[619] Venter J., Remington K., Heidelberg J. *et al.* (2004). Environmental genome shotgun sequencing of the Sargasso Sea. *Science* **304**, 66–74.
[620] Venter, J. C. *et al.* (2001). The sequence of the human genome. *Science* **291**, 5507, 1304–1351.
[621] Versteeg R., van Schaik B., van Batenburg M. *et al.* (2003). The human transcriptome map reveals extremes in gene density, intron length, GC content, and repeat pattern for domains of highly and weakly expressed genes. *Genome Research* **13**, 1998–2004.
[622] Vinson C., Myakishev M., Acharya A., Mir A., Moll J., Bonovich M. (2002). Classification of human b-ZIP proteins based on dimerization properties. *Molecular and Cellular Biology* **22**, 6321–6335.
[623] Vojtek A., Hollenberg S., Cooper J. (1993). Mammalian ras interacts directly with the serine/threonine kinase raf. *Cell 74,* 1, 205–214.
[624] Vollert C., Uetz P. (2004). The phox homology (px) domain protein interaction network in yeast. *Molecular & Cellular Proteomics* **3**, 11, 1053–1064.
[625] von Mering C., Krause R., Snel B. *et al.* (2002). Comparative assessment of large-scale data sets of protein–protein interactions. *Nature* **417**, 399–403.
[626] Wagner A. (2000). Mutational robustness in genetic networks of yeast. *Nature Genetics* **24**, 355–361.
[627] Wagner A., Fell D. (2001). The small world inside large metabolic networks. *Proceedings of the Royal Society of London Series B* **268**, 1803–1810.
[628] Wajant H., Pfizenmaier K., Scheurich P. (2003). Tumor necrosis factor signaling. *Cell Death Differentiation* **10**, 45–65.
[629] Walhout A., Boulton S., Vidal M. (2000). Yeast two-hybrid systems and protein interaction mapping projects for yeast and worm. *Yeast* **17**, 2, 88–94.
[630] Walhout A., Sordella R., Lu X. *et al.* (2000). Protein interaction mapping in *C. elegans* using proteins involved in vulval development. *Science* **287**, 116–122.
[631] Wang Y. (2007). Chromatin structure of repeating CTG/CAG and CGG/CCG sequences in human disease. *Frontiers in Biosciences* **12**, 4731–4741.
[632] Ward C., Lawrence M., Streltsov V., Adams T., McKern N. (2007). The insulin and EGF receptor structures: new insights into ligand-induced receptor activation. *Trends in Biochemical Science* **32**, 129–137.
[633] Wasserman S., Faust K. (1994). *Social Network Analysis.* Cambridge University Press, Cambridge.
[634] Watanabe T., Yoshida N., Satake M. (2005). Biological implications of filamin a-bound PEBP2β/CBFβ retention in the cytoplasm. *Critical Reviews in Eukaryotic Gene Expression* **15**, 197–206.
[635] Watnick P., Kolter R. (2000). Biofilm, city of microbes. *Journal of Bacteriology* **182**, 2675–2679.
[636] Watts D. (2003). *Six Degrees: The Science of a Connected Age.* Norton, New York.
[637] Watts D., Strogatz S. (1998). Collective dynamics of small-world networks. *Nature* **393**, 440–442.
[638] Wehling M., Lösel R. (2006). Non-genomic steroid hormone effects: membrane or intracellular receptors? *Journal of Steroid Biochemistry and Molecular Biology* **102**, 180–183.

[639] Wehr M., Laage R., Bolz U. *et al.* (2006 Dec). Monitoring regulated protein-protein interactions using split TEV. *Nature Methods* **3**, 12, 985–993.
[640] Wei W., Jin J., Schlisio S., Harper J., Kaelin Jr. W. (2005). The v-Jun point mutation allows c-Jun to escape GSK3-dependent recognition and destruction by the fbw7 ubiquitin ligase. *Cancer Cell* **8**, 25–33.
[641] Weickert M., Adhya S. (1992). A family of bacterial regulators homologous to gal and lac repressors. *Journal of Biological Chemistry* **267**, 15869–15874.
[642] Weigel N., Moore N. (2007). Kinases and protein phosphorylation as regulators of steroid hormone action. *Nuclear Receptor Signaling* **5**, 1–13.
[643] Weinzierl R. (1999). *Mechanisms of Gene Expression.* Imperial College Press, London.
[644] Weiss C., Bohmann D. (2004). Deregulated repression of c-Jun provides a potential link to its role in tumorigenesis. *Cell Cycle* **3**, 111–113.
[645] Weiss C., Schneider S., Wagner E., Zhang X., Seto E., Bohmann D. (2003). Jnk phosphorylation relieves HDAC3-dependent suppression of the transcriptional activity of c-Jun. *EMBO Journal* **22**, 3686–3695.
[646] Weiss M., Ellenberger T., Wobbe C., Lee J., Harrison S., Struhl K. (1990). Folding, transition in the DNA-binding domain of GCN4 on specific binding to DNA. *Nature* **347**, 575–578.
[647] Wellbrock C., Karasarides M., Marais R. (2004). The raf proteins take centre stage. *Nature Review Molecular Cell Biology* **5**, 875–885.
[648] Wera S., Hemmings B. (1995). Serine/threonine protein phosphatases. *Biochemistry Journal* **311**, 17–29.
[649] Wetzker R., Böhmer F.-D. (2003). Transactivation joins multiple tracks to the ERK/MAPK cascade. *Nature Reviews Molecular Cell Biology* **4**, 651–657.
[650] White R. (2001). *Gene Transcription: Mechanisms and Control.* Blackwell Science, Malden, MA.
[651] Wiback S., Mahadevan R., Palsson B. (2003). Reconstructing metabolic flux vectors from extreme pathways: defining the alpha-spectrum. *Journal of Theoretical Biology* **224**, 313–324.
[652] Wiback S., Palsson B. (2002). Extreme pathway analysis of human red blood cell metabolism. *Biophysical Journal* **83**, 808–818.
[653] Wilkins A. (2002). *The Evolution of Developmental Pathways.* Sinauer Associates, Sunderland, Mass.
[654] Wilson D., Madera M., Vogel C., Chothia C., Gough J. (2007). The superfamily database in 2007: families and functions. *Nucleic Acids Research* **35**, D308–313.
[655] Wilson K., Shewchuk L., Brennan R., Otsuka A., Matthews B. (1992). *Escherichia coli* biotin holoenzyme synthetase/bio repressor crystal structure delineates the biotin- and DNA-binding domains. *Proceedings of the National Academy of Sciences of the United States of America* **89**, 9257–9261.
[656] Wingender E., Chen X., Hehl R. *et al.* (2000). TRANSFAC: an integrated system for gene expression regulation. *Nucleic Acids Research* **28**, 316–319.
[657] Wodak S., Méndez R. (2004). Prediction of protein–protein interactions: the capri experiment, its evaluation and implications. *Current Opinion in Structural Biology* **14**, 2, 242–249.
[658] Wold, S. *et al.* (2001). Pls-regression: a basic tool of chemometrics. *Chemometrics and Intelligent Laboratory Systems* **58**, 109–130.
[659] Wolf D., Arkin A. (2003). Motifs, modules and games in bacteria. *Current Opinion in Microbiology* **6**, 125–134.

[660] Wolf Y., Karev G., Koonin E. (2002). Scale-free networks in biology: new insights into the fundamentals of evolution? *Bioessays* **24**, 105–109.
[661] Woodley L., Valcarcel J. (2002). Regulation of alternative pre-mRNA splicing. *Briefings in Functional Genomics and Proteomics* **1**, 266–277.
[662] Wray G., Hahn M., Abouheif E. *et al.* (2003). The evolution of transcriptional regulation in eukaryotes. *Molecular Biology and Evolution* **20**, 1377–1419.
[663] Wright S. (1982). Character change, speciation, and the higher taxa. *Evolution* **36**, 427–443.
[664] Wu F., Huberman B. (2004). Finding communities in linear time: a physics approach. *European Physical Journal B* **38**, 331–338.
[665] Wu W., Li W., Chen B. (2006). Computational reconstruction of transcriptional regulatory modules of the yeast cell cycle. *BMC Bioinformatics* **7**, 421.
[666] Wuchty S. (2001). Scale-free behavior in protein domain networks. *Molecular Biology and Evolution* **18**, 1694–1702.
[667] Wuchty S. (2002). Interaction and domain networks of yeast. *Proteomics* **2**, 1715–1723.
[668] Wullaert A., van Loo G., Heyninck K., Beyaert R. (2007). Hepatic tumor necrosis factor signaling and nuclear factor-kB: effects on liver hoemeostasis and beyond. *Endocrinology Review* **28**, 365–386.
[669] Xenarios I., Fernandez E., Salwinski L. *et al.* (2001). DIP: the database of interacting proteins: 2001 update. *Nucleic Acids Research* **29**, 239–241.
[670] Xenarios I., Rice D., Salwinski L., Baron M., Marcotte E., Eisenberg D. (2000). DIP: the database of interacting proteins. *Nucleic Acids Research* **28**, 289.
[671] Xi H., Yu Y., Fu Y., Foley J., Halees A., Weng Z. (2007). Analysis of overrepresented motifs in human core promoters reveals dual regulatory roles of YY1. *Genome Research* **17**, 798–806.
[672] Xu L. (2003). Regulation of SMAD activities. *Biochimica et Biophysica Acta* **1759**, 503–513.
[673] Xu M., Paulus H., Chong S. (2000). Fusions to self-splicing inteins for protein purification. *Methods in Enzymology* **326**, 376–418.
[674] Yaffe M. (2002). Phosphotyrosine-binding domains in signal transduction. *Nature Reviews Molecular Cell Biology* **3**, 177–186.
[675] Yamada L., Kobayashi K., Degnan B., Satoh N., Satou Y. (2003). A genomewide survey of developmentally relevant genes in ciona intestinalis. iv. genes for hmg transcriptional regulators, bZIP and gata/gli/zic/snail. *Development Genes and Evolution* **213**, 245–253.
[676] Yeh P., Kishony R. (2007). Networks from drug–drug surfaces. *Molecular Systems Biology* **3**, 85.
[677] Yeh P., Tschumi A., Kishony R. (2006). Functional classification of drugs by properties of their pairwise interactions. *Nature Genetics* **38**, 489–494.
[678] Youle R., Strasser A. (2008). The BCL-2 protein family: opposing activities that mediate cell death. *Nature Reviews Molecular Cell Biology* **9**, 47–59.
[679] Youngren J. (2007). Regulation of insulin receptor function. *Cellular Molecular Life Science* **64**, 873–891.
[680] Yu H., Gerstein M. (2006). Genomic analysis of the hierarchical structure of regulatory networks. *Proceedings of the National Academy of Sciences USA* **103**, 14724–14731.
[681] Yu H., Luscombe N., Lu H. *et al.* (2004). Annotation transfer between genomes: protein–protein interologs and protein–DNA regulogs. *Genome Research* **14**, 1107–1118.

[682] Yura K., Yamaguchi A., Go M. (2006). Coverage of whole proteome by structural genomics observed through protein homology modeling database. *Journal of Structural and Functional Genomics 7,* 2, 65–76.

[683] Zakeri Z., Lockshin R. (2001). Programmed cell death and apoptosis: origins of the theory. *Nature Revew Molecular Cell Biology* **2**, 545–550.

[684] Zaman Z., Ansari A., Gaudreau L., Nevado J., Ptashne M. (1998). Gene transcription by recruitment. *Cold Spring Harb. Symp. Quant. Biol.* **63**, 167–171.

[685] Zaslaver A., Mayo A., Rosenberg R. *et al.* (2004). Just-in-time transcription program in metabolic pathways. *Nature Genetics* **36**, 486–491.

[686] Zeggini E., Scott L., Saxena R. *et al.* (2008). Meta-analysis of genome-wide association data and large-scale replication identifies additional susceptibility loci for type 2 diabetes. *Nature Genetics* **40**, 638–645.

[687] Zhang G., Campbell E., Minakhin L., Richter C., Severinov K., Darst S. (1999). Crystal structure of thermus aquaticus core RNA polymerase at 3.3 Å resolution. *Cell* **98**, 6, 811–824.

[688] Zhang J. (2003). Evolution by gene duplication: an update. *Trends in Ecology & Evolution* **18**, 6, 292–298.

[689] Zhang Z.-Y. (1998). Protein-tyrosine phosphatases: biological function, structural characteristics, and mechanism of catalysis. *Critical Reviews in Biochemistry and Molecular Biology* **33**, 1–52.

[690] Zhivotovsky B., Kroemer G. (2004). Apoptosis and genomic instability. *Nature Reviews Molecular Cell Biology* **5**, 752–762.

[691] Zhong J., Zhang H., Stanyon C., Tromp G., Finley R. J. (2003). A strategy for constructing large protein interaction maps using the yeast two-hybrid system: regulated expression arrays and two-phase mating. *Genome Research 13,* 12, 2691–2699.

[692] Zhou H. (2003a). Distance, dissimilarity index, and network community structure. *Physical Review E* **67**, 061901.

[693] Zhou H. (2003b). Network landscape from a brownian particle's perspective. *Physical Review E* **67**, 041908.

[694] Zhu H., Bilgin M., Bangham R. *et al.* (2001). Global analysis of protein activities using proteome chips. *Science* **293**, 5537, 2101–2105.

Index

Printed in the United States
by Baker & Taylor Publisher Services